CALCULUS
SEVERAL VARIABLES

THE WILEY BICENTENNIAL—KNOWLEDGE FOR GENERATIONS

*E*ach generation has its unique needs and aspirations. When Charles Wiley first opened his small printing shop in lower Manhattan in 1807, it was a generation of boundless potential searching for an identity. And we were there, helping to define a new American literary tradition. Over half a century later, in the midst of the Second Industrial Revolution, it was a generation focused on building the future. Once again, we were there, supplying the critical scientific, technical, and engineering knowledge that helped frame the world. Throughout the 20th Century, and into the new millennium, nations began to reach out beyond their own borders and a new international community was born. Wiley was there, expanding its operations around the world to enable a global exchange of ideas, opinions, and know-how.

For 200 years, Wiley has been an integral part of each generation's journey, enabling the flow of information and understanding necessary to meet their needs and fulfill their aspirations. Today, bold new technologies are changing the way we live and learn. Wiley will be there, providing you the must-have knowledge you need to imagine new worlds, new possibilities, and new opportunities.

Generations come and go, but you can always count on Wiley to provide you the knowledge you need, when and where you need it!

WILLIAM J. PESCE
PRESIDENT AND CHIEF EXECUTIVE OFFICER

PETER BOOTH WILEY
CHAIRMAN OF THE BOARD

STUDENT SOLUTIONS MANUAL

Garret Etgen
University of Houston

to accompany

CALCULUS
SEVERAL VARIABLES

10th Edition

Saturnino Salas
Einar Hille
Garret Etgen
University of Houston

BICENTENNIAL
1807
WILEY
2007
BICENTENNIAL

John Wiley & Sons, Inc.

SKY10020817_082720

CONTENTS

CALCULUS
SEVERAL VARIABLES

CHAPTER 13

SECTION 13.1

1.

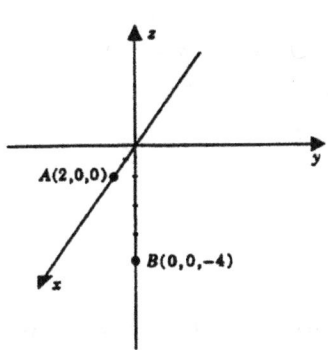

midpoint: $(1, 0, -2)$

3.

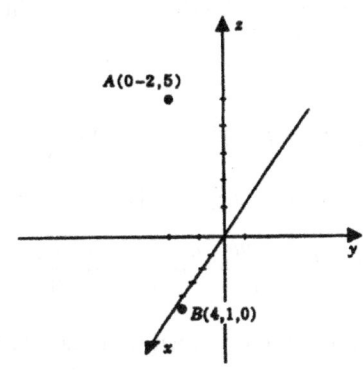

length $\overline{AB}$: $5\sqrt{2}$

midpoint: $\left(2, -\frac{1}{2}, \frac{5}{2}\right)$

5. $z = -2$ **7.** $y = 1$ **9.** $x = 3$

11. $x^2 + (y-2)^2 + (z+1)^2 = 9$ **13.** $(x-2)^2 + (y-4)^2 + (z+4)^2 = 36$

15. $(x-3)^2 + (y-2)^2 + (z-2)^2 = 13$

17.
$$x^2 + y^2 + z^2 + 4x - 8y - 2z + 5 = 0$$
$$x^2 + 4x + 4 + y^2 - 8y + 16 + z^2 - 2z + 1 = -5 + 4 + 16 + 1$$
$$(x+2)^2 + (y-4)^2 + (z-1)^2 = 16$$

center: $(-2, 4, 1)$,   radius: 4

19. $(2, 3, -5)$ **21.** $(-2, 3, 5)$ **23.** $(-2, 3, -5)$

25. $(-2, -3, -5)$ **27.** $(2, -5, 5)$ **29.** $(-2, 1, -3)$

31. $d(PR) = \sqrt{14}$, $d(QR) = \sqrt{45}$, $d(PQ) = \sqrt{59}$; $[d(PR)]^2 + [d(QR)]^2 = [d(PQ)]^2$

33. The sphere of radius 2 centered at the origin, together with its interior.

35. A rectangular box in the first octant with sides on the coordinate planes and dimensions $1 \times 2 \times 3$, together with its interior.

37. A circular cylinder with base the circle $x^2 + y^2 = 4$ and height 4, together with its interior.

39. Let $B = (x, y, z)$. Then

$$\frac{x+2}{2} = 1 \Longrightarrow x = 0, \quad \frac{y+3}{2} = 2 \Longrightarrow y = 1, \quad \frac{z+4}{2} = 3 \Longrightarrow z = 2.$$

Therefore $B = (0, 1, 2)$.

41. Let $P_1 = (x, y, z)$ be the trisection point closest to A. Then

$$\overrightarrow{AP_1} = \tfrac{1}{3}\,\overrightarrow{AB} \Longrightarrow (x - a_1, y - a_2, z - a_3) = \tfrac{1}{3}(b_1 - a_1, b_2 - a_2, b_3 - a_3).$$

Solving for x, y, z gives $(x, y, z) = \left(\dfrac{2a_1 + b_1}{3}, \dfrac{2a_2 + b_2}{3}, \dfrac{2a_3 + b_3}{3} \right)$.

Similarly, if $P_2 = (x, y, z)$ is the trisection point closest to B, then

$$(x, y, z) = \left(\frac{a_1 + 2b_1}{3}, \frac{a_2 + 2b_2}{3}, \frac{a_3 + 2b_3}{3} \right).$$

43. Substituting the coordinates of the points into the equation $Ax + By + Cz + D = 0$, we get the equations

$$Ax_0 + D = 0, \ By_0 + D = 0, \ Cz_0 + D = 0 \quad \text{which implies} \quad Ax_0 = By_0 = Cz_0.$$

Therefore, we have

$$Ax + \frac{Ax_0}{y_0}y + \frac{Ax_0}{z_0}z + D = 0 \quad \text{or} \quad \frac{x}{x_0} + \frac{y}{y_0} + \frac{z}{z_0} + \frac{D}{Ax_0} = 0.$$

Substituting the point $(x_0, 0, 0)$ into this equation gives

$$\frac{x}{x_0} + \frac{y}{y_0} + \frac{z}{z_0} = 1.$$

45. (i) $a_3 \neq 0$ The line through the origin and (a_1, a_2, a_3) is given by $x = a_1 t, \ y = a_2 t, \ z = a_3 t, \ t$ any real number. The line intersects the plane $z = z_0$ at the point Q where $t = z_0/a_3$. The coordinates of Q are: $\frac{a_1}{a_3}z_0, \ \frac{a_2}{a_3}z_0, \ z_0$.

(ii) $a_3 = 0$ If $z_0 \neq 0$, the line does not intersect the plane. If $z_0 = 0$, the line lies in the plane.

47. The ray that emanates from the origin and passes through the point (a_1, a_2, a_3) is given by $x = a_1 t,$ $y = a_2 t, \ z = a_3 t, \ t \geq 0$. The ray intersects the sphere $x^2 + y^2 + z^2 = 1$ at the point Q where

$$a_1^2 t^2 + a_2^2 t^2 + a_3^2 t^2 = 1 \Longrightarrow t = \frac{1}{\sqrt{a_1^2 + a_2^2 + a_3^2}}.$$

The coordinates of Q are: $\dfrac{a_1}{\sqrt{a_1^2 + a_2^2 + a_3^2}}, \ \dfrac{a_2}{\sqrt{a_1^2 + a_2^2 + a_3^2}}, \ \dfrac{a_3}{\sqrt{a_1^2 + a_2^2 + a_3^2}}.$

SECTION 13.2

1. $\overrightarrow{PQ} = (3, 4, -2); \quad \| \overrightarrow{PQ} \| = \sqrt{29}$ **3.** $\overrightarrow{PQ} = (0, -2, -1); \quad \| \overrightarrow{PQ} \| = \sqrt{5}$

5. $2\mathbf{a} - \mathbf{b} = (2 \cdot 1 - 3, 2 \cdot [-2] - 0, 2 \cdot 3 + 1) = (-1, -4, 7)$

7. $-2\mathbf{a} + \mathbf{b} - \mathbf{c} = [-(2\mathbf{a} - \mathbf{b})] - \mathbf{c} = (1 + 4, 4 - 2, -7 - 1) = (5, 2, -8)$

9. $3\mathbf{i} - 4\mathbf{j} + 6\mathbf{k}$ **11.** $-3\mathbf{i} - \mathbf{j} + 8\mathbf{k}$

13. 5 **15.** 3 **17.** $\sqrt{6}$

19. (a) $\mathbf{a}$, $\mathbf{c}$, and $\mathbf{d}$ since $\mathbf{a} = \frac{1}{3}\mathbf{c} = -\frac{1}{2}\mathbf{d}$
  (b) $\mathbf{a}$ and $\mathbf{c}$ since $\mathbf{a} = \frac{1}{3}\mathbf{c}$
  (c) $\mathbf{a}$ and $\mathbf{c}$ both have direction opposite to $\mathbf{d}$

21. $\overrightarrow{RQ} = (3 - x, -1 - y, 1 - z)$ and $\overrightarrow{OP} = (1, 4, -2)$.
$$\overrightarrow{RQ} = \overrightarrow{OP} \Longrightarrow 3 - x = 1, \ -1 - y = 4, \ 1 - z = -2 \Longrightarrow x = 2, \ y = -5, \ z = 3.$$

23. $\overrightarrow{RQ} = (3 - x, -1 - y, 1 - z) = -2\overrightarrow{OP} = (-2, -8, 4) \Longrightarrow 3 - x = -2, \ -1 - y = -8, \ 1 - z = 4$
$\Longrightarrow x = 5, \ y = 7, \ z = -3.$

25. $\|\mathbf{a}\| = 5$; $\dfrac{\mathbf{a}}{\|\mathbf{a}\|} = \left(\dfrac{3}{5}, -\dfrac{4}{5}, 0\right)$ **27.** $\|\mathbf{a}\| = 3$; $\dfrac{\mathbf{a}}{\|\mathbf{a}\|} = \dfrac{1}{3}\mathbf{i} - \dfrac{2}{3}\mathbf{j} + \dfrac{2}{3}\mathbf{k}$

29. $\|\mathbf{a}\| = \sqrt{14}$; $-\dfrac{\mathbf{a}}{\|\mathbf{a}\|} = \dfrac{1}{\sqrt{14}}\mathbf{i} - \dfrac{3}{\sqrt{14}}\mathbf{j} - \dfrac{2}{\sqrt{14}}\mathbf{k}$

31. (i) $\mathbf{a} - \mathbf{b}$ (ii) $-(\mathbf{a} + \mathbf{b})$ (iii) $\mathbf{a} - \mathbf{b}$ (iv) $\mathbf{b} - \mathbf{a}$

33. (a) $\mathbf{a} - 3\mathbf{b} + 2\mathbf{c} + 4\mathbf{d} = (2\mathbf{i} - \mathbf{k}) - 3(\mathbf{i} + 3\mathbf{j} + 5\mathbf{k}) + 2(-\mathbf{i} + \mathbf{j} + \mathbf{k}) + 4(\mathbf{i} + \mathbf{j} + 6\mathbf{k})$
$$= \mathbf{i} - 3\mathbf{j} + 10\mathbf{k}$$
  (b) The vector equation
$$(1, 1, 6) = A(2, 0, -1) + B(1, 3, 5) + C(-1, 1, 1)$$
  implies
$$1 = 2A + B - C,$$
$$1 = 3B + C,$$
$$6 = -A + 5B + C.$$
  Simultaneous solution gives $A = -2, \quad B = \frac{3}{2}, \quad C = -\frac{7}{2}.$

35. $\|3\mathbf{i} + \mathbf{j}\| = \|\alpha\mathbf{j} - \mathbf{k}\| \implies 10 = \alpha^2 + 1$ so $\alpha = \pm 3$

37.
$$\|\alpha\mathbf{i} + (\alpha - 1)\mathbf{j} + (\alpha + 1)\mathbf{k}\| = 2 \implies \alpha^2 + (\alpha - 1)^2 + (\alpha + 1)^2 = 4$$
$$\implies 3\alpha^2 = 2 \ \text{ so } \ \alpha = \pm\frac{1}{3}\sqrt{6}$$

39. $\pm\frac{2}{13}\sqrt{13}\,(3\mathbf{j} + 2\mathbf{k})$ since $\|\alpha(3\mathbf{j} + 2\mathbf{k})\| = 2 \implies \alpha = \pm\frac{2}{13}\sqrt{13}$

41. (a) Since $\|\mathbf{a} - \mathbf{b}\|$ and $\|\mathbf{a} + \mathbf{b}\|$ are the lengths of the diagonals of the parallelogram, the parallelogram must be a rectangle.

(b) Simplify
$$\sqrt{(a_1 - b_1)^2 + (a_2 - b_2)^2 + (a_3 - b_3)^2} = \sqrt{(a_1 + b_1)^2 + (a_2 + b_2)^2 + (a_3 + b_3)^2}.$$
the result is $a_1 b_1 + a_2 b_2 + a_3 b_3 = 0$.

43. Let $P = (x_1, y_1, z_1)$, $Q = (x_2, y_2, z_2)$, and $M = (x_m, y_m, z_m)$. Then
$$(x_m, y_m, z_m) = (x_1, y_1, z_1) + \frac{1}{2}(x_2 - x_1, y_2 - y_1, z_2 - z_1) \Longrightarrow \mathbf{m} = \mathbf{p} + \frac{1}{2}(\mathbf{q} - \mathbf{p}).$$

SECTION 13.3

1. $\mathbf{a} \cdot \mathbf{b} = (2)(-2) + (-3)(0) + (1)(3) = -1$ **3.** $\mathbf{a} \cdot \mathbf{b} = (2)(1) + (-4)(1/2) + (0)(0) = 0$

5. $\mathbf{a} \cdot \mathbf{b} = (2)(1) + (1)(1) - (2)(2) = -1$ **7.** $\mathbf{a} \cdot \mathbf{b}$

9. $(\mathbf{a} - \mathbf{b}) \cdot \mathbf{c} + \mathbf{b} \cdot (\mathbf{c} + \mathbf{a}) = \mathbf{a} \cdot \mathbf{c} - \mathbf{b} \cdot \mathbf{c} + \mathbf{b} \cdot \mathbf{c} + \mathbf{b} \cdot \mathbf{a} = \mathbf{a} \cdot (\mathbf{b} + \mathbf{c})$

11. (a) $\mathbf{a} \cdot \mathbf{b} = (2)(3) + (1)(-1) + (0)(2) = 5$

$\mathbf{a} \cdot \mathbf{c} = (2)(4) + (1)(0) + (0)(3) = 8$

$\mathbf{b} \cdot \mathbf{c} = (3)(4) + (-1)(0) + (2)(3) = 18$

(b) $\|\mathbf{a}\| = \sqrt{5}$, $\|\mathbf{b}\| = \sqrt{14}$, $\|\mathbf{c}\| = 5$. Then,
$$\cos \sphericalangle (\mathbf{a}, \mathbf{b}) = \frac{\mathbf{a} \cdot \mathbf{b}}{\|\mathbf{a}\| \, \|\mathbf{b}\|} = \frac{5}{\left(\sqrt{5}\right)\left(\sqrt{14}\right)} = \frac{1}{14}\sqrt{70},$$
$$\cos \sphericalangle (\mathbf{a}, \mathbf{c}) = \frac{8}{\left(\sqrt{5}\right)(5)} = \frac{8}{25}\sqrt{5},$$
$$\cos \sphericalangle (\mathbf{b}, \mathbf{c}) = \frac{18}{\left(\sqrt{14}\right)(5)} = \frac{9}{35}\sqrt{14}.$$

(c) $\mathbf{u_b} = \frac{1}{\sqrt{14}}(3\mathbf{i} - \mathbf{j} + 2\mathbf{k})$, $\text{comp}_\mathbf{b}\, \mathbf{a} = \mathbf{a} \cdot \mathbf{u_b} = \frac{1}{\sqrt{14}}(6 - 1) = \frac{5}{14}\sqrt{14}$,

$\mathbf{u_c} = \frac{1}{5}(4\mathbf{i} + 3\mathbf{k})$, $\text{comp}_\mathbf{c}\, \mathbf{a} = \mathbf{a} \cdot \mathbf{u_c} = \frac{8}{5}$

(d) $\mathbf{proj_b}\, \mathbf{a} = (\text{comp}_\mathbf{b}\, \mathbf{a})\, \mathbf{u_b} = \frac{5}{14}(3\mathbf{i} - \mathbf{j} + 2\mathbf{k})$, $\mathbf{proj_c}\, \mathbf{a} = (\text{comp}_\mathbf{c}\, \mathbf{a})\, \mathbf{u_c} = \frac{8}{25}(4\mathbf{i} + 3\mathbf{k})$

13. $\mathbf{u} = \cos\frac{\pi}{3}\mathbf{i} + \cos\frac{\pi}{4}\mathbf{j} + \cos\frac{2\pi}{3}\mathbf{k} = \frac{1}{2}\mathbf{i} + \frac{1}{2}\sqrt{2}\mathbf{j} - \frac{1}{2}\mathbf{k}$

15. $\cos\theta = \dfrac{(3\mathbf{i} - \mathbf{j} - 2\mathbf{k}) \cdot (\mathbf{i} + 2\mathbf{j} - 3\mathbf{k})}{\|3\mathbf{i} - \mathbf{j} - 2\mathbf{k}\| \, \|\mathbf{i} + 2\mathbf{j} - 3\mathbf{k}\|} = \dfrac{7}{\sqrt{14}\sqrt{14}} = \dfrac{1}{2}, \quad \theta = \dfrac{\pi}{3}$

17. Since $\|\mathbf{i} - \mathbf{j} + \sqrt{2}\,\mathbf{k}\| = 2$, we have $\cos\alpha = \frac{1}{2}$, $\cos\beta = -\frac{1}{2}$, $\cos\gamma = \frac{1}{2}\sqrt{2}$. The direction angles are $\frac{1}{3}\pi$, $\frac{2}{3}\pi$, $\frac{1}{4}\pi$.

19. $\theta = \arccos\dfrac{\mathbf{a}\cdot\mathbf{b}}{\|\mathbf{a}\|\|\mathbf{b}\|} = \arccos\left(\dfrac{-9}{\sqrt{231}}\right) \cong 2.2$ radians   or $126.3°$

21. $\theta = \arccos\dfrac{\mathbf{a}\cdot\mathbf{b}}{\|\mathbf{a}\|\|\mathbf{b}\|} = \arccos\left(\dfrac{-13}{5\sqrt{10}}\right) \cong 2.54$ radians   or $145.3°$

23. angles: $38.51°$, $95.52°$, $45.97°$;   perimeter $: \cong 15.924$

25. $\|\mathbf{a}\| = \sqrt{1^2 + 2^2 + 2^2} = 3$;

$$\cos\alpha = \frac{1}{3}, \quad \cos\beta = \frac{2}{3}, \quad \cos\gamma = \frac{2}{3}$$
$$\alpha \cong 70.5° \quad \beta \cong 48.2°, \quad \gamma \cong 48.2°$$

27. $\|\mathbf{a}\| = \sqrt{3^2 + (12)^2 + 4^2} = 13$;

$$\cos\alpha = \frac{3}{13}, \quad \cos\beta = \frac{12}{13} \quad \cos\gamma = \frac{4}{13}$$
$$\alpha \cong 76.7° \quad \beta \cong 22.6°, \quad \gamma \cong 72.1°$$

29. $2\mathbf{i} + 5\mathbf{j} + 2x\,\mathbf{k} \perp 6\mathbf{i} + 4\mathbf{j} - x\,\mathbf{k} \Longrightarrow 12 + 20 - 2x^2 = 0 \Longrightarrow x^2 = 16 \Longrightarrow x = \pm 4$

31. $\cos\dfrac{\pi}{3} = \dfrac{\mathbf{c}\cdot\mathbf{d}}{\|\mathbf{c}\|\,\|\mathbf{d}\|}$, $\dfrac{1}{2} = \dfrac{2x+1}{x^2+2}$, $x^2 = 4x$; $x = 0$, $x = 4$

33. (a) The direction angles of a vector always satisfy

$$\cos^2\alpha + \cos^2\beta + \cos^2\gamma = 1$$

and, as you can check,

$$\cos^2\tfrac{1}{4}\pi + \cos^2\tfrac{1}{6}\pi + \cos^2\tfrac{2}{3}\pi \neq 1.$$

(b) The relation

$$\cos^2\alpha + \cos^2\tfrac{1}{4}\pi + \cos^2\tfrac{1}{4}\pi = 1$$

gives

$$\cos^2\alpha + \tfrac{1}{2} + \tfrac{1}{2} = 1, \quad \cos\alpha = 0, \quad a_1 = \|\mathbf{a}\|\cos\alpha = 0.$$

35. Let $\theta_1, \theta_2, \theta_3$ be the direction angles of $-\mathbf{a}$. Then

$$\theta_1 = \arccos\left[\frac{(-\mathbf{a}\cdot\mathbf{i})}{\|-\mathbf{a}\|}\right] = \arccos\left[-\frac{(\mathbf{a}\cdot\mathbf{i})}{\|\mathbf{a}\|}\right] = \arccos\left(-\cos\alpha\right) = \pi - \arccos\left(\cos\alpha\right) = \pi - \alpha.$$

Similarly $\theta_2 = \pi - \beta$ and $\theta_3 = \pi - \gamma$.

37. Set $\mathbf{u} = a\,\mathbf{i} + b\,\mathbf{j} + c\,\mathbf{k}$.   The relations

$$(a\,\mathbf{i} + b\,\mathbf{j} + c\,\mathbf{k})\cdot(\mathbf{i} + 2\mathbf{j} + \mathbf{k}) = 0 \quad\text{and}\quad (a\,\mathbf{i} + b\,\mathbf{j} + c\,\mathbf{k})\cdot(3\mathbf{i} - 4\mathbf{j} + 2\mathbf{k}) = 0$$

give

$$a + 2b + c = 0 \qquad 3a - 4b + 2c = 0$$

so that $b = \frac{1}{8}a$ and $c = -\frac{5}{4}a$.

Then, since $\mathbf{u}$ is a unit vector,

$$a^2 + b^2 + c^2 = 1, \quad a^2 + \left(\frac{a}{8}\right)^2 + \left(\frac{-5a}{4}\right)^2 = 1, \quad \frac{165}{64}a^2 = 1.$$

Thus, $a = \pm\dfrac{8}{165}\sqrt{165}$ and $\mathbf{u} = \pm\dfrac{\sqrt{165}}{165}(8\,\mathbf{i} + \mathbf{j} - 10\,\mathbf{k}).$

39.

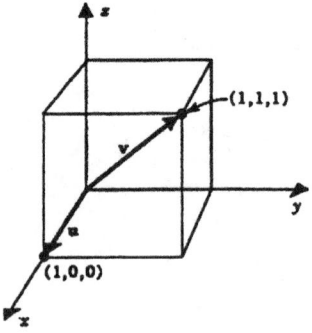

We take $\mathbf{u} = \mathbf{i}$ as an edge and $\mathbf{v} = \mathbf{i} + \mathbf{j} + \mathbf{k}$ as a diagonal of a cube. Then,

$$\cos\theta = \frac{\mathbf{u} \cdot \mathbf{v}}{\|\mathbf{u}\|\,\|\mathbf{v}\|} = \frac{1}{3}\sqrt{3},$$

$$\theta = \cos^{-1}\left(\tfrac{1}{3}\sqrt{3}\right) \cong 0.96 \text{ radians.}$$

41. (a) $\mathbf{proj_b}\,\alpha\mathbf{a} = (\alpha\mathbf{a} \cdot \mathbf{u_b})\mathbf{u_b} = \alpha(\mathbf{a} \cdot \mathbf{u_b})\mathbf{u_b} = \alpha\,\mathbf{proj_b}\,\mathbf{a}$

(b) $$\mathbf{proj_b}\,(\mathbf{a} + \mathbf{c}) = [(\mathbf{a} + \mathbf{c}) \cdot \mathbf{u_b}]\,\mathbf{u_b}$$
$$= (\mathbf{a} \cdot \mathbf{u_b} + \mathbf{c} \cdot \mathbf{u_b})\mathbf{u_b}$$
$$= (\mathbf{a} \cdot \mathbf{u_b})\mathbf{u_b} + (\mathbf{c} \cdot \mathbf{u_b})\mathbf{u_b} = \mathbf{proj_b}\,\mathbf{a} + \mathbf{proj_b}\,\mathbf{c}$$

43. (a) $\mathbf{a} \cdot \mathbf{b} = \mathbf{a} \cdot \mathbf{c} \Longrightarrow \mathbf{a}(\mathbf{b} - \mathbf{c}) = 0 \Longrightarrow \mathbf{a} \perp (\mathbf{b} - \mathbf{c}).$

For $\mathbf{a} \neq \mathbf{0}$ the following statements are equivalent:

$$\mathbf{a} \cdot \mathbf{b} = \mathbf{a} \cdot \mathbf{c}, \quad \mathbf{b} \cdot \mathbf{a} = \mathbf{c} \cdot \mathbf{a},$$
$$\mathbf{b} \cdot \frac{\mathbf{a}}{\|\mathbf{a}\|} = \mathbf{c} \cdot \frac{\mathbf{a}}{\|\mathbf{a}\|}, \quad \mathbf{b} \cdot \mathbf{u_a} = \mathbf{c} \cdot \mathbf{u_a}$$
$$(\mathbf{b} \cdot \mathbf{u_a})\mathbf{u_a} = (\mathbf{c} \cdot \mathbf{u_a})\mathbf{u_a},$$
$$\mathbf{proj_a}\,\mathbf{b} = \mathbf{proj_a}\,\mathbf{c}.$$

Thus, $\mathbf{a} \cdot \mathbf{b} = \mathbf{a} \cdot \mathbf{c}$ implies only that the projection of $\mathbf{b}$ on $\mathbf{a}$ equals the projection of $\mathbf{c}$ on $\mathbf{a}$.

(b) $\mathbf{b} = (\mathbf{b} \cdot \mathbf{i})\mathbf{i} + (\mathbf{b} \cdot \mathbf{j})\mathbf{j} + (\mathbf{b} \cdot \mathbf{k})\mathbf{k} = (\mathbf{c} \cdot \mathbf{i})\mathbf{i} + (\mathbf{c} \cdot \mathbf{j})\mathbf{j} + (\mathbf{c} \cdot \mathbf{k})\mathbf{k} = \mathbf{c}$

$\qquad\quad\uparrow\underline{\quad}$ (13.3.14) $\qquad\qquad\qquad\qquad\qquad$ (13.3.14)$\underline{\quad}\uparrow$

45. (a) $\|\mathbf{a} + \mathbf{b}\|^2 - \|\mathbf{a} - \mathbf{b}\|^2 = (\mathbf{a} + \mathbf{b}) \cdot (\mathbf{a} + \mathbf{b}) - (\mathbf{a} - \mathbf{b}) \cdot (\mathbf{a} - \mathbf{b})$
$$= [(\mathbf{a} \cdot \mathbf{a}) + 2(\mathbf{a} \cdot \mathbf{b}) + (\mathbf{b} \cdot \mathbf{b})] - [(\mathbf{a} \cdot \mathbf{a}) - 2(\mathbf{a} \cdot \mathbf{b}) + (\mathbf{b} \cdot \mathbf{b})] = 4(\mathbf{a} \cdot \mathbf{b})$$

(b) By (b), the relation $\|\mathbf{a} + \mathbf{b}\| = \|\mathbf{a} - \mathbf{b}\|$ gives $\mathbf{a} \perp \mathbf{b}$. The relation $\mathbf{a} + \mathbf{b} \perp \mathbf{a} - \mathbf{b}$ gives

$$0 = (\mathbf{a} + \mathbf{b}) \cdot (\mathbf{a} - \mathbf{b}) = \|\mathbf{a}\|^2 - \|\mathbf{b}\|^2 \quad \text{and thus} \quad \|\mathbf{a}\| = \|\mathbf{b}\|.$$

The parallelogram is a square since it has two adjacent sides of equal length and these meet at right angles.

47. $\|\mathbf{a} + \mathbf{b}\|^2 = (\mathbf{a} + \mathbf{b}) \cdot (\mathbf{a} + \mathbf{b}) = \mathbf{a} \cdot \mathbf{a} + 2\,\mathbf{a} \cdot \mathbf{b} + \mathbf{b} \cdot \mathbf{b} = \|\mathbf{a}\|^2 + 2\,\mathbf{a} \cdot \mathbf{b} + \|\mathbf{b}\|^2$

$\|\mathbf{a} - \mathbf{b}\|^2 = (\mathbf{a} - \mathbf{b}) \cdot (\mathbf{a} - \mathbf{b}) = \mathbf{a} \cdot \mathbf{a} - 2\,\mathbf{a} \cdot \mathbf{b} + \mathbf{b} \cdot \mathbf{b} = \|\mathbf{a}\|^2 - 2\,\mathbf{a} \cdot \mathbf{b} + \|\mathbf{b}\|^2$

Add the two equations and the result follows.

49. Let $\mathbf{c} = \|\mathbf{b}\|\mathbf{a} + \|\mathbf{a}\|\mathbf{b}$. Then

$$\frac{\mathbf{a} \cdot \mathbf{c}}{\|\mathbf{a}\|\,\|\mathbf{c}\|} = \|\mathbf{a}\|\,\|\mathbf{b}\| + \mathbf{a} \cdot \mathbf{b} = \frac{\mathbf{b} \cdot \mathbf{c}}{\|\mathbf{b}\|\,\|\mathbf{c}\|}$$

51. Existence of decomposition:

$$\mathbf{a} = (\mathbf{a} \cdot \mathbf{u_b})\mathbf{u_b} + [\mathbf{a} - (\mathbf{a} \cdot \mathbf{u_b})\mathbf{u_b}].$$

Uniqueness of decomposition: suppose that

$$\mathbf{a} = \mathbf{a}_\| + \mathbf{a}_\perp = \mathbf{A}_\| + \mathbf{A}_\perp.$$

Then the vector $\mathbf{a}_\| - \mathbf{A}_\| = \mathbf{A}_\perp - \mathbf{a}_\perp$ is both parallel to $\mathbf{b}$ and perpendicular to $\mathbf{b}$. Therefore it is zero. Consequently $\mathbf{A}_\| = \mathbf{a}_\|$ and $\mathbf{A}_\perp = \mathbf{a}_\perp$.

53. Place center of sphere at the origin.

$$\overrightarrow{P_1Q} \;\cdot\; \overrightarrow{P_2Q} = (-\mathbf{a} + \mathbf{b}) \cdot (\mathbf{a} + \mathbf{b})$$

$$= -\|\mathbf{a}\|^2 + \|\mathbf{b}\|^2$$

$$= 0.$$

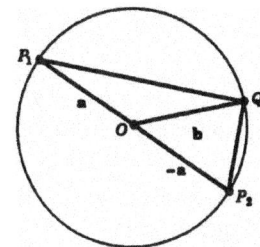

PROJECT 13.3

1. (a) $W = \mathbf{F} \cdot \mathbf{r}$ (b) 0   (c) $\|\mathbf{F}\|\,\mathbf{i} \cdot (b - a)\,\mathbf{i} = \|\mathbf{F}\|(b - a)$

3. (a) $W_1 = \mathbf{F}_1 \cdot \mathbf{r} = \|\mathbf{F}_1\|\|\mathbf{r}\|\cos\theta;\quad W_2 = \mathbf{F}_2 \cdot \mathbf{r} = \|\mathbf{F}_2\|\|\mathbf{r}\|\cos(-\theta) = \|\mathbf{F}_2\|\|\mathbf{r}\|\cos\theta$

Therefore $W_2 = \dfrac{\|\mathbf{F}_2\|}{\|\mathbf{F}_1\|}W_1$.

(b) $W_1 = \|\mathbf{F}_1\|\|\mathbf{r}\|\cos\pi/3 = \frac{1}{2}\|\mathbf{F}_1\|\|\mathbf{r}\|;\quad W_2 = \|\mathbf{F}_2\|\|\mathbf{r}\|\cos\pi/6 = \frac{1}{2}\sqrt{3}\|\mathbf{F}_2\|\|\mathbf{r}\|$

Therefore $W_2 = \sqrt{3}\dfrac{\|\mathbf{F}_2\|}{\|\mathbf{F}_1\|}W_1$.

SECTION 13.4

1. $(\mathbf{i}+\mathbf{j}) \times (\mathbf{i}-\mathbf{j}) = [\mathbf{i} \times (\mathbf{i}-\mathbf{j})] + [\mathbf{j} \times (\mathbf{i}-\mathbf{j})] = (\mathbf{0}-\mathbf{k}) + (-\mathbf{k}-\mathbf{0}) = -2\mathbf{k}$

3. $(\mathbf{i}-\mathbf{j}) \times (\mathbf{j}-\mathbf{k}) = [\mathbf{i} \times (\mathbf{j}-\mathbf{k})] - [\mathbf{j} \times (\mathbf{j}-\mathbf{k})] = (\mathbf{j}+\mathbf{k}) - (\mathbf{0}-\mathbf{i}) = \mathbf{i}+\mathbf{j}+\mathbf{k}$

5. $(2\mathbf{j}-\mathbf{k}) \times (\mathbf{i}-3\mathbf{j}) = [2\mathbf{j} \times (\mathbf{i}-3\mathbf{j})] - [\mathbf{k} \times (\mathbf{i}-3\mathbf{j})] = (-2\mathbf{k}) - (\mathbf{j}+3\mathbf{i}) = -3\mathbf{i}-\mathbf{j}-2\mathbf{k}$

or

$$(2\mathbf{j}-\mathbf{k}) \times (\mathbf{i}-3\mathbf{j}) = \begin{vmatrix} \mathbf{i} & \mathbf{j} & \mathbf{k} \\ 0 & 2 & -1 \\ 1 & -3 & 0 \end{vmatrix} = \mathbf{i}\begin{vmatrix} 2 & -1 \\ -3 & 0 \end{vmatrix} - \mathbf{j}\begin{vmatrix} 0 & -1 \\ 1 & -3 \end{vmatrix} + \mathbf{k}\begin{vmatrix} 0 & 2 \\ 1 & -3 \end{vmatrix} = -3\mathbf{i}-\mathbf{j}-2\mathbf{k}$$

7. $\mathbf{j} \cdot (\mathbf{i} \times \mathbf{k}) = \mathbf{j} \cdot (-\mathbf{j}) = -1$ **9.** $(\mathbf{i} \times \mathbf{j}) \times \mathbf{k} = \mathbf{k} \times \mathbf{k} = \mathbf{0}$ **11.** $\mathbf{j} \cdot (\mathbf{k} \times \mathbf{i}) = \mathbf{j} \cdot (\mathbf{j}) = 1$

13. $(\mathbf{i}+3\mathbf{j}-\mathbf{k}) \times (\mathbf{i}+\mathbf{k}) = \begin{vmatrix} \mathbf{i} & \mathbf{j} & \mathbf{k} \\ 1 & 3 & -1 \\ 1 & 0 & 1 \end{vmatrix} = [(3)(1)-(-1)(0)]\mathbf{i} - [(1)(1)-(-1)(1)]\mathbf{j} + [(1)0-(3)(1)]\mathbf{k}$

$$= 3\mathbf{i}-2\mathbf{j}-3\mathbf{k}$$

15. $(\mathbf{i}+\mathbf{j}+\mathbf{k}) \times (2\mathbf{i}+\mathbf{k}) = \begin{vmatrix} \mathbf{i} & \mathbf{j} & \mathbf{k} \\ 1 & 1 & 1 \\ 2 & 0 & 1 \end{vmatrix} = [(1)(1)-(1)(0)]\mathbf{i} - [(1)(1)-(1)(2)]\mathbf{j} + [(1)(0)-(1)(2)]\mathbf{k}$

$$= \mathbf{i}+\mathbf{j}-2\mathbf{k}$$

17. $[2\mathbf{i}+\mathbf{j}] \cdot [(\mathbf{i}-3\mathbf{j}+\mathbf{k}) \times (4\mathbf{i}+\mathbf{k})] = \begin{vmatrix} 1 & -3 & 1 \\ 4 & 0 & 1 \\ 2 & 1 & 0 \end{vmatrix} =$

$$[(0)(0)-(1)(1)] - (-3)[(4)(0)-(1)(2)] + [(4)(1)-(0)(2)] = -3$$

19.
$$[(\mathbf{i}-\mathbf{j}) \times (\mathbf{j}-\mathbf{k})] \times [\mathbf{i}+5\mathbf{k}] = \{[\mathbf{i} \times (\mathbf{j}-\mathbf{k})] - [\mathbf{j} \times (\mathbf{j}-\mathbf{k})]\} \times [\mathbf{i}+5\mathbf{k}]$$

$$= [(\mathbf{k}+\mathbf{j}) - (-\mathbf{i})] \times [\mathbf{i}+5\mathbf{k}]$$

$$= (\mathbf{i}+\mathbf{j}+\mathbf{k}) \times (\mathbf{i}+5\mathbf{k})$$

$$= [(\mathbf{i}+\mathbf{j}+\mathbf{k}) \times \mathbf{i}] + [(\mathbf{i}+\mathbf{j}+\mathbf{k}) \times 5\mathbf{k}]$$

$$= (-\mathbf{k}+\mathbf{j}) + (-5\mathbf{j}+5\mathbf{i})$$

$$= 5\mathbf{i}-4\mathbf{j}-\mathbf{k}$$

21. $\mathbf{a} \times \mathbf{b} = \begin{vmatrix} \mathbf{i} & \mathbf{j} & \mathbf{k} \\ 1 & 3 & -1 \\ 2 & 0 & 1 \end{vmatrix} = 3\mathbf{i} - 3\mathbf{j} - 6\mathbf{k}$

$$\frac{\mathbf{a} \times \mathbf{b}}{\|\mathbf{a} \times \mathbf{b}\|} = \frac{1}{\sqrt{6}}\mathbf{i} - \frac{1}{\sqrt{6}}\mathbf{j} - \frac{2}{\sqrt{6}}\mathbf{k}; \quad \frac{\mathbf{b} \times \mathbf{a}}{\|\mathbf{b} \times \mathbf{a}\|} = -\frac{1}{\sqrt{6}}\mathbf{i} + \frac{1}{\sqrt{6}}\mathbf{j} + \frac{2}{\sqrt{6}}\mathbf{k}$$

23. Set $\quad \mathbf{a} = \overrightarrow{PQ} = -\mathbf{i} + 2\mathbf{k} \quad$ and $\quad \mathbf{b} = \overrightarrow{PR} = 2\mathbf{i} - \mathbf{k}.$ Then

$$\mathbf{a} \times \mathbf{b} = \begin{vmatrix} \mathbf{i} & \mathbf{j} & \mathbf{k} \\ -1 & 0 & 2 \\ 2 & 0 & -1 \end{vmatrix} = 3\mathbf{j}; \quad \frac{\mathbf{a} \times \mathbf{b}}{\|\mathbf{a} \times \mathbf{b}\|} = \mathbf{j}$$

and $A = \frac{1}{2} \| \mathbf{a} \times \mathbf{b} \| = \frac{1}{2} \| 3\mathbf{j} \| = \frac{3}{2}.$

25. Set $\quad \mathbf{a} = \overrightarrow{PQ} = \mathbf{i} + \mathbf{j} - 3\mathbf{k} \quad$ and $\quad \mathbf{b} = \overrightarrow{PR} = -\mathbf{i} + 3\mathbf{j} - \mathbf{k}.$ Then

$$\mathbf{a} \times \mathbf{b} = \begin{vmatrix} \mathbf{i} & \mathbf{j} & \mathbf{k} \\ 1 & 1 & -3 \\ -1 & 3 & -1 \end{vmatrix} = 8\mathbf{j} + 4\mathbf{j} + 4\mathbf{k}; \quad \frac{\mathbf{a} \times \mathbf{b}}{\|\mathbf{a} \times \mathbf{b}\|} = \frac{2}{\sqrt{6}}\mathbf{i} + \frac{1}{\sqrt{6}}\mathbf{j} + \frac{1}{\sqrt{6}}\mathbf{k}$$

and $A = \frac{1}{2} \| \mathbf{a} \times \mathbf{b} \| = \frac{1}{2} \| 8\mathbf{i} + 4\mathbf{j} + 4\mathbf{k} \| = \frac{1}{2}\sqrt{8^2 + 4^2 + 4^2} = 2\sqrt{6}.$

27. $V = \left| [(\mathbf{i} + \mathbf{j}) \times (2\mathbf{i} - \mathbf{k})] \cdot (3\mathbf{j} + \mathbf{k}) \right| = \left| (-\mathbf{i} + \mathbf{j} - 2\mathbf{k}) \cdot (3\mathbf{j} + \mathbf{k}) \right| = 1$

29. $V = \left| \overrightarrow{OP} \cdot \left(\overrightarrow{OQ} \times \overrightarrow{OR} \right) \right| = \left\| \begin{vmatrix} 1 & 2 & 3 \\ 1 & 1 & 2 \\ 2 & 1 & 1 \end{vmatrix} \right\| = 2$

31. $$(\mathbf{a} + \mathbf{b}) \times (\mathbf{a} - \mathbf{b}) = [\mathbf{a} \times (\mathbf{a} - \mathbf{b})] + [\mathbf{b} \times (\mathbf{a} - \mathbf{b})]$$

$$= [\mathbf{a} \times (-\mathbf{b})] + [\mathbf{b} \times \mathbf{a}]$$

$$= -(\mathbf{a} \times \mathbf{b}) - (\mathbf{a} \times \mathbf{b}) = -2(\mathbf{a} \times \mathbf{b})$$

33. $\mathbf{a} \times \mathbf{i} = \mathbf{0}, \quad \mathbf{a} \times \mathbf{j} = \mathbf{0} \quad \Longrightarrow \quad \mathbf{a} \| \mathbf{i} \quad$ and $\quad \mathbf{a} \| \mathbf{j} \quad \Longrightarrow \quad \mathbf{a} = \mathbf{0}$

35. By (13.4.4) $\alpha \mathbf{a} \times \beta \mathbf{b} = (\alpha\beta)\mathbf{a} \times \mathbf{b}.$ Therefore, $\| \alpha \mathbf{a} \times \beta \mathbf{b} \| = \| (\alpha\beta)\mathbf{a} \times \mathbf{b} \|.$

37. (a) $\mathbf{a} \cdot (\mathbf{b} \times \mathbf{c})$: makes sense – this is the dot product of two vectors.

     (b) $\mathbf{a} \times (\mathbf{b} \cdot \mathbf{c})$: does not make sense – this is the cross product of a vector with a number.

     (c) $\mathbf{a} \cdot (\mathbf{b} \cdot \mathbf{c})$: does not make sense – this is the dot product of a vector with a number.

     (d) $\mathbf{a} \times (\mathbf{b} \times \mathbf{c})$: makes sense – this is the cross product of two vectors.

39. $\mathbf{d} \cdot \mathbf{a} = \mathbf{d} \cdot \mathbf{b} \Longrightarrow \mathbf{d} \perp (\mathbf{a} - \mathbf{b}); \quad \mathbf{d} \cdot \mathbf{a} = \mathbf{d} \cdot \mathbf{c} \Longrightarrow \mathbf{d} \perp (\mathbf{a} - \mathbf{c})$

Therefore, $\mathbf{d} = \lambda[(\mathbf{a} - \mathbf{b}) \times (\mathbf{a} - \mathbf{c})]$ for some number λ.

41. $\mathbf{a} \cdot \mathbf{b} = \mathbf{a} \cdot \mathbf{c} \implies \mathbf{a} \cdot (\mathbf{b} - \mathbf{c}) = 0; \quad \mathbf{a}$ is perpendicular to $\mathbf{b} - \mathbf{c}$.

$\mathbf{a} \times \mathbf{b} = \mathbf{a} \times \mathbf{c} \implies \mathbf{a} \times (\mathbf{b} - \mathbf{c}) = \mathbf{0}; \quad \mathbf{a}$ is parallel to $\mathbf{b} - \mathbf{c}$.

Since $\mathbf{a} \neq \mathbf{0}$ it follows that $\mathbf{b} - \mathbf{c} = \mathbf{0}$ or $\mathbf{b} = \mathbf{c}$.

43. $\mathbf{c} \times \mathbf{a} = (\mathbf{a} \times \mathbf{b}) \times \mathbf{a} = (\mathbf{a} \cdot \mathbf{a})\mathbf{b} - (\mathbf{a} \cdot \mathbf{b})\mathbf{a} = (\mathbf{a} \cdot \mathbf{a})\mathbf{b} = \|\mathbf{a}\|^2 \mathbf{b}$

Exercise 42(a) $\mathbf{a} \cdot \mathbf{b} = 0$

45. Suppose $\mathbf{a} \neq \mathbf{0}$. Then

$$\mathbf{a} \cdot \mathbf{b} = 0 \Longrightarrow \mathbf{b} \perp \mathbf{a}; \quad \mathbf{a} \times \mathbf{b} = \mathbf{0} \Longrightarrow \mathbf{b} \| \mathbf{a}$$

Thus $\mathbf{b}$ is simultaneously perpendicular to, and parallel to $\mathbf{a}$. It follows that $\mathbf{b} = \mathbf{0}$.

47. The result is an immediate consequence of Exercise 46.

SECTION 13.5

1. P (when $t = 0$) and Q (when $t = -1$)

3. Take $\mathbf{r}_0 = \overrightarrow{OP} = 3\mathbf{i} + \mathbf{j}$ and $\mathbf{d} = \mathbf{k}$. Then, $\mathbf{r}(t) = (3\mathbf{i} + \mathbf{j}) + t\mathbf{k}$.

5. Take $\mathbf{r}_0 = \mathbf{0}$ and $\mathbf{d} = \overrightarrow{OQ}$. Then, $\mathbf{r}(t) = t(x_1\mathbf{i} + y_1\mathbf{j} + z_1\mathbf{k})$.

7. $\overrightarrow{PQ} = \mathbf{i} - \mathbf{j} + \mathbf{k}$ so direction numbers are $1, -1, 1$. Using P as a point on the line, we have

$$x(t) = 1 + t, \quad y(t) = -t, \quad z(t) = 3 + t.$$

9. The line is parallel to the y-axis so we can take $0, 1, 0$ as direction numbers. Therefore

$$x(t) = 2, \quad y(t) = -2 + t, \quad z(t) = 3.$$

11. Since the line $2(x + 1) = 4(y - 3) = z$ can be written

$$\frac{x + 1}{2} = \frac{y - 3}{1} = \frac{z}{4},$$

it has direction numbers $2, 1, 4$. The line through $P(-1, 2, -3)$ with direction vector $2\mathbf{i} + \mathbf{j} + 4\mathbf{k}$ can be parameterized

$$\mathbf{r}(t) = (-\mathbf{i} + 2\mathbf{j} - 3\mathbf{k}) + t(2\mathbf{i} + \mathbf{j} + 4\mathbf{k}).$$

13. $\mathbf{r}(t) = (3\mathbf{i} + \mathbf{j} + 5\mathbf{k}) + t(\mathbf{i} - \mathbf{j} + 2\mathbf{k}) = (3 + t)\mathbf{i} + (1 - t)\mathbf{j} + (5 + 2t)\mathbf{k}$

$\mathbf{R}(u) = (\mathbf{i} + 4\mathbf{j} + 2\mathbf{k}) + u(\mathbf{j} + \mathbf{k}) = \mathbf{i} + (4 + u)\mathbf{j} + (2 + u)\mathbf{k}$

$\mathbf{d} = \mathbf{i} - \mathbf{j} + 2\mathbf{k}$ is a direction vector for l_1; $\mathbf{D} = \mathbf{j} + \mathbf{k}$ is a direction vector for l_2. Since $\mathbf{d}$ is not a multiple of $\mathbf{D}$, the lines either intersect or are skew. Setting $\mathbf{r}(t) = \mathbf{R}(u)$ we get the system of equations:

$$3 + t = 1, \quad 1 - t = 4 + u, \quad 5 + 2t = 2 + u$$

This system has the solution $t = -2$, $u = -1$. The point of intersection is: $(1, 3, 1)$.

15. $\mathbf{d} = 2\mathbf{i} + 4\mathbf{j} - \mathbf{k}$ is a direction vector for l_1; $\mathbf{D} = 2\mathbf{i} + \mathbf{j} + 2\mathbf{k}$ is a direction vector for l_2. Since $\mathbf{d}$ is not a multiple of $\mathbf{D}$, the lines either intersect or are skew. Equating coordinates, we get the system of equations:

$$3 + 2t = 3 + 2u, \quad -1 + 4t = 2 + u, \quad 2 - t = -2 + 2u$$

From the first two equations, we get $t = u = 1$. Since these values of t and u do not satisfy the third equation, the lines are skew.

17. $\mathbf{d} = -6\mathbf{i} + 9\mathbf{j} - 3\mathbf{k}$ is a direction vector for l_1; $\mathbf{D} = 2\mathbf{i} - 3\mathbf{j} + \mathbf{k}$ is a direction vector for l_2. Since $\mathbf{d} = -3\mathbf{D}$, we conclude that l_1 and l_2 are either parallel or coincident. The point $(1, 2, 0)$ lies on l_1 but does not line on l_2. Therefore, the lines are parallel.

19. $\mathbf{d} = 2\mathbf{i} + 4\mathbf{j} + 3\mathbf{k}$ is a direction vector for l_1; $\mathbf{D} = \mathbf{i} + 3\mathbf{j} + 2\mathbf{k}$ is a direction vector for l_2. Since $\mathbf{d}$ is not a multiple of $\mathbf{D}$, the lines either intersect or are skew. The system of equations

$$4 + 2t = 2 + u, \quad -5 + 4t = -1 + 3u, \quad 1 + 3t = 2u$$

does not have a solution. Therefore the lines are skew.

21. We set $\mathbf{r}_1(t) = \mathbf{r}_2(u)$ and solve for t and u:

$$\mathbf{i} + t\mathbf{j} = \mathbf{j} + u(\mathbf{i} + \mathbf{j}),$$

$$(1 - u)\mathbf{i} + (-1 - u + t)\mathbf{j} = \mathbf{0}.$$

Thus,

$$1 - u = 0 \quad \text{and} \quad -1 - u + t = 0.$$

These equations give $u = 1$, $t = 2$.   The point of intersection is $P(1, 2, 0)$. As direction vectors for the lines we can take $\mathbf{u} = \mathbf{j}$ and $\mathbf{v} = \mathbf{i} + \mathbf{j}$.   Thus

$$\cos \theta = \frac{\mathbf{u} \cdot \mathbf{v}}{\|\mathbf{u}\| \, \|\mathbf{v}\|} = \frac{1}{(1)(\sqrt{2})} = \frac{1}{2}\sqrt{2}.$$

The angle of intersection is $\frac{1}{4}\pi$ radians.

23. $\left(x_0 - \dfrac{d_1}{d_3} z_0, \quad y_0 - \dfrac{d_2}{d_3} z_0, \quad 0 \right)$

25. The lines are parallel.

27. $\mathbf{r}(t) = (2\,\mathbf{i} + 7\,\mathbf{j} - \mathbf{k}) + t(2\,\mathbf{i} - 5\,\mathbf{j} + 4\,\mathbf{k}), \quad 0 \le t \le 1$

29. Set
$$\mathbf{u} = \frac{\overrightarrow{PQ}}{\|\overrightarrow{PQ}\|} = \frac{-4\,\mathbf{i} + 2\,\mathbf{j} + 4\,\mathbf{k}}{\| - 4\,\mathbf{i} + 2\,\mathbf{j} + 4\,\mathbf{k}\|} = -\frac{2}{3}\,\mathbf{i} + \frac{1}{3}\,\mathbf{j} + \frac{2}{3}\,\mathbf{k}.$$

Then $\mathbf{r}(t) = (6\,\mathbf{i} - 5\,\mathbf{j} + \mathbf{k}) + t\mathbf{u}$ is $\overrightarrow{OP}$ at $t = 9$ and it is $\overrightarrow{OQ}$ at $t = 15$. (Check this.)

Answer: $\mathbf{u} = -\frac{2}{3}\,\mathbf{i} + \frac{1}{3}\,\mathbf{j} + \frac{2}{3}\,\mathbf{k}, \quad 9 \le t \le 15.$

31. The given line, call it l, has direction vector $2\,\mathbf{i} - 4\,\mathbf{j} + 6\,\mathbf{k}$.

If $a\,\mathbf{i} + b\,\mathbf{j} + c\,\mathbf{k}$ is a direction vector for a line perpendicular to l, then

$$(2\,\mathbf{i} - 4\,\mathbf{j} + 6\,\mathbf{k}) \cdot (a\,\mathbf{i} + b\,\mathbf{j} + c\,\mathbf{k}) = 2a - 4b + 6c = 0.$$

The lines through $P(3, -1, 8)$ perpendicular to l can be parameterized

$$X(u) = 3 + au, \quad Y(u) = -1 + bu, \quad Z(u) = 8 + cu$$

with $2a - 4b + 6c = 0.$

33. $d(P, l) = \dfrac{\|(\mathbf{i} + 2\,\mathbf{k}) \times (2\,\mathbf{i} - \mathbf{j} + 2\,\mathbf{k})\|}{\|2\,\mathbf{i} - \mathbf{j} + 2\,\mathbf{k}\|} = 1$

35. The line contains the point $P_0(1, 0, 2)$. Therefore
$$d(P, l) = \frac{\|(2\,\mathbf{j} + \mathbf{k}) \times (\mathbf{i} - 2\,\mathbf{j} + 3\,\mathbf{k})\|}{\|\mathbf{i} - 2\,\mathbf{j} + 3\,\mathbf{k}\|} = \sqrt{\frac{69}{14}} \cong 2.22$$

37. The line contains the point $P_0(2, -1, 0)$. Therefore
$$d(P, l) = \frac{\|(\mathbf{i} - \mathbf{j} - \mathbf{k}) \times (\mathbf{i} + \mathbf{j})\|}{\|\mathbf{i} + \mathbf{j}\|} = \sqrt{3} \cong 1.73.$$

39. (a) The line passes through $P(1, 1, 1)$ with direction vector $\mathbf{i} + \mathbf{j}$. Therefore
$$d(0, l) = \frac{\|(\mathbf{i} + \mathbf{j} + \mathbf{k}) \times (\mathbf{i} + \mathbf{j})\|}{\|\mathbf{i} + \mathbf{j}\|} = 1.$$

(b) The distance from the origin to the line segment is $\sqrt{3}$.

Solution. The line segment can be parameterized
$$\mathbf{r}(t) = \mathbf{i} + \mathbf{j} + \mathbf{k} + t(\mathbf{i} + \mathbf{j}), \quad t \in [0, 1].$$

This is the set of all points $P(1 + t, 1 + t, 1)$ with $t \in [0, 1]$.

The distance from the origin to such a point is
$$f(t) = \sqrt{2\,(1 + t)^2 + 1}.$$

The minimum value of this function is $f(0) = \sqrt{3}$.

Explanation. The point on the line through P and Q closest to the origin is not on the line segment $\overline{PQ}$.

41. We begin with $\mathbf{r}(t) = \mathbf{j} - 2\mathbf{k} + t(\mathbf{i} - \mathbf{j} + 3\mathbf{k})$. The scalar t_0 for which $\mathbf{r}(t_0) \perp l$ can be found by solving the equation

$$[\mathbf{j} - 2\mathbf{k} + t_0(\mathbf{i} - \mathbf{j} + 3\mathbf{k})] \cdot [\mathbf{i} - \mathbf{j} + 3\mathbf{k}] = 0.$$

This equation gives $-7 + 11t_0 = 0$ and thus $t_0 = 7/11$. Therefore

$$\mathbf{r}(t_0) = \mathbf{j} - 2\mathbf{k} + \tfrac{7}{11}(\mathbf{i} - \mathbf{j} + 3\mathbf{k}) = \tfrac{7}{11}\mathbf{i} + \tfrac{4}{11}\mathbf{j} - \tfrac{1}{11}\mathbf{k}.$$

The vectors of norm 1 parallel to $\mathbf{i} - \mathbf{j} + 3\mathbf{k}$ are

$$\pm \frac{1}{\sqrt{11}}(\mathbf{i} - \mathbf{j} + 3\mathbf{k}).$$

The standard parameterizations are

$$\mathbf{R}(t) = \frac{7}{11}\mathbf{i} + \frac{4}{11}\mathbf{j} - \frac{1}{11}\mathbf{k} \pm \frac{t}{\sqrt{11}}(\mathbf{i} - \mathbf{j} + 3\mathbf{k})$$

$$= \frac{1}{11}(7\mathbf{i} + 4\mathbf{j} - \mathbf{k}) \pm t\left[\frac{\sqrt{11}}{11}(\mathbf{i} - \mathbf{j} + 3\mathbf{k})\right].$$

43. $0 < t < s$

By similar triangles, if $0 < s < 1$, the tip of $\overrightarrow{OA} + s\,\overrightarrow{AB} + s\,\overrightarrow{BC}$ falls on $\overline{AC}$. If $0 < t < s$, then the tip of $\overrightarrow{OA} + s\,\overrightarrow{AB} + t\,\overrightarrow{BC}$ falls short of $\overline{AC}$ and stays within the triangle. Clearly all points in the interior of the triangle can be reached in this manner.

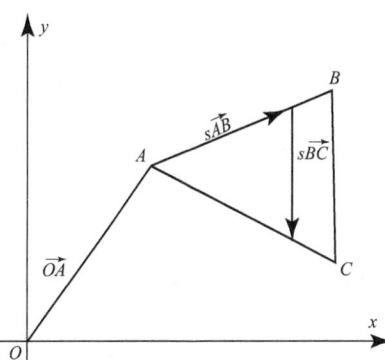

45. $\mathbf{d} = \mathbf{i} + 3\mathbf{j} - 2\mathbf{k}$ is a direction vector for l_1; $\mathbf{D} = 4\mathbf{i} - \mathbf{j} + 2\mathbf{k}$ is a direction vector for l_2. Since $\mathbf{d}$ is not a multiple of $\mathbf{D}$, the lines either intersect or are skew. Equating coordinates, we get the system of equations:

$$2 + t = -1 + 4u, \quad -1 + 3t = 2 - u, \quad 1 - 2t = -3 + 2u$$

This system does not have a solution. Therefore the lines are skew. The point $P(2, -1, 1)$ is on l_1 and the point $Q(-1, 2, -3)$ is on l_2, and $\overrightarrow{PQ} = -3\mathbf{i} + 3\mathbf{j} - 4\mathbf{k}$. By Exercise 44, the distance between l_1 and l_2 is:

$$\frac{\left|\overrightarrow{PQ} \cdot (\mathbf{d} \times \mathbf{D})\right|}{\|\mathbf{d} \times \mathbf{D}\|} = \frac{|(-3\mathbf{i} + 3\mathbf{j} - 4\mathbf{k}) \cdot (4\mathbf{i} - 10\mathbf{j} - 13\mathbf{k})|}{\sqrt{285}} = \frac{10}{\sqrt{285}}$$

SECTION 13.6

1. Q

3. Since $\mathbf{i} - 4\mathbf{j} + 3\mathbf{k}$ is normal to the plane, we have

$$(x - 2) - 4(y - 3) + 3(z - 4) = 0 \quad \text{and thus} \quad x - 4y + 3z - 2 = 0.$$

5. The vector $3\mathbf{i} - 2\mathbf{j} + 5\mathbf{k}$ is normal to the given plane and thus to every parallel plane: the equation we want can be written

$$3(x - 2) - 2(y - 1) + 5(z - 1) = 0, \quad 3x - 2y + 5z - 9 = 0.$$

7. The point $Q\,(0, 0, -2)$ lies on the line l; and $\mathbf{d} = \mathbf{i} + \mathbf{j} + \mathbf{k}$ is a direction vector for l. We want an equation for the plane which has the vector

$$\mathbf{N} = \overrightarrow{PQ} \times \mathbf{d} = (\mathbf{i} + 3\mathbf{j} + 3\mathbf{k}) \times (\mathbf{i} + \mathbf{j} + \mathbf{k})$$

as a normal vector:

$$\mathbf{N} = \begin{vmatrix} \mathbf{i} & \mathbf{j} & \mathbf{k} \\ -1 & -3 & -3 \\ 1 & 1 & 1 \end{vmatrix} = -2\mathbf{j} + 2\mathbf{k}$$

An equation for the plane is: $-2(y - 3) + 2(z - 1) = 0$ or $y - z - 2 = 0$

9. $\overrightarrow{OP_0} = x_0\,\mathbf{i} + y_0\,\mathbf{j} + z_0\,\mathbf{k}$ An equation for the plane is:

$$x_0\,(x - x_0) + y_0\,(y - y_0) + z_0\,(z - z_0) = 0$$

11. The vector $\mathbf{N} = 2\mathbf{i} - \mathbf{j} + 5\mathbf{k}$ is normal to the plane $2x - y + 5z - 10 = 0$. The unit normals are:

$$\frac{\mathbf{N}}{\|\mathbf{N}\|} = \frac{1}{\sqrt{30}}(2\mathbf{i} - \mathbf{j} + 5\mathbf{k}) \quad \text{and} \quad -\frac{\mathbf{N}}{\|\mathbf{N}\|} = -\frac{1}{\sqrt{30}}(2\mathbf{i} - \mathbf{j} + 5\mathbf{k})$$

13. Intercept form: $\dfrac{x}{15} + \dfrac{y}{12} - \dfrac{z}{10} = 1$ x-intercept: $(15, 0, 0)$

     y-intercept: $(0, 12, 0)$

     z-intercept: $(0, 0, -10)$

15. $\mathbf{u_{N_1}} = \dfrac{\sqrt{38}}{38}(5\mathbf{i} - 3\mathbf{j} + 2\mathbf{k}), \quad \mathbf{u_{N_2}} = \dfrac{\sqrt{14}}{14}(\mathbf{i} + 3\mathbf{j} + 2\mathbf{k}), \quad \cos\theta = \left| \mathbf{u_{N_1}} \cdot \mathbf{u_{N_2}} \right| = 0.$
    Therefore $\theta = \pi/2$ radians.

17. $\mathbf{u_{N_1}} = \dfrac{\sqrt{3}}{3}(\mathbf{i} - \mathbf{j} + \mathbf{k}), \quad \mathbf{u_{N_2}} = \dfrac{\sqrt{14}}{14}(2\mathbf{i} + \mathbf{j} + 3\mathbf{k}), \quad \cos\theta = \left| \mathbf{u_{N_1}} \cdot \mathbf{u_{N_2}} \right| = \dfrac{2}{21}\sqrt{42} \cong 0.617.$
    Therefore $\theta \cong 0.91$ radians.

19. coplanar since $0(4\mathbf{j} - \mathbf{k}) + 0(3\mathbf{i} + \mathbf{j} + 2\mathbf{k}) + 1(\mathbf{0}) = 0$

21. We need to determine whether there exist scalars s, t, u not all zero such that

$$s(\mathbf{i} + \mathbf{j} + \mathbf{k}) + t(2\mathbf{i} - \mathbf{j}) + u(3\mathbf{i} - \mathbf{j} - \mathbf{k}) = \mathbf{0}$$

$$(s + 2t + 3u)\mathbf{i} + (s - t - u)\mathbf{j} + (s - u)\mathbf{k} = \mathbf{0}.$$

The only solution of the system

$$s + 2t + 3u = 0, \quad s - t - u = 0, \quad s - u = 0$$

is $s = t = u = 0$.   Thus, the vectors are not coplanar.

23. By (13.6.5), $d(P, p) = \dfrac{|2(2) + 4(-1) - (3) + 1|}{\sqrt{4 + 16 + 1}} = \dfrac{2}{\sqrt{21}} = \dfrac{2}{21}\sqrt{21}.$

25. By (13.6.5), $d(P, p) = \dfrac{|(-3)(1) + 0(-3) + 4(5) + 5|}{\sqrt{9 + 16}} = \dfrac{22}{5}.$

27. $\overrightarrow{P_1P} = (x - 1)\mathbf{i} + y\mathbf{j} + (z - 1)\mathbf{k}, \quad \overrightarrow{P_1P_2} = \mathbf{i} + \mathbf{j} - \mathbf{k}, \quad \overrightarrow{P_1P_3} = \mathbf{j}.$
Therefore

$$(\overrightarrow{P_1P_2} \times \overrightarrow{P_1P_3}) = (\mathbf{i} + \mathbf{j} - \mathbf{k}) \times \mathbf{j} = \mathbf{i} + \mathbf{k}$$

and

$$\overrightarrow{P_1P} \cdot (\overrightarrow{P_1P_2} \times \overrightarrow{P_1P_3}) = [(x - 1)\mathbf{i} + y\mathbf{j} + (z - 1)\mathbf{k}] \cdot [\mathbf{i} + \mathbf{k}] = x - 1 + z - 1.$$

An equation for the plane can be written $x + z = 2$.

29. $\overrightarrow{P_1P} = (x - 3)\mathbf{i} + (y + 4)\mathbf{j} + (z - 1)\mathbf{k}, \quad \overrightarrow{P_1P_2} = 6\mathbf{j}, \quad \overrightarrow{P_1P_3} = -2\mathbf{i} + 5\mathbf{j} - 3\mathbf{k}.$
Therefore

$$(\overrightarrow{P_1P_2} \times \overrightarrow{P_1P_3}) = 6\mathbf{j} \times (-4\mathbf{i} + 5\mathbf{j} - 3\mathbf{k}) = -18\mathbf{i} + 12\mathbf{k}$$

and

$$\overrightarrow{P_1P} \cdot (\overrightarrow{P_1P_2} \times \overrightarrow{P_1P_3}) = [(x - 3)\mathbf{i} + (y + 4)\mathbf{j} + (z - 1)\mathbf{k}] \cdot [-18\mathbf{i} + 12\mathbf{k}]$$

$$= -18(x - 3) + 12(z - 1)$$

An equation for the plane can be written $-18(x - 3) + 12(z - 1) = 0$ or $3x - 2z - 7 = 0$.

31. The line passes through the point $P_0(x_0, y_0, z_0)$ with direction numbers: A, B, C.
Equations for the line written in symmetric form are:

$$\frac{x - x_0}{A} = \frac{y - y_0}{B} = \frac{z - z_0}{C}, \quad \text{provided}   A \neq 0, \ B \neq 0, \ C \neq 0.$$

33.
$$\frac{x - x_0}{d_1} = \frac{y - y_0}{d_2}, \qquad \frac{y - y_0}{d_2} = \frac{z - z_0}{d_3}$$

35. We set $x = 0$ and find that $P_0(0,0,0)$ lies on the line of intersection. As normals to the plane we use

$$\mathbf{N}_1 = \mathbf{i} + 2\mathbf{j} + 3\mathbf{k} \quad \text{and} \quad \mathbf{N}_2 = -3\mathbf{i} + 4\mathbf{j} + \mathbf{k}.$$

Note that

$$\mathbf{N}_1 \times \mathbf{N}_2 = (\mathbf{i} + 2\mathbf{j} + 3\mathbf{k}) \times (-3\mathbf{i} + 4\mathbf{j} + \mathbf{k}) = -10\mathbf{i} - 10\mathbf{j} + 10\mathbf{k}.$$

We take $-\frac{1}{10}(\mathbf{N}_1 \times \mathbf{N}_2) = \mathbf{i} + \mathbf{j} - \mathbf{k}$ as a direction vector for the line through $P_0(0,0,0)$. Then

$$x(t) = t, \quad y(t) = t, \quad z(t) = -t.$$

37. Straightforward computations give us

$$l: x(t) = 1 - 3t, \quad y(t) = -1 + 4t, \quad z(t) = 2 - t$$

and

$$p: x + 4y - z = 6.$$

Substitution of the scalar parametric equations for l in the equation for p gives

$$(1 - 3t) + 4(-1 + 4t) - (2 - t) = 6 \quad \text{and thus} \quad t = 11/14.$$

Using $t = 11/14$, we get $x = -19/14, \quad y = 15/7, \quad z = 17/14.$

39. Let $\mathbf{N} = A\mathbf{i} + B\mathbf{j} + C\mathbf{k}$ be normal to the plane. Then
$$\mathbf{N} \cdot \mathbf{d} = (\mathbf{i} + B\mathbf{j} + C\mathbf{k}) \cdot (\mathbf{i} + 2\mathbf{j} + 4\mathbf{k}) = 1 + 2B + 4C = 0$$
and
$$\mathbf{N} \cdot \mathbf{D} = (\mathbf{i} + B\mathbf{j} + C\mathbf{k}) \cdot (-\mathbf{i} - \mathbf{j} + 3\mathbf{k}) = -1 - B + 3C = 0.$$
This gives $B = -7/10$ and $C = 1/10$. The equation for the plane can be written
$$1(x - 0) - \frac{7}{10}(y - 0) + \frac{1}{10}(z - 0) = 0, \quad \text{which simplifies to} \quad 10x - 7y + z = 0.$$

41. $\mathbf{N} + \overrightarrow{PQ}$ and $\mathbf{N} - \overrightarrow{PQ}$ are the diagonals of a rectangle with sides $\mathbf{N}$ and $\overrightarrow{PQ}$. Since the diagonals are perpendicular, the rectangle is a square; that is $\|\mathbf{N}\| = \|\overrightarrow{PQ}\|$. Thus, the points Q form a circle centered at P with radius $\|\mathbf{N}\|$.

43. If $\alpha > 0$, then P_0 lies on the same side of the plane as the tip of $\mathbf{N}$; if $\alpha < 0$, then P_0 and the tip of $\mathbf{N}$ lie on opposite sides of the plane.

To see this, suppose that $\mathbf{N}$ emanates from the point $P_1(x_1, y_1, z_1)$ on the plane. Then

$$\mathbf{N} \cdot \overrightarrow{P_1 P_0} = A(x_0 - x_1) + B(y_0 - y_1) + C(z_0 - z_1) = Ax_0 + By_0 + Cz_0 + D = \alpha.$$

If $\alpha > 0$, $0 \le \sphericalangle \left(\mathbf{N}, \overrightarrow{P_0 P_1}\right) < \pi/2$; if $\alpha < 0$, then $\pi/2 < \sphericalangle \left(\mathbf{N}, \overrightarrow{P_0 P_1}\right) < \pi$. Since $\mathbf{N}$ is perpendicular to the plane, the result follows.

45. (a) intercepts:

$(4, 0, 0), \ (0, 5, 0), \ (0, 0, 2)$

(b) traces:

in the x, y-plane: $5x + 4y = 20$

in the x, z-plane: $x + 2z = 4$

in the y, z-plane: $2y + 5z = 10$

(c) unit normals: $\pm \dfrac{1}{\sqrt{141}} \left(5\,\mathbf{i} + 4\,\mathbf{j} + 10\,\mathbf{k}\right)$

(d)

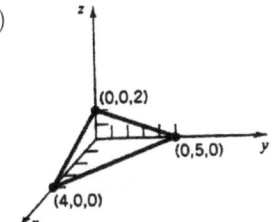

47. (a) intercepts:

$(4, 0, 0), \ $ no y-intercept, $ \ (0, 0, 6)$

(b) traces:

in the x, y-plane: $x = 4$

in the x, z-plane: $3x + 2z = 12$

in the y, z-plane: $z = 6$

(c) unit normals: $\pm \dfrac{1}{\sqrt{13}} \left(3\,\mathbf{i} + 2\,\mathbf{k}\right)$

(d)

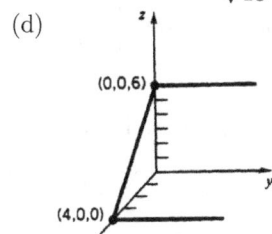

49. The normal vectors to the planes are: $\mathbf{N}_1 = 2\,\mathbf{i} + \mathbf{j} + 3\,\mathbf{k}, \ \mathbf{N}_2 = \mathbf{i} + 5\,\mathbf{j} - 2\,\mathbf{k}.$

The cosine of the angle θ between the planes is: $\cos \theta = \dfrac{|\mathbf{N}_1 \cdot \mathbf{N}_2|}{\|\mathbf{N}_1\| \, \|\mathbf{N}_2\|} = \dfrac{1}{2\sqrt{105}};$

$\theta \cong 1.52$ radians $\cong 87.2°.$

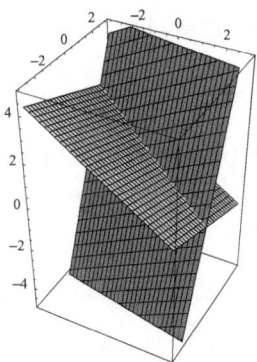

51. An equation of the plane that passes through $(2, 7, -3)$ with normal vector $\mathbf{N} = 3\,\mathbf{i} + \mathbf{j} + 4\,\mathbf{k}$ is:

$3(x - 2) + (y - 7) + 4(z + 3) = 0$ or $3x + y + 4z = 1.$

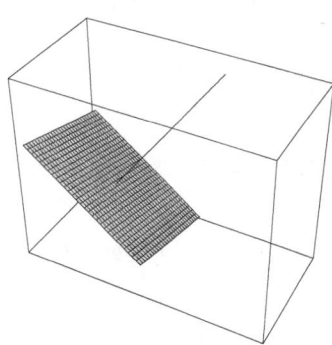

53. $\dfrac{x}{2} + \dfrac{y}{5} + \dfrac{z}{4} = 1$

$10x + 4y + 5z = 20$

55. $\dfrac{x}{3} + \dfrac{y}{5} = 1$

$5x + 3y = 15$

REVIEW EXERCISES

1. (a) $\overline{PQ} = \sqrt{(7-3)^2 + (-5-2)^2 + (-4-[-1])^2} = \sqrt{16 + 49 + 25} = 3\sqrt{10}$

(b) Midpoint of $\overline{QR}$: $\left(\dfrac{7+5}{2}, \dfrac{-5+6}{2}, \dfrac{4-3}{2}\right) = \left(6, \dfrac{1}{2}, \dfrac{1}{2}\right)$

(c) Let $X = (x, y, z)$. Then

$$(7, -5, 4) = \left(\dfrac{3+x}{2}, \dfrac{2+y}{2}, \dfrac{-1+z}{2}\right) \implies (x, y, z) = (11, -12, 9).$$

(d) Midpoint of $\overline{PR}$: $(4, 4, -2)$; radius of the sphere:

$$r = \dfrac{1}{2}|\overline{PR}| = \dfrac{1}{2}\sqrt{(5-3)^2 + (6-2)^2 + (-3+1)^2} = \sqrt{6}.$$

The equation of the sphere is: $(x-4)^2 + (y-4)^2 + (z+2)^2 = 6$.

3. radius: $\sqrt{2^2 + (-3)^2 + 1^2} = \sqrt{14}$

equation: $(x-2)^2 + (y+3)^2 + (z-1)^2 = 14$

5. By completing the square, the equation can be written as

$$(x+1)^2 + (y+2)^2 + (z-4)^2 = 4 = 2^2.$$

center: $(-1, -2, 4)$. radius: 2

7. $\frac{3}{2}\mathbf{i} + \mathbf{j} - \frac{1}{2}\mathbf{k}$

9. $\mathbf{b} + \mathbf{c} = 3\mathbf{i} + 7\mathbf{j} + \mathbf{k}$, $\mathbf{a} \cdot (\mathbf{b} + \mathbf{c}) = (3\mathbf{i} + 2\mathbf{j} - \mathbf{k}) \cdot (3\mathbf{i} + 7\mathbf{j} + \mathbf{k}) = 22$

9. $\|\mathbf{c}\|^2 = (-2)^2 + 4^2 + 1^2 = 21$

13. $2\mathbf{a} - \mathbf{b} = \mathbf{i} + \mathbf{j} - 2\mathbf{k}$; $(2\mathbf{a} - \mathbf{b}) \cdot \mathbf{c} = (\mathbf{i} + \mathbf{j} - 2\mathbf{k}) \cdot (-2\mathbf{i} + 4\mathbf{j} + \mathbf{k}) = 0$

15. $\|\mathbf{a}\| = \sqrt{14}$; $\mathbf{u_a} = \dfrac{1}{\sqrt{14}}(3\mathbf{i} + 2\mathbf{j} - \mathbf{k})$

17. $\cos\theta = \dfrac{\mathbf{a} \cdot \mathbf{c}}{\|\mathbf{a}\|\|\mathbf{c}\|} = \dfrac{1}{\sqrt{14}\sqrt{21}} = \dfrac{1}{7\sqrt{6}}$; $\theta \cong 1.51$ radians

19. $\|\mathbf{a}\| = \sqrt{14}$; $\cos\alpha = \dfrac{3}{\sqrt{14}}$, $\alpha \cong 0.64$ radians, $\cos\beta = \dfrac{2}{\sqrt{14}}$, $\beta \cong 1.01$, $\cos\gamma = \dfrac{-1}{\sqrt{14}}$, $\gamma \approx 1.84$

21. $\mathbf{b} \times \mathbf{c} = \begin{vmatrix} \mathbf{i} & \mathbf{j} & \mathbf{k} \\ 5 & 3 & 0 \\ -2 & 4 & 1 \end{vmatrix} = 3\,\mathbf{i} - 5\,\mathbf{j} + 26\,\mathbf{k};$

$\operatorname{comp}_{\mathbf{a}}(\mathbf{b} \times \mathbf{c}) = (\mathbf{b} \times \mathbf{c}) \cdot \mathbf{u_a} = (3\,\mathbf{i} - 5\,\mathbf{j} + 26\,\mathbf{k}) \cdot \dfrac{1}{\sqrt{14}}(3\,\mathbf{i} + 2\,\mathbf{j} - \mathbf{k}) = -\dfrac{27}{\sqrt{14}}$

23. $V = |(\mathbf{a} \times \mathbf{b}) \cdot \mathbf{c}|; \quad (\mathbf{a} \times \mathbf{b}) \cdot \mathbf{c} = \begin{vmatrix} 3 & 2 & -1 \\ 5 & 3 & 0 \\ -2 & 4 & 1 \end{vmatrix} = -27; \quad V = |-27| = 27$

25. (a) Direction vector: $\overrightarrow{QR} = (6, -3, 3);$ scalar parametric equations for the line:

$$x = 1 + 6t, \quad y = 1 - 3t, \quad z = 1 + 3t.$$

(b) Normal vector: $\overrightarrow{PR} = (3, -3, 2);$ equation of the plane:

$$3(x - 1) + (-3)(y - 1) + 2(z - 1) = 0.$$

(c) A normal vector for the plane is: $\overrightarrow{QR} \times \overrightarrow{PR} = (3, -3, -9)$ or $\mathbf{N} = (1, -1, -3);$

an equation for the plane: $\quad (x - 1) - (y - 1) - 3(z - 1) = 0$

27. Solve, if possible, the system of equations: $t = 1 - u, \quad -t = 1 + 3u, \quad -6 + 2t = 2u.$ In this case, the solution is $t = 2, u = -1.$ The lines intersect at the point $(2, -2, -2).$

29. The lines l_1 and l_2 written in scalar parametric form are:

$$l_1 : x = 1 + 2t, \quad y = -2 - t, \quad z = 3 + 4t; \quad l_2 : x = -2 + u, \quad y = 3 + 3u, \quad z = u.$$

Solve, if possible, the system of equations: $1 + 2t = -2 + u, \quad -2 - t = 3 + 3u, \quad 3 + 4t = u.$ In this case there is no solution; the lines are skew.

31. (a) No. $\overrightarrow{PQ} = (4, -7, 5), \quad \overrightarrow{PR} = (2, -3, 2);$ the vectors are not parallel; the points are not collinear.

(b) $\overrightarrow{PQ} = (4, -7, 5), \quad \overrightarrow{PR} = (2, -3, 2), \quad \overrightarrow{PS} = (-2, 0, 1)$

$$(\overrightarrow{PQ} \times \overrightarrow{PR}) \cdot \overrightarrow{PS} = \begin{vmatrix} 4 & -7 & 5 \\ 2 & -3 & 2 \\ -2 & 0 & 1 \end{vmatrix} = 0.$$

The points are coplanar.

33. $\overrightarrow{PQ} \times \overrightarrow{PR} = -10\,\mathbf{i} + 5\,\mathbf{k}$ is a normal vector for the plane; so is $\mathbf{N} = 2\,\mathbf{i} - \mathbf{k}.$

An equation for the plane is: $2(x - 1) - (z - 1) = 0$ or $2x - z = 1$

35. $\mathbf{N} = 3\,\mathbf{i} + 2\,\mathbf{j} - 1\,\mathbf{k}$ is a normal vector for the plane. An equation for the plane is:

$$3(x - 1) + 2(y + 2) - (z + 1) = 0 \quad \text{or} \quad 3x + 2y - z = 0.$$

37. Let P be the plane that satisfies the conditions. A direction vector for the given line is $\mathbf{d} = (3, 2, 4)$; a normal vector for the given plane is $\mathbf{N} = (2, 1, -3)$. The cross product $\mathbf{d} \times \mathbf{N}$ is a normal vector for P.

$$\mathbf{d} \times \mathbf{N} = \begin{vmatrix} \mathbf{i} & \mathbf{j} & \mathbf{k} \\ 3 & 2 & 4 \\ 2 & 1 & -3 \end{vmatrix} = -10\,\mathbf{i} + 17\,\mathbf{j} - \mathbf{k}.$$

The point $Q\,(-1, 1, 2)$ is on the plane. An equation for P is:

$$-10(x + 1) + 17(y - 1) - (z - 2) = 0 \quad \text{or} \quad 10x - 17y + z + 25 = 0.$$

39. The line l which passes through Q and R has direction vector $\mathbf{d} = \overrightarrow{QR} = (2, 1, -2)$. By (13.5.6), the distance from P to l is given by

$$d(P, l) = \frac{\|\overrightarrow{QP} \times \mathbf{d}\|}{\|\mathbf{d}\|} = \frac{\|(2, 4, -5) \times (2, 1, -2)\|}{3} = \frac{9}{3} = 3.$$

41. The normals are: $\mathbf{N}_1 = (2, 1, 1)$, $\mathbf{N}_2 = (2, 2, -1)$. The cosine of the angle between the planes is:

$$\cos \theta = \frac{|\mathbf{N}_1 \cdot \mathbf{N}_2|}{\|\mathbf{N}_1\|\|\mathbf{N}_2\|} = \frac{5}{\sqrt{54}} \quad \text{and} \quad \theta \cong 0.822 \text{ radians.}$$

43. The normal vectors to the two planes are: $\mathbf{N}_1 = 3\,\mathbf{i} + 5\,\mathbf{j} + 2\,\mathbf{k}$, $\mathbf{N}_2 = \mathbf{i} + 2\,\mathbf{j} - \mathbf{k}$. A direction vector for the line of intersection is:

$$\mathbf{N}_1 \times \mathbf{N}_2 = \begin{vmatrix} \mathbf{i} & \mathbf{j} & \mathbf{k} \\ 3 & 5 & 2 \\ 1 & 2 & -1 \end{vmatrix} = -9\,\mathbf{i} + 5\,\mathbf{j} + \mathbf{k}.$$

A solution of the pair of equations $3x + 5y + 2z - 4 = 0 \quad x + 2y - z - 2 = 0$ is $x = -2$, $y = 2$, $z = 0$ (set $z = 0$ and solve for x and y). Scalar parametric equations for the line of intersection are:

$$x = -2 - 9t, \quad y = 2 + 5t, \quad z = t.$$

45. $\mathbf{a} \times \mathbf{b} = -5\,\mathbf{i} + 11\,\mathbf{j} + 7\,\mathbf{k}$ is perpendicular to both $\mathbf{a}$ and $\mathbf{b}$; $\|\mathbf{a} \times \mathbf{b}\| = \sqrt{195}$. The vectors are:

$$\pm \frac{4}{\sqrt{195}}(-5\,\mathbf{i} + 11\,\mathbf{j} + 7\,\mathbf{k}).$$

47. $(\|\mathbf{b}\|\mathbf{a} - \|\mathbf{a}\|\mathbf{b}) \cdot (\|\mathbf{b}\|\mathbf{a} + \|\mathbf{a}\|\mathbf{b}) = \|\mathbf{a}\|^2\|\mathbf{b}\|^2 + \|\mathbf{a}\|\|\mathbf{b}\|\mathbf{a} \cdot \mathbf{b} - \|\mathbf{a}\|\|\mathbf{b}\|\mathbf{a} \cdot \mathbf{b} - \|\mathbf{a}\|^2\|\mathbf{b}\|^2 = 0.$
Therefore, $(\|\mathbf{b}\|\mathbf{a} - \|\mathbf{a}\|\mathbf{b}) \perp (\|\mathbf{b}\|\mathbf{a} + \|\mathbf{a}\|\mathbf{b})$

49. Let $\mathbf{a}$ and $\mathbf{b}$ be adjacent sides of a parallelogram. Then the diagonals of the parallelogram are $\mathbf{a} + \mathbf{b}$ and $\mathbf{a} - \mathbf{b}$. By Exercise 48, the diagonals have equal length iff $\mathbf{a} \perp \mathbf{b}$, which means that the parallelogram is a rectangle.

51. Let A, B, C be the vertices of a triangle. Without loss of generality, assume that $A(0,0)$, $B(x_1, y_1)$, $C(x_2, 0)$. Let D and E be the midpoints of $\overline{AB}$ and $\overline{BC}$, respectively. Then $D\left(\dfrac{x_1}{2}, \dfrac{y_1}{2}\right)$ and $E\left(\dfrac{x_1 + x_2}{2}, \dfrac{y_1}{2}\right)$. Now

$$\overrightarrow{DE} = \left(\frac{x_2}{2}, 0\right), \quad \text{and} \quad \overrightarrow{AC} = (x_2, 0).$$

Therefore $\overrightarrow{DE} \| \overrightarrow{AC}$ and $\| \overrightarrow{DE} \| = \frac{1}{2} \| \overrightarrow{AC} \|$.

CHAPTER 14

SECTION 14.1

1. $\mathbf{f}'(t) = 2\mathbf{i} - \mathbf{j} + 3\mathbf{k}$

3. $\mathbf{f}'(t) = -\dfrac{1}{2\sqrt{1-t}}\,\mathbf{i}\dfrac{1}{2\sqrt{1+t}}\,\mathbf{j} + \dfrac{1}{(1-t)^2}\,\mathbf{k}$

5. $\mathbf{f}'(t) = \cos t\,\mathbf{i} - \sin t\,\mathbf{j} + \sec^2 t\,\mathbf{k}$

7. $\mathbf{f}'(t) = \dfrac{-1}{1-t}\,\mathbf{i} - \sin t\,\mathbf{j} + 2t\,\mathbf{k}$

9. $\mathbf{f}'(t) = 4\mathbf{i} + 6t^2\,\mathbf{j} + (2t+2)\,\mathbf{k};\qquad \mathbf{f}''(t) = 12t\,\mathbf{j} + 2\,\mathbf{k}$

11. $\mathbf{f}'(t) = -2\sin 2t\,\mathbf{i} + 2\cos 2t\,\mathbf{j} + 4t\,\mathbf{k};\quad \mathbf{f}''(t) = -4\cos 2t\,\mathbf{i} - 4\sin 2t\,\mathbf{j}$

13. (a) $\mathbf{r}'(t) = -2te^{-t^2}\,\mathbf{i} - e^{-t}\,\mathbf{j};\quad \mathbf{r}'(0) = -\mathbf{j}$

    (b) $\mathbf{r}'(t) = \cot t\,\mathbf{i} - \tan t\,\mathbf{j} + (2\cos t + 3\sin t)\,\mathbf{k};\quad \mathbf{r}'(\pi/4) = \mathbf{i} - \mathbf{j} + \dfrac{5}{\sqrt{2}}\,\mathbf{k}$

15. $\displaystyle\int_1^2 (\mathbf{i} + 2t\,\mathbf{j})\,dt = \left[t\,\mathbf{i} + t^2\,\mathbf{j}\right]_1^2 = \mathbf{i} + 3\,\mathbf{j}$

17. $\displaystyle\int_0^1 (e^t\,\mathbf{i} + e^{-t}\,\mathbf{k})\,dt = \left[e^t\,\mathbf{i} - e^{-t}\,\mathbf{k}\right]_0^1 = (e-1)\,\mathbf{i} + \left(1 - \dfrac{1}{e}\right)\mathbf{k}$

19. $\displaystyle\int_0^1 \left(\dfrac{1}{1+t^2}\,\mathbf{i} + \sec^2 t\,\mathbf{j}\right)dt = \left[\tan^{-1} t\,\mathbf{i} + \tan t\,\mathbf{j}\right]_0^1 = \dfrac{\pi}{4}\,\mathbf{i} + \tan(1)\,\mathbf{j}$

21. $\displaystyle\lim_{t\to 0}\mathbf{f}(t) = \left(\lim_{t\to 0}\dfrac{\sin t}{2t}\right)\mathbf{i} + \left(\lim_{t\to 0} e^{2t}\right)\mathbf{j} + \left(\lim_{t\to 0}\dfrac{t^2}{e^t}\right)\mathbf{k} = \dfrac{1}{2}\,\mathbf{i} + \mathbf{j}$

23. $\displaystyle\lim_{t\to 0}\mathbf{f}(t) = \left(\lim_{t\to 0} t^2\right)\mathbf{i} + \left(\lim_{t\to 0}\dfrac{1-\cos t}{3t}\right)\mathbf{j} + \left(\lim_{t\to 0}\dfrac{t}{t+1}\right)\mathbf{k} = 0\mathbf{i} + \dfrac{1}{3}\left(\lim_{t\to 0}\dfrac{1-\cos t}{t}\right)\mathbf{j} + 0\mathbf{k} = \mathbf{0}$

25. (a) $\displaystyle\int_0^1 \left(te^t\,\mathbf{i} + te^{t^2}\,\mathbf{j}\right)dt = \mathbf{i} + \dfrac{e-1}{2}\,\mathbf{j}$

    (b) $\displaystyle\int_3^8 \left(\dfrac{t}{t+1}\,\mathbf{i} + \dfrac{t}{(t+1)^2}\,\mathbf{j} + \dfrac{t}{(t+1)^3}\,\mathbf{k}\right)dt = \left[5 + \ln\left(\tfrac{4}{9}\right)\right]\mathbf{i} + \left[-\tfrac{5}{36} + \ln\left(\tfrac{4}{9}\right)\right]\mathbf{j} + \tfrac{295}{2592}\,\mathbf{k}$

27.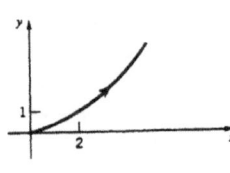

29.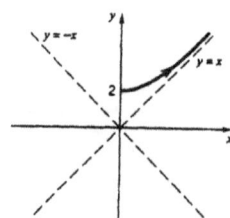

31.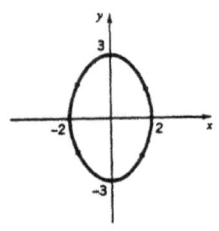

33. 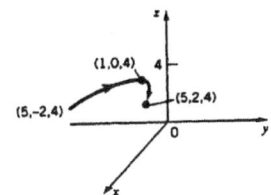 **35.** 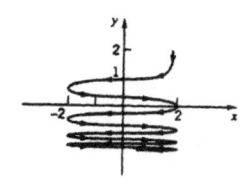 **37.**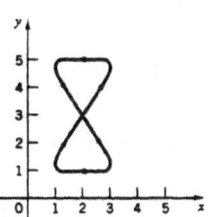

39. (a) $\mathbf{f}(t) = 3\cos t\,\mathbf{i} + 2\sin t\,\mathbf{j}$ (b) $\mathbf{f}(t) = 3\cos t\,\mathbf{i} - 2\sin t\,\mathbf{j}$

41. (a) $\mathbf{f}(t) = t\,\mathbf{i} + t^2\,\mathbf{j}$ (b) $\mathbf{f}(t) = -t\,\mathbf{i} + t^2\,\mathbf{j}$

43. $\mathbf{f}(t) = (1 + 2t)\,\mathbf{i} + (4 + 5t)\,\mathbf{j} + (-2 + 8t)\,\mathbf{k}, \quad 0 \le t \le 1$

45. $\mathbf{f}'(t_0) = \mathbf{i} + m\,\mathbf{j},$

$$\int_a^b \mathbf{f}(t)\,dt = \left[\frac{1}{2}t^2\,\mathbf{i}\right]_a^b + \left[\int_a^b f(t)\,dt\right]\mathbf{j} = \frac{1}{2}\left(b^2 - a^2\right)\mathbf{i} + A\,\mathbf{j},$$

$$\int_a^b \mathbf{f}'(t)\,dt = [t\,\mathbf{i} + f(t)\,\mathbf{j}]_a^b = (b - a)\,\mathbf{i} + (d - c)\,\mathbf{j}$$

47.
$$\mathbf{f}'(t) = \mathbf{i} + t^2\,\mathbf{j}$$
$$\mathbf{f}(t) = (t + C_1)\,\mathbf{i} + \left(\tfrac{1}{3}t^3 + C_2\right)\mathbf{j} + C_3\,\mathbf{k}$$
$$\mathbf{f}(0) = \mathbf{j} - \mathbf{k} \implies C_1 = 0, \quad C_2 = 1, \quad C_3 = -1$$
$$\mathbf{f}(t) = t\,\mathbf{i} + \left(\tfrac{1}{3}t^3 + 1\right)\mathbf{j} - \mathbf{k}$$

49. $\mathbf{f}'(t) = \alpha\,\mathbf{f}(t) \implies \mathbf{f}(t) = e^{\alpha t}\,\mathbf{f}(0) = e^{\alpha t}\,\mathbf{c}$

51. (a) If $\mathbf{f}'(t) = \mathbf{0}$ on an interval, then the derivative of each component is 0 on that interval, each component is constant on that interval, and therefore $\mathbf{f}$ itself is constant on that interval.

    (b) Set $\mathbf{h}(t) = \mathbf{f}(t) - \mathbf{g}(t)$ and apply part (a).

53. If $\mathbf{f}$ is differentiable at t, then each component is differentiable at t, each component is continuous at t, and therefore $\mathbf{f}$ is continuous at t.

55. no; as a counter-example, set $\mathbf{f}(t) = \mathbf{i} = \mathbf{g}(t)$.

57. Suppose $\mathbf{f}(t) = f_1(t)\,\mathbf{i} + f_2(t)\,\mathbf{j} + f_3(t)\,\mathbf{k}$. Then $\|\mathbf{f}(t)\| = \sqrt{f_1^2(t) + f_2^2(t) + f_3^2(t)}$ and

$$\frac{d}{dt}\left(\|\mathbf{f}\|\right) = \frac{1}{2}\left[f_1^2 + f_2^2 + f_3^2\right]^{-1/2}\left(2f_1 \cdot f_1' + 2f_2 \cdot f_2' + 2f_3 \cdot f_3'\right) = \frac{\mathbf{f}(t) \cdot \mathbf{f}'(t)}{\|\mathbf{f}(t)\|}$$

The Answer Section of the text gives an alternative approach.

SECTION 14.2

1. $\mathbf{f}'t = \mathbf{b}, \quad \mathbf{f}''(t) = \mathbf{0}$

3. $\mathbf{f}'(t) = 2e^{2t}\,\mathbf{i} - \cos t\,\mathbf{j}, \quad \mathbf{f}''(t) = 4e^{2t}\,\mathbf{i} + \sin t\,\mathbf{j}$

5. $\mathbf{f}'t = [(t^2\,\mathbf{i} - 2t\,\mathbf{j}) \cdot (\mathbf{i} + 3t^2\,\mathbf{j}) + (2t\,\mathbf{i} - 2\,\mathbf{j}) \cdot (t\,\mathbf{i} + t^3\,\mathbf{j}]\,\mathbf{j} = [3t^2 - 8t^3]\,\mathbf{j}$
    $\mathbf{f}''(t) = (6t - 24t^2)\,\mathbf{j}$

7.
$$\mathbf{f}'(t) = \left[(e^t\,\mathbf{i} + t\,\mathbf{k}) \times \frac{d}{dt}\,(t\,\mathbf{j} + e^{-t}\,\mathbf{k}) \right] + \left[\frac{d}{dt}\,(e^t\,\mathbf{i} + t\,\mathbf{k}) \times (t\,\mathbf{j} + e^{-t}\,\mathbf{k}) \right]$$
$$= [(e^t\,\mathbf{i} + t\,\mathbf{k}) \times (\mathbf{j} - e^{-t}\,\mathbf{k})] + [(e^t\,\mathbf{i} + \mathbf{k}) \times (t\,\mathbf{j} + e^{-t}\,\mathbf{k})]$$
$$= (-t\,\mathbf{i} + \mathbf{j} + e^t\,\mathbf{k}) + (-t\,\mathbf{i} - \mathbf{j} + te^t\,\mathbf{k})$$
$$= -2t\,\mathbf{i} + e^t(t + 1)\,\mathbf{k}$$
$$\mathbf{f}''(t) = -2\,\mathbf{i} + e^t(t + 2)\,\mathbf{k}$$

9. $\mathbf{f}'(t) = (\mathbf{a} \times t\,\mathbf{b}) \times 2t\,\mathbf{b} + (\mathbf{a} \times \mathbf{b}) \times (\mathbf{a} + t^2\,\mathbf{b}), \qquad \mathbf{f}''(t) = (\mathbf{a} \times t\,\mathbf{b}) \times 2\mathbf{b} + 4t(\mathbf{a} \times \mathbf{b}) \times \mathbf{b}$

11. $\mathbf{f}'(t) = \dfrac{1}{2}\,\sqrt{t}\,\mathbf{g}'\left(\sqrt{t}\right) + \mathbf{g}\left(\sqrt{t}\right), \quad \mathbf{f}''(t) = \dfrac{1}{4}\,\mathbf{g}''\left(\sqrt{t}\right) + \dfrac{3}{4\sqrt{t}}\,\mathbf{g}'\left(\sqrt{t}\right)$

13. $-(\sin t)\,e^{\cos t}\,\mathbf{i} + (\cos t)\,e^{\sin t}\,\mathbf{j}$

15. $(e^t\,\mathbf{i} + e^{-t}\,\mathbf{j}) \cdot (e^t\,\mathbf{i} - e^{-t}\,\mathbf{j}) = e^{2t} - e^{-2t};$ therefore
$$\frac{d^2}{dt^2}\,[(e^t\,\mathbf{i} + e^{-t}\,\mathbf{j}) \cdot (e^t\,\mathbf{i} - e^{-t}\,\mathbf{j})] = \frac{d^2}{dt^2}\,[e^{2t} - e^{-2t}] = \frac{d}{dt}\,[2e^{2t} + 2e^{-2t}] = 4e^{2t} - 4e^{-2t}$$

17. $\dfrac{d}{dt}\,[(a + t\,\mathbf{b}) \times (c + t\,\mathbf{d})] = [(a + t\,\mathbf{b}) \times \mathbf{d}] + [\mathbf{b} \times (c + t\,\mathbf{d})] = (\mathbf{a} \times \mathbf{d}) + (\mathbf{b} \times \mathbf{c}) + 2t\,(\mathbf{b} \times \mathbf{d})$

19. $\dfrac{d}{dt}\,[(a + t\,\mathbf{b}) \cdot (c + t\,\mathbf{d})] = [(a + t\,\mathbf{b}) \cdot \mathbf{d}] + [\mathbf{b} \cdot (c + t\,\mathbf{d})] = (\mathbf{a} \cdot \mathbf{d}) + (\mathbf{b} \cdot \mathbf{c}) + 2t\,(\mathbf{b} \cdot \mathbf{d})$

21. $\mathbf{r}(t) = \mathbf{a} + t\,\mathbf{b}$

23. $\mathbf{r}(t) = \dfrac{1}{2}t^2\,\mathbf{a} + \dfrac{1}{6}t^3\,\mathbf{b} + t\,\mathbf{c} + \mathbf{d}$ 24. $\mathbf{r}(t) = \left(1 + 2t - \dfrac{1}{4}\cos 2t\right)\,\mathbf{i} + \left(1 - \dfrac{1}{4}\sin 2t\right)\,\mathbf{j}$

25. $\mathbf{r}(t) = \sin t\,\mathbf{i} + \cos t\,\mathbf{j}, \quad \mathbf{r}'(t) = \cos t\,\mathbf{i} - \sin t\,\mathbf{j}, \quad \mathbf{r}''(t) = -\sin t\,\mathbf{i} - \cos t\,\mathbf{j} = -\mathbf{r}(t).$

    Thus $\mathbf{r}(t)$ and $\mathbf{r}''(t)$ are parallel, and they always point in opposite directions.

27.
$$\mathbf{r}(t) \cdot \mathbf{r}'(t) = (\cos t\,\mathbf{i} + \sin t\,\mathbf{j}) \cdot (-\sin t\,\mathbf{i} + \cos t\,\mathbf{j}) = 0$$
$$\mathbf{r}(t) \times \mathbf{r}'(t) = (\cos t\,\mathbf{i} + \sin t\,\mathbf{j}) \times (-\sin t\,\mathbf{i} + \cos t\,\mathbf{j})$$
$$= \cos^2 t\,\mathbf{k} + \sin^2 t\,\mathbf{k} = (\cos^2 t + \sin^2 t)\,\mathbf{k} = \mathbf{k}$$

29. $\dfrac{d}{dt}\left[\mathbf{f}\left(t\right)\times\mathbf{f}'(t)\right]=\left[\mathbf{f}\left(t\right)\times\mathbf{f}''(t)\right]+\underbrace{\left[\mathbf{f}'(t)\times\mathbf{f}'(t)\right]}_{0}=\mathbf{f}\left(t\right)\times\mathbf{f}''(t).$

31. $\left[\mathbf{f}\cdot\mathbf{g}\times\mathbf{h}\right]'=\mathbf{f}'\cdot\left(\mathbf{g}\times\mathbf{h}\right)+\mathbf{f}\cdot\left(\mathbf{g}\times\mathbf{h}\right)'=\mathbf{f}'\cdot\left(\mathbf{g}\times\mathbf{h}\right)+\mathbf{f}\cdot\left[\mathbf{g}'\times\mathbf{h}+\mathbf{g}\times\mathbf{h}'\right]$

and the result follows.

33. $\qquad\|\mathbf{r}\left(t\right)\|$ is constant $\quad\Longleftrightarrow\quad\|\mathbf{r}\left(t\right)\|^2=\mathbf{r}\left(t\right)\cdot\mathbf{r}\left(t\right)$ is constant

$$\Longleftrightarrow\quad\dfrac{d}{dt}\left[\mathbf{r}\left(t\right)\cdot\mathbf{r}\left(t\right)\right]=2\left[\mathbf{r}\left(t\right)\cdot\mathbf{r}'(t)\right]=0\text{ identically}$$

$$\Longleftrightarrow\quad\mathbf{r}\left(t\right)\cdot\mathbf{r}'(t)=0\text{ identically}$$

35. Write

$$\dfrac{\left[\mathbf{f}\left(t+h\right)\times\mathbf{g}\left(t+h\right)\right]-\left[\mathbf{f}\left(t\right)\times\mathbf{g}\left(t\right)\right]}{h}$$

as

$$\left(\mathbf{f}\left(t+h\right)\times\left[\dfrac{\mathbf{g}\left(t+h\right)-\mathbf{g}\left(t\right)}{h}\right]\right)+\left(\left[\dfrac{\mathbf{f}\left(t+h\right)-\mathbf{f}\left(t\right)}{h}\right]\times\mathbf{g}\left(t\right)\right)$$

and take the limit as $h\to 0$. (Appeal to Theorem 13.1.3.)

SECTION 14.3

1. $\mathbf{r}'\left(t\right)=-\pi\sin\pi t\,\mathbf{i}+\pi\cos\pi t\,\mathbf{j}+\mathbf{k},\quad\mathbf{r}'(2)=\pi\,\mathbf{j}+\mathbf{k}$

$\mathbf{R}\left(u\right)=\left(\mathbf{i}+2\,\mathbf{k}\right)+u(\pi\,\mathbf{j}+\mathbf{k})$

3. $\mathbf{r}'(t)=\mathbf{b}+2t\,\mathbf{c},\quad\mathbf{r}'(-1)=\mathbf{b}-2\,\mathbf{c},\quad\mathbf{R}\left(u\right)=\left(\mathbf{a}-\mathbf{b}+\mathbf{c}\right)+u(\mathbf{b}-2\,\mathbf{c})$

5. $\mathbf{r}'(t)=4t\,\mathbf{i}-\mathbf{j}+4t\,\mathbf{k},\quad P$ is tip of $\mathbf{r}\left(1\right),\quad\mathbf{r}'(1)=4\,\mathbf{i}-\mathbf{j}+4\,\mathbf{k}$

$\mathbf{R}\left(u\right)=\left(2\,\mathbf{i}+5\,\mathbf{k}\right)+u\left(4\,\mathbf{i}-\mathbf{j}+4\,\mathbf{k}\right)$

7. $\mathbf{r}'(t)=-2\,\sin t\,\mathbf{i}+3\,\cos t\,\mathbf{j}+\mathbf{k},\quad\mathbf{r}'(\pi/4)=-\sqrt{2}\,\mathbf{i}+\frac{3}{2}\,\sqrt{2}\,\mathbf{j}+\mathbf{k}$

$\mathbf{R}\left(u\right)=\left(\sqrt{2}\,\mathbf{i}+\frac{3}{2}\,\sqrt{2}\,\mathbf{j}+\frac{\pi}{4}\,\mathbf{k}\right)+u\left(-\sqrt{2}\,\mathbf{i}+\frac{3}{2}\,\sqrt{2}\,\mathbf{j}+\mathbf{k}\right)$

9. The scalar components $x(t)=at$ and $y(t)=bt^2$ satisfy the equation

$$a^2y(t)=a^2(bt^2)=b\left(a^2t^2\right)=b\left[x(t)\right]^2$$

and generate the parabola $\quad a^2y=bx^2$.

11. $\mathbf{r}\left(t\right)=t\,\mathbf{i}+\left(1+t^2\right)\mathbf{j},\qquad\mathbf{r}'(t)=\mathbf{i}+2t\,\mathbf{j}$

(a) $\mathbf{r}\left(t\right)\perp\mathbf{r}'(t)\quad\Longrightarrow\quad\mathbf{r}\left(t\right)\cdot\mathbf{r}'(t)=\left[t\,\mathbf{i}+\left(1+t^2\right)\mathbf{j}\right]\cdot\left(\mathbf{i}+2t\,\mathbf{j}\right)$

$$=t\left(2t^2+3\right)=0\quad\Longrightarrow\quad t=0$$

$\mathbf{r}\left(t\right)$ and $\mathbf{r}'(t)$ are perpendicular at $(0,1)$.

(b) and (c) $\mathbf{r}(t) = \alpha\,\mathbf{r}'(t)$ with $\alpha \neq 0 \implies t = \alpha$ and $1 + t^2 = 2t\alpha \implies t = \pm 1$.

If $\alpha > 0$, then $t = 1$. $\mathbf{r}(t)$ and $\mathbf{r}'(t)$ have the same direction at $(1, 2)$.

If $\alpha < 0$, then $t = -1$. $\mathbf{r}(t)$ and $\mathbf{r}'(t)$ have opposite directions at $(-1, 2)$.

13. The tangent line at $t = t_0$ has the form $\mathbf{R}(u) = \mathbf{r}(t_0) + u\,\mathbf{r}'(t_0)$. If $\mathbf{r}'(t_0) = \alpha\,\mathbf{r}(t_0)$, then
$$\mathbf{R}(u) = \mathbf{r}(t_0) + u\,\alpha\,\mathbf{r}(t_0) = (1 + u\alpha)\,\mathbf{r}(t_0).$$

The tangent line passes through the origin at $u = -1/\alpha$.

15. $\mathbf{r}_1(t)$ passes through $P(0, 0, 0)$ at $t = 0$; $\quad$ $\mathbf{r}_2(u)$ passes through $P(0, 0, 0)$ at $u = -1$.

$$\mathbf{r}_1'(t) = e^t\,\mathbf{i} + 2\cos t\,\mathbf{j} + \frac{1}{t+1}\,\mathbf{k}; \quad \mathbf{r}_1'(0) = \mathbf{i} + 2\mathbf{j} + \mathbf{k}$$

$$\mathbf{r}_2'(u) = \mathbf{i} + 2u\,\mathbf{j} + 3u^2\,\mathbf{k}; \quad \mathbf{r}_2'(-1) = \mathbf{i} - 2\mathbf{j} + 3\mathbf{k}$$

$$\cos\theta = \frac{\mathbf{r}_1'(0) \cdot \mathbf{r}_2'(1)}{\|\mathbf{r}_1'(0)\|\,\|\mathbf{r}_2'(1)\|} = 0; \quad \theta = \frac{\pi}{2} \cong 1.57, \text{ or } 90°.$$

17. $\mathbf{r}_1(t) = \mathbf{r}_2(u)$ implies

$$\left.\begin{cases} e^t = u \\ 2\sin\left(t + \tfrac{1}{2}\pi\right) = 2 \\ t^2 - 2 = u^2 - 3 \end{cases}\right\} \quad \text{so that} \quad t = 0, \quad u = 1.$$

The point of intersection is $(1, 2, -2)$.

$$\mathbf{r}_1'(t) = e^t\,\mathbf{i} + 2\cos\left(t + \frac{\pi}{2}\right)\mathbf{j} + 2t\,\mathbf{k}, \quad \mathbf{r}_1'(0) = \mathbf{i}$$

$$\mathbf{r}_2'(u) = \mathbf{i} + 2u\,\mathbf{k}, \quad \mathbf{r}_2'(1) = \mathbf{i} + 2\,\mathbf{k}$$

$$\cos\theta = \frac{\mathbf{r}_1'(0) \cdot \mathbf{r}_2'(1)}{\|\mathbf{r}_1'(0)\|\,\|\mathbf{r}_2'(1)\|} = \frac{1}{5}\sqrt{5} \cong 0.447, \quad \theta \cong 1.11 \text{ radians}$$

19. (a) $\mathbf{r}(t) = a\cos t\,\mathbf{i} + b\sin t\,\mathbf{j}$ $\qquad\qquad$ (b) $\mathbf{r}(t) = a\cos t\,\mathbf{i} - b\sin t\,\mathbf{j}$

(c) $\mathbf{r}(t) = a\cos 2t\,\mathbf{i} + b\sin 2t\,\mathbf{j}$ $\qquad\qquad$ (d) $\mathbf{r}(t) = a\cos 3t\,\mathbf{i} - b\sin 3t\,\mathbf{j}$

21. $\mathbf{r}'(t) = t^3\,\mathbf{i} + 2t\,\mathbf{j}$ $\qquad$ **23.** $\mathbf{r}'(t) = 2e^{2t}\,\mathbf{i} - 4e^{-4t}\,\mathbf{j}$ $\qquad$ **25.** $\mathbf{r}'(t) = -2\sin t\,\mathbf{i} + 3\cos t\,\mathbf{j}$

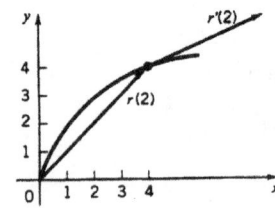 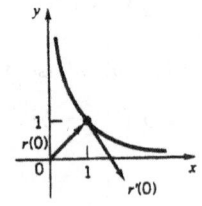 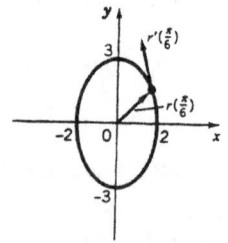

27. $\mathbf{r}(t) = (t^2 + 1)\,\mathbf{i} + t\,\mathbf{j}, \quad t \geq 1;$ or, $\mathbf{r}(t) = \sec^2 t\,\mathbf{i} + \tan t\,\mathbf{j}, \quad t \in \left[\tfrac{1}{4}\pi, \tfrac{1}{2}\pi\right)$

29. $\mathbf{r}(t) = \cos t \sin 3t\,\mathbf{i} + \sin t \sin 3t\,\mathbf{j}, \quad t \in [0, \pi]$

31. $y^3 = x^2$

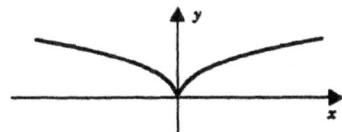

There is no tangent vector at the origin.

33. We substitute $x = t$, $y = t^2$, $z = t^3$ in the plane equation to obtain

$$4t + 2t^2 + t^3 = 24, \quad (t-2)\left(t^2 + 4t + 12\right) = 0, \quad t = 2.$$

The twisted cubic intersects the plane at the tip of $\mathbf{r}(2)$, the point $(2, 4, 8)$.

The angle between the curve and the normal line at the point of intersection is the angle between the tangent vector $\mathbf{r}'(2) = \mathbf{i} + 4\mathbf{j} + 12\mathbf{k}$ and the normal $\mathbf{N} = 4\mathbf{i} + 2\mathbf{j} + \mathbf{k}$:

$$\cos\theta = \frac{(\mathbf{i} + 4\mathbf{j} + 12\mathbf{k}) \cdot (4\mathbf{i} + 2\mathbf{j} + \mathbf{k})}{\|\mathbf{i} + 4\mathbf{j} + 12\mathbf{k}\| \, \|4\mathbf{i} + 2\mathbf{j} + \mathbf{k}\|} = \frac{24}{\sqrt{161}\sqrt{21}} \cong 0.412, \quad \theta \cong 1.15 \text{ radians.}$$

35. $\mathbf{r}'(t) = 2\mathbf{j} + 2t\mathbf{k}, \quad \|\mathbf{r}'(t)\| = 2\sqrt{1 + t^2}$

$$\mathbf{T}(t) = \frac{\mathbf{r}'(t)}{\|\mathbf{r}'(t)\|} = \frac{1}{\sqrt{1 + t^2}}(\mathbf{j} + t\mathbf{k}),$$

$$\mathbf{T}'(t) = \frac{1}{(1 + t^2)^{3/2}}[-t\mathbf{j} + \mathbf{k}]$$

at $t = 1$: tip of $\mathbf{r} = (1, 2, 1)$, $\quad \mathbf{T} = \mathbf{T}(1) = \frac{1}{\sqrt{2}}\mathbf{j} + \frac{1}{\sqrt{2}}\mathbf{k}$;

$$\mathbf{T}'(1) = -\frac{1}{2\sqrt{2}}\mathbf{j} + \frac{1}{2\sqrt{2}}\mathbf{k}; \quad \|\mathbf{T}'(1)\| = \frac{1}{2}; \quad \mathbf{N} = \mathbf{N}(1) = \frac{\mathbf{T}'(1)}{\|\mathbf{T}'(1)\|} = -\frac{1}{\sqrt{2}}\mathbf{j} + \frac{1}{\sqrt{2}}\mathbf{k}$$

normal for osculating plane:

$$\mathbf{T} \times \mathbf{N} = \left(\frac{1}{\sqrt{2}}\mathbf{j} + \frac{1}{\sqrt{2}}\mathbf{k}\right) \times \left(-\frac{1}{\sqrt{2}}\mathbf{j} + \frac{1}{\sqrt{2}}\mathbf{k}\right) = \frac{1}{2}\mathbf{i}$$

equation for osculating plane:

$$\frac{1}{2}(x - 1) + 0(y - 2) + 0(z - 1) = 0, \quad \text{which gives} \quad x - 1 = 0$$

37.

$$\mathbf{r}'(t) = -2\sin 2t\,\mathbf{i} + 2\cos 2t\,\mathbf{j} + \mathbf{k}, \quad \|\mathbf{r}'(t)\| = \sqrt{5}$$

$$\mathbf{T}(t) = \frac{\mathbf{r}'(t)}{\|\mathbf{r}'(t)\|} = \frac{1}{5}\sqrt{5}\,(-2\sin 2t\,\mathbf{i} + 2\cos 2t\,\mathbf{j} + \mathbf{k})$$

$$\mathbf{T}'(t) = -\frac{4}{5}\sqrt{5}\,(\cos 2t\,\mathbf{i} + \sin 2t\,\mathbf{j}), \quad \|\mathbf{T}'(t)\| = \frac{4}{5}\sqrt{5}$$

$$\mathbf{N}(t) = \frac{\mathbf{T}'(t)}{\|\mathbf{T}'(t)\|} = -(\cos 2t\,\mathbf{i} + \sin 2t\,\mathbf{j})$$

at $t = \pi/4$:   tip of $\mathbf{r} = (0, 1, \pi/4)$, $\mathbf{T} = \frac{1}{5}\sqrt{5}\,(-2\,\mathbf{i} + \mathbf{k})$, $\mathbf{N} = -\mathbf{j}$

normal for osculating plane:

$$\mathbf{T} \times \mathbf{N} = \frac{1}{5}\sqrt{5}\,(-2\,\mathbf{i} + \mathbf{k}) \times (-\mathbf{j}) = \frac{1}{5}\sqrt{5}\,\mathbf{i} + \frac{2}{5}\sqrt{5}\,\mathbf{k}$$

equation for osculating plane:

$$\frac{1}{5}\sqrt{5}\,(x - 0) + \frac{2}{5}\sqrt{5}\left(z - \frac{\pi}{4}\right) = 0, \quad \text{which gives} \quad x + 2z = \frac{\pi}{2}$$

39.
$$\mathbf{r}'(t) = \cosh t\,\mathbf{i} + \sinh t\,\mathbf{j} + \mathbf{k}, \quad \|\mathbf{r}'(t)\| = \sqrt{\cosh^2 t + \sinh^2 t + 1} = \sqrt{2}\,\cosh t$$

$$\mathbf{T}(t) = \frac{\mathbf{r}'(t)}{\|\mathbf{r}'(t)\|} = \frac{1}{\sqrt{2}}\,(\mathbf{i} + \tanh t\,\mathbf{j} + \operatorname{sech} t\,\mathbf{k}),$$

$$\mathbf{T}'(t) = \frac{1}{\sqrt{2}}\left(\operatorname{sech}^2 t\,\mathbf{j} - \operatorname{sech} t\,\tanh t\,\mathbf{k}\right)$$

at $t = 0$:   tip of $\mathbf{r} = (0, 1, 0)$, $\mathbf{T} = \frac{1}{\sqrt{2}}\,(\mathbf{i} + \mathbf{k})$, $\mathbf{T}'(0) = \frac{1}{\sqrt{2}}\,\mathbf{j}$; $\mathbf{N} = \mathbf{j}$

normal for osculating plane:

$$\mathbf{T} \times \mathbf{N} = \left(\frac{1}{\sqrt{2}}\,(-\mathbf{i} + \mathbf{k})\right) \times \mathbf{j} = \frac{1}{\sqrt{2}}\,(-\mathbf{i} + \mathbf{k}) \quad \text{or} \quad \mathbf{i} - \mathbf{k}$$

equation for osculating plane: $x - z = 0$

41.
$$\mathbf{r}'(t) = e^t\left[(\sin t + \cos t)\,\mathbf{i} + (\cos t - \sin t)\,\mathbf{j} + \mathbf{k}\right], \quad \|\mathbf{r}'(t)\| = e^t\sqrt{3}$$

$$\mathbf{T}(t) = \frac{\mathbf{r}'(t)}{\|\mathbf{r}'(t)\|} = \frac{1}{\sqrt{3}}\left[(\sin t + \cos t)\,\mathbf{i} + (\cos t - \sin t)\,\mathbf{j} + \mathbf{k}\right],$$

$$\mathbf{T}'(t) = \frac{1}{\sqrt{3}}\left[(\cos t - \sin t)\,\mathbf{i} - (\sin t + \cos t)\,\mathbf{j}\right]$$

at $t = 0$:   tip of $\mathbf{r} = (0, 1, 1)$, $\mathbf{T} = \mathbf{T}(0) = \frac{1}{\sqrt{3}}\,(\mathbf{i} + \mathbf{j} + \mathbf{k})$;

$$\mathbf{T}'(0) = \frac{1}{\sqrt{3}}\,(\mathbf{i} - \mathbf{j}); \quad \|\mathbf{T}'(0)\| = \frac{\sqrt{2}}{\sqrt{3}}; \quad \mathbf{N} = \mathbf{N}(0) = \frac{\mathbf{T}'(0)}{\|\mathbf{T}'(0)\|} = \frac{1}{\sqrt{2}}\,(\mathbf{i} - \mathbf{j})$$

normal for osculating plane:

$$\mathbf{T} \times \mathbf{N} = \frac{1}{\sqrt{3}}\,(\mathbf{i} + \mathbf{j} + \mathbf{k}) \times \frac{1}{\sqrt{2}}\,(\mathbf{i} - \mathbf{j}) = \frac{1}{\sqrt{6}}\,(\mathbf{i} + \mathbf{j} - 2\,\mathbf{k})$$

equation for osculating plane:

$$\frac{1}{\sqrt{6}}\,(x - 0) + \frac{1}{\sqrt{6}}\,(y - 1) - \frac{2}{\sqrt{6}}\,(z - 1) = 0, \quad \text{or} \quad x + y - 2z + 1 = 0$$

43. $\mathbf{T}_1 = \dfrac{\mathbf{R}'(u)}{\|\mathbf{R}'(u)\|} = -\dfrac{\mathbf{r}'(a + b - u)}{\|\mathbf{r}'(a + b - u)\|} = -\mathbf{T}.$

Therefore $\mathbf{T}_1'(u) = \mathbf{T}'(a + b - u)$ and $\mathbf{N}_1 = \mathbf{N}.$

45. (a) $\mathbf{r}'(t) = -\sqrt{2}\,\sin t\,\mathbf{i} + \sqrt{2}\,\cos t\,\mathbf{j} + 5\cos t\,\mathbf{k}$

tangent line $(t = \pi/4):\quad x = 1 - t,\quad y = 1 + t,\quad z = -\dfrac{\sqrt{2}}{2} - \dfrac{5\sqrt{2}}{2}t$

(c) The tangent line is parallel to the x, y-plane at the points where $t = \dfrac{(2n+1)\pi}{10}$, $n = 0, 1, 2, \ldots, 9$.

SECTION 14.4

1. $\mathbf{r}'(t) = \mathbf{i} + t^{1/2}\,\mathbf{j},\quad \|\mathbf{r}'(t)\| = \sqrt{1+t}$

$$L = \int_0^8 \sqrt{1+t}\,dt = \left[\frac{2}{3}(1+t)^{3/2}\right]_0^8 = \frac{52}{3} \qquad\qquad L = \int_0^2 (t^2 + 1)\,dt = \frac{14}{3}$$

3. $\mathbf{r}'(t) = -a\sin t\,\mathbf{i} + a\cos t\,\mathbf{j} + b\,\mathbf{k},\quad \|\mathbf{r}'(t)\| = \sqrt{a^2 + b^2}$

$$L = \int_0^{2\pi} \sqrt{a^2 + b^2}\,dt = 2\pi\sqrt{a^2 + b^2}$$

5. $\mathbf{r}'(t) = \mathbf{i} + \tan t\,\mathbf{j},\quad \|\mathbf{r}'(t)\| = \sqrt{1 + \tan^2 t} = |\sec t|$

$$L = \int_0^{\pi/4} |\sec t|\,dt = \int_0^{\pi/4} \sec t\,dt = \left[\ln|\sec t + \tan t|\right]_0^{\pi/4} = \ln\left(1 + \sqrt{2}\right)$$

7. $\mathbf{r}'(t) = 3t^2\,\mathbf{i} + 2t\,\mathbf{j},\quad \|\mathbf{r}'(t)\| = \sqrt{9t^4 + 4t^2} = |t|\sqrt{4 + 9t^2}$

$$L = \int_0^1 \left| t\sqrt{4 + 9t^2} \right| dt = \int_0^1 t\sqrt{4 + 9t^2}\,dt = \left[\frac{1}{27}(4 + 9t^2)^{3/2}\right]_0^1 = \frac{1}{27}\left(13\sqrt{13} - 8\right)$$

9. $\mathbf{r}'(t) = (\cos t - \sin t)e^t\,\mathbf{i} + (\sin t + \cos t)e^t\,\mathbf{j},\quad \|\mathbf{r}'(t)\| = \sqrt{2}\,e^t$

$$L = \int_0^\pi \sqrt{2}\,e^t\,dt = \sqrt{2}\left(e^\pi - 1\right)$$

11. $\mathbf{r}'(t) = \dfrac{1}{t}\,\mathbf{i} + 2\,\mathbf{j} + 2t\,\mathbf{k},\quad \|\mathbf{r}'(t)\| = \sqrt{\dfrac{1}{t^2} + 4 + 4t^2}$

$$L = \int_1^e \sqrt{\frac{1}{t^2} + 4 + 4t^2}\,dt = \int_1^e \left(\frac{1}{t} + 2t\right) dt = \left[\ln|t| + t^2\right]_1^e = e^2$$

13. $\mathbf{r}'(t) = t\cos t\,\mathbf{i} + t\sin t\,\mathbf{j} + \sqrt{3}\,t\,\mathbf{k},\quad \|\mathbf{r}'(t)\| = \sqrt{t^2\cos^2 t + t^2\sin^2 t + 3t^2} = \sqrt{4t^2} = 2t$

$$L = \int_0^{2\pi} 2t\,dt = \left[t^2\right]_0^{2\pi} = 4\pi^2$$

15. $\mathbf{r}'(t) = 2\,\mathbf{i} + 2t\,\mathbf{j} - 2t\,\mathbf{k},\quad \|\mathbf{r}'(t)\| = 2\sqrt{1 + 2t^2}$

$$L = \int_0^2 2\sqrt{1 + 2t^2}\,dt = \sqrt{2}\int_0^{\tan^{-1}(2\sqrt{2})} \sec^3 u\,du$$

$$\underset{\displaystyle (t\sqrt{2} = \tan u)}{\big\llcorner}$$

$$= \tfrac{1}{2}\sqrt{2}\left[\sec u\tan u + \ln|\sec u + \tan u|\right]_0^{\tan^{-1}(2\sqrt{2})} = 6 + \tfrac{1}{2}\sqrt{2}\,\ln\left(3 + 2\sqrt{2}\right)$$

17.
$$s = s(t) = \int_a^t \|\mathbf{r}'(u)\| \, du$$
$$s'(t) = \|\mathbf{r}'(t)\| = \|x'(t)\,\mathbf{i} + y'(t)\,\mathbf{j} + z'(t)\,\mathbf{k}\|$$
$$= \sqrt{[x'(t)]^2 + [y'(t)]^2 + [z'(t)]^2}.$$

In the Leibniz notation this translates to
$$\frac{ds}{dt} = \sqrt{\left(\frac{dx}{dt}\right)^2 + \left(\frac{dy}{dt}\right)^2 + \left(\frac{dz}{dt}\right)^2}.$$

19.
$$s = s(x) = \int_a^x \sqrt{1 + [f'(t)]^2} \, dt$$
$$s'(x) = \sqrt{1 + [f'(x)]^2}.$$

In the Leibniz notation this translates to
$$\frac{ds}{dx} = \sqrt{1 + \left(\frac{dy}{dx}\right)^2}.$$

21. $\mathbf{r}'(t) = -\sin t\,\mathbf{i} + \cos t\,\mathbf{j}$. Since $\|\mathbf{r}'\| \equiv 1$, the parametrization is by arc length.

23. $\mathbf{r}'(t) = t \sin t\,\mathbf{i} + t \cos t\,\mathbf{j} + t\,\mathbf{k}$; $\|\mathbf{r}'\| = t\sqrt{2}$.
$$s = \int_0^t u\sqrt{2}\,du = \frac{\sqrt{2}}{2}t^2; \quad t = 2^{1/4}\sqrt{s}.$$
$$\mathbf{R}(s) = \left(\sin 2^{1/4}\sqrt{s} - 2^{1/4}\sqrt{s}\cos 2^{1/4}\sqrt{s}\right)\mathbf{i} + \left(\cos 2^{1/4}\sqrt{s} + 2^{1/4}\sqrt{s}\sin 2^{1/4}\sqrt{s}\right)\mathbf{j} + \frac{1}{\sqrt{2}}s\,\mathbf{k}.$$

25. $\mathbf{r}'(t) = t^{3/2}\,\mathbf{j} + \mathbf{k}$, $\|\mathbf{r}'(t)\| = \sqrt{\left(t^{3/2}\right)^2 + 1} = \sqrt{t^3 + 1}$
$$s = \int_0^{1/2} \sqrt{t^3 + 1}\,dt \cong 0.5077$$

27. $\mathbf{r}'(t) = -3\sin t\,\mathbf{i} + 4\cos t\,\mathbf{j}$, $\|\mathbf{r}'(t)\| = \sqrt{9\sin^2 t + 16\cos^2 t}\,dt$
$$s = \int_0^{2\pi} \sqrt{9\sin^2 t + 16\cos^2 t}\,dt \cong 22.0939$$

29. (a)   (b) $s = \int_0^{2\pi} \sqrt{1 + 16\cos^2 4t}\,dt \cong 17.6286$

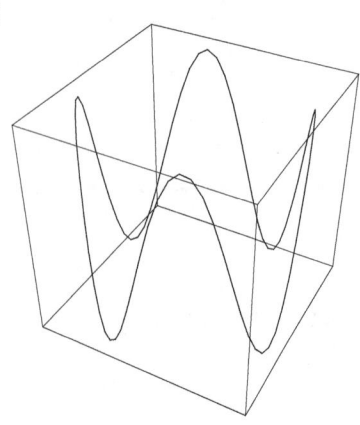

PROJECT 14.4

1. Given the differentiable curve $\mathbf{r} = \mathbf{r}(t)$, $t \in I$. Let $t = \phi(u)$ be a continuously differentiable one-to-one function that maps the interval J onto the interval I, and let $\mathbf{R}(u) = \mathbf{r}(\phi(u))$, $u \in J$. Suppose that $\phi'(u) > 0$ on J. Then, as u increases across J, $t = \phi(u)$ increases across I. As a result, $\mathbf{R}(u)$ takes on exactly the same values in exactly the same order as $\mathbf{r}$. If $\phi'(u) < 0$ on J, then $\mathbf{R}(u)$ takes on exactly the same values as $\mathbf{r}$ but in the reverse order.

3. Suppose that $\phi'(u) < 0$ on J. Then

$$\mathbf{R}'(u) = [\mathbf{r}(\phi(u))]' = \mathbf{r}'(\phi(u))\phi'(u); \quad \frac{\mathbf{R}'(u)}{\mathbf{R}'(u)} = \frac{\mathbf{r}'(\phi(u))\phi'(u)}{\|\mathbf{r}'(\phi(u))\phi'(u)\|} = -\frac{\mathbf{r}'(\phi(u))\phi'(u)}{\phi'(u)\|\mathbf{r}'(\phi(u))\|} = -\frac{\mathbf{r}'(t)}{\|\mathbf{r}'(t)\|}.$$

$$\text{since } \phi'(u) < 0 \longrightarrow$$

Thus, the unit tangent is reversed by a sense-reversing change of parameter. That is, if $T_{\mathbf{R}}$ and $T_{\mathbf{r}}$ are the respective unit tangents, then $T_{\mathbf{R}} = -T_{\mathbf{r}}$.

Now consider the principal normals:

$$\frac{T'_{\mathbf{R}}}{\|T'_{\mathbf{R}}\|} = -\frac{T'_{\mathbf{r}}(\phi(u)\phi'(u)}{\|T'_{\mathbf{r}}(\phi(u)\phi'(u)\|} = \frac{T'_{\mathbf{r}}(\phi(u)\phi'(u)}{\phi'(u)\|T'_{\mathbf{r}}(\phi(u)\phi'(u)\|} = \frac{T'_{\mathbf{r}}(t)}{\|T'_{\mathbf{r}}(t)\|}.$$

$$\text{since } \phi'(u) < 0 \longrightarrow$$

Thus the principal normal is unchanged by a sense-reversing change of parameter. The osculating plane is also unchanged since $\mathbf{T} \times \mathbf{N}$ and $-\mathbf{T} \times \mathbf{N}$ are each normal to the osculating plane.

5. Let L be the length as computed from $\mathbf{r}$ and L^* the length as computed from $\mathbf{R}$. Then

$$L^* = \int_c^d \|\mathbf{R}'(u)\| \, du = \int_c^d \|\mathbf{r}'(\phi(u))\| \, \phi'(u) \, du = \int_a^b \|\mathbf{r}'(t)\| \, dt = L.$$

$$t = \phi(u) \longrightarrow$$

SECTION 14.5

1. $\mathbf{r}(t) = a[\cos\theta(t)\,\mathbf{i} + \sin\theta(t)\,\mathbf{j}], \quad \mathbf{r}'(t) = a[-\sin\theta(t)\,\mathbf{i} + \cos\theta(t)\,\mathbf{j}]\theta'(t)$

$\|\mathbf{r}'(t)\| = v \implies a|\theta'(t)| = v \implies |\theta'(t)| = v/a$

$\mathbf{r}''(t) = a[-\cos\theta(t)\,\mathbf{i} - \sin\theta(t)\,\mathbf{j}][\theta'(t)]^2, \quad \|\mathbf{r}''(t)\| = a[\theta'(t)]^2 = v^2/a$

3. $\mathbf{r}(t) = at\,\mathbf{i} + b\sin at\,\mathbf{j}, \quad = a\,\mathbf{i} + ab\cos at\,\mathbf{j}$

$\mathbf{r}'(t) = -a^2 b\sin at\,\mathbf{j}, \quad \|\mathbf{r}'(t)\| = a^2|b\sin at| = a^2|y(t)|$

5. $y = \cos \pi x, \quad 0 \le x \le 2$

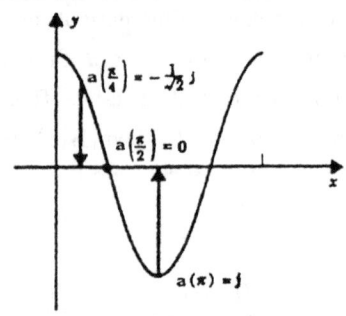

7. $x = \sqrt{1 + y^2}, \quad y \ge -1$

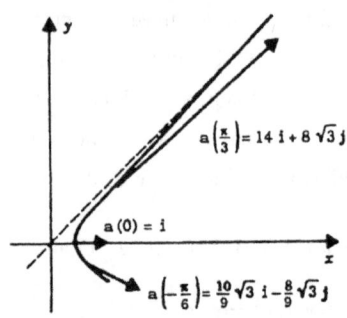

9. (a) initial position is tip of $\mathbf{r}(0) = x_0 \mathbf{i} + y_0 \mathbf{j} + z_0 \mathbf{k}$

(b) $\mathbf{r}'(t) = (\alpha \cos \theta) \mathbf{j} + (\alpha \sin \theta - 32t) \mathbf{k}, \quad \mathbf{r}'(0) = (\alpha \cos \theta) \mathbf{j} + (\alpha \sin \theta) \mathbf{k}$

(c) $|\mathbf{r}'(0)| = |\alpha|$ (d) $\mathbf{r}''(t) = -32 \mathbf{k}$

(e) a parabolic arc from the parabola

$$z = z_0 + (\tan \theta)(y - y_0) - 16 \frac{(y - y_0)^2}{\alpha^2 \cos^2 \theta}$$

in the plane $x = x_0$

11.
$$\|\mathbf{r}(t)\| = C \text{ iff } \|\mathbf{r}(t)\|^2 = \mathbf{r}(t) \cdot \mathbf{r}(t) = C$$
$$\text{iff } \frac{d}{dt}\|\mathbf{r}(t)\| = 2\mathbf{r}(t) \cdot \mathbf{r}'(t) = 0$$
$$\text{iff } \mathbf{r}(t) \perp \mathbf{r}'(t)$$

13. $\kappa = \dfrac{e^{-x}}{(1 + e^{-2x})^{3/2}}$

15. $y' = \dfrac{1}{2x^{1/2}}; \quad y'' = \dfrac{-1}{4x^{3/2}} \quad \kappa = \dfrac{\left|-1/4x^{3/2}\right|}{\left[1 + (1/2x^{1/2})^2\right]^{3/2}} = \dfrac{2}{(1 + 4x)^{3/2}}$

17. $\kappa = \dfrac{\sec^2 x}{(1 + \tan^2 x)^{3/2}} = |\cos x|$

19. $\kappa = \dfrac{|\sin x|}{(1 + \cos^2 x)^{3/2}}$

21. $\kappa = \dfrac{|x|}{(1 + x^4/4)^{3/2}}; \quad$ at $\left(2, \dfrac{4}{3}\right), \quad \kappa = \dfrac{2}{5\sqrt{5}}$

23. $\kappa = \dfrac{\left|-1/y^3\right|}{(1 + 1/y^2)^{3/2}} = \dfrac{1}{(1 + y^2)^{3/2}}; \quad$ at $(2, 2), \quad \kappa = \dfrac{1}{5\sqrt{5}}$

25. $y'(x) = \dfrac{1}{x+1}$, $\quad y'(2) = \dfrac{1}{3}$; $\qquad y''(x) = \dfrac{-1}{(x+1)^2}$, $\quad y''(2) = -\dfrac{1}{9}$.

At $\ x = 2, \ \ \kappa = \dfrac{\left| -\dfrac{1}{9} \right|}{\left[1 + \left(\dfrac{1}{3} \right)^2 \right]^{3/2}} = \dfrac{3}{10\sqrt{10}}$

27. $\kappa(x) = \dfrac{\left| -1/x^2 \right|}{(1 + 1/x^2)^{3/2}} = \dfrac{x}{(x^2 + 1)^{3/2}}$, $\quad x > 0$

$\kappa'(x) = \dfrac{(1 - 2x^2)}{(x^2 + 1)^{5/2}}$, $\qquad \kappa'(x) = 0 \implies x = \dfrac{1}{2}\sqrt{2}$

Since κ increases on $\ (0, \frac{1}{2}\sqrt{2}\,]$ and decreases on $\ [\frac{1}{2}\sqrt{2}, \infty)$, κ is maximal at $\ (\frac{1}{2}\sqrt{2}, \frac{1}{2}\ln\frac{1}{2})$.

29. $x(t) = t$, $\quad x'(t) = 1$, $\quad x''(t) = 0$; $\qquad y(t) = \frac{1}{2}t^2$, $\quad y'(t) = t$, $\quad y''(t) = 1 \quad \kappa = \dfrac{1}{(1 + t^2)^{3/2}}$

31. $x(t) = 2t$, $\quad x'(t) = 2$, $\quad x''(t) = 0$; $\qquad y(t) = t^3$, $\quad y'(t) = 3t^2$, $\quad y''(t) = 6t$; $\quad \kappa = \dfrac{12|t|}{(4 + 9t^4)^{3/2}}$

33. $x(t) = e^t \cos t$, $\quad x'(t) = e^t(\cos t - \sin t)$, $\quad x''(t) = -2\,e^t \sin t$

$y(t) = e^t \sin t$, $\quad y'(t) = e^t(\sin t + \cos t)$, $\quad y''(t) = 2\,e^t \cos t$

$\kappa = \dfrac{\left| 2e^{2t}\cos t\,(\cos t - \sin t) + 2e^{2t}\sin t\,(\cos t + \sin t) \right|}{\left[e^{2t}(\cos t - \sin t)^2 + e^{2t}(\cos t + \sin t)^2 \right]^{3/2}} = \dfrac{2e^{2t}}{(2e^{2t})^{3/2}} = \dfrac{1}{2}\sqrt{2}\,e^{-t}$

35. $x(t) = t \cos t$, $\quad x'(t) = \cos t - t \sin t$, $\quad x''(t) = -2\sin t - t \cos t$

$y(t) = t \sin t$, $\quad y'(t) = \sin t + t \cos t$, $\quad y''(t) = 2\cos t - t \sin t$

$\kappa = \dfrac{\left| (\cos t - t\sin t)(2\cos t - t\sin t) - (\sin t + t\cos t)(-2\sin t - t\cos t) \right|}{\left[(\cos t - t\sin t)^2 + (\sin t + t\cos t)^2 \right]^{3/2}} = \dfrac{2 + t^2}{[1 + t^2]^{3/2}}$

37. $\kappa = \dfrac{|2/x^3|}{[1 + 1/x^4]^{3/2}} = \dfrac{2|x^3|}{(x^4 + 1)^{3/2}}$; $\quad$ at $x = \pm 1$, $\quad \kappa = \dfrac{\sqrt{2}}{2}$

39. We use (14.5.3) and the hint to obtain

$$\kappa = \dfrac{\left| ab\sinh^2 t - ab\cosh^2 t \right|}{\left[a^2\sinh^2 t + b^2\cosh^2 t \right]^{3/2}} = \dfrac{\left| \dfrac{a}{b}y^2 - \dfrac{b}{a}x^2 \right|}{\left[\left(\dfrac{ay}{b} \right)^2 + \left(\dfrac{bx}{a} \right)^2 \right]^{3/2}}$$

$$= \dfrac{a^3 b^3 \left| \dfrac{a}{b}y^2 - \dfrac{b}{a}x^2 \right|}{[a^4 y^2 + b^4 x^2]^{3/2}} = \dfrac{a^4 b^4}{[a^4 y^2 + b^4 x^2]^{3/2}}.$$

41. $\mathbf{r}'(t) = e^t(\cos t - \sin t)\,\mathbf{i} + e^t(\sin t + \cos t)\,\mathbf{j} + e^t\,\mathbf{k}$

$$\frac{ds}{dt} = \|\mathbf{r}'(t)\| = \sqrt{3}\,e^t, \quad \frac{d^2s}{dt^2} = \sqrt{3}\,e^t$$

$$\mathbf{T}(t) = \frac{\mathbf{r}'(t)}{\|\mathbf{r}'(t)\|} = \frac{1}{\sqrt{3}}\left[(\cos t - \sin t)\,\mathbf{i} + (\sin t + \cos t)\,\mathbf{j} + \mathbf{k}\right]$$

$$\mathbf{T}'(t) = \frac{1}{\sqrt{3}}\left[(-\sin t - \cos t)\,\mathbf{i} + (\cos t - \sin t)\,\mathbf{j}\right]; \quad \|\mathbf{T}'(t)\| = \sqrt{2/3}$$

$$\kappa = \frac{\|\mathbf{T}'(t)\|}{ds/dt} = \frac{\sqrt{2/3}}{\sqrt{3}\,e^t} = \frac{1}{3}\sqrt{2}\,e^{-t}; \qquad \mathbf{a_T} = \frac{d^2s}{dt^2} = \sqrt{3}\,e^t, \quad \mathbf{a_N} = \kappa\left(\frac{ds}{dt}\right)^2 = \sqrt{2}\,e^t$$

43. $\mathbf{r}'(t) = -2\sin 2t\,\mathbf{i} + 2\cos 2t\,\mathbf{j}; \quad \dfrac{ds}{dt} = \|\mathbf{r}'(t)\| = 2, \quad \dfrac{d^2s}{dt^2} = 0$

$$\mathbf{T}(t) = \frac{\mathbf{r}'(t)}{\|\mathbf{r}'(t)\|} = -\sin 2t\,\mathbf{i} + \cos 2t\,\mathbf{j}$$

$$\mathbf{T}'(t) = -2\left(\cos 2t\,\mathbf{i} + \sin 2t\,\mathbf{j}\right); \quad \|\mathbf{T}'(t)\| = 2$$

$$\kappa = \frac{\|\mathbf{T}'(t)\|}{ds/dt} = \frac{2}{2} = 1; \qquad \mathbf{a_T} = \frac{d^2s}{dt^2} = 0, \quad \mathbf{a_N} = \kappa\left(\frac{ds}{dt}\right)^2 = 1\cdot 4 = 4.$$

45. $\mathbf{r}'(t) = -3\sin 3t\,\mathbf{i} + 4\,\mathbf{j} - 3\cos 3t\,\mathbf{k}; \quad \dfrac{ds}{dt} = \|\mathbf{r}'\| = 5; \quad \dfrac{d^2s}{dt^2} = 0.$

$$\mathbf{T}(t) = \frac{\mathbf{r}'(t)}{\|\mathbf{r}'(t)\|} = -\frac{3}{5}\sin 3t\,\mathbf{i} + \frac{4}{5}\,\mathbf{j} - \frac{3}{5}\cos 3t\,\mathbf{k}$$

$$\mathbf{T}'(t) = -\frac{9}{5}\cos 3t\,\mathbf{i} + \frac{9}{5}\sin 3t\,\mathbf{k}; \quad \|\mathbf{T}'(t)\| = \frac{9}{5}$$

$$\kappa = \frac{\|\mathbf{T}'(t)\|}{ds/dt} = \frac{9/5}{5} = \frac{9}{25}; \qquad \mathbf{a_T} = \frac{d^2s}{dt^2} = 0, \quad \mathbf{a_N} = \kappa\left(\frac{ds}{dt}\right)^2 = \frac{9}{25}\cdot 25 = 9.$$

47. $\mathbf{r}'(t) = \sqrt{1+t}\,\mathbf{i} - \sqrt{1-t}\,\mathbf{j} + \sqrt{2}\,\mathbf{k}, \quad \dfrac{ds}{dt} = \|\mathbf{r}'(t)\| = \sqrt{(1+t) + (1-t) + 2} = 2, \quad \dfrac{d^2s}{dt^2} = 0$

$$\mathbf{T}(t) = \frac{\mathbf{r}'(t)}{\|\mathbf{r}'(t)\|} = \frac{\sqrt{1+t}}{2}\,\mathbf{i} - \frac{\sqrt{1-t}}{2}\,\mathbf{j} + \frac{\sqrt{2}}{2}\,\mathbf{k}$$

$$\mathbf{T}'(t) = \frac{1}{4\sqrt{1+t}}\,\mathbf{i} + \frac{1}{4\sqrt{1-t}}\,\mathbf{j}.$$

$$\|\mathbf{T}'(t)\| = \sqrt{\frac{1}{16(1+t)} + \frac{1}{16(1-t)}} = \frac{1}{4}\sqrt{\frac{2}{1-t^2}}$$

Then, $\quad \kappa = \dfrac{\|\mathbf{T}'(t)\|}{ds/dt} = \dfrac{1}{8}\sqrt{\dfrac{2}{1-t^2}}$

$$\mathbf{a_T} = \frac{d^2s}{dt^2} = 0, \quad \mathbf{a_N} = \kappa\left(\frac{ds}{dt}\right)^2 = \frac{1}{2}\sqrt{\frac{2}{1-t^2}}.$$

49. tangential component: $\quad \mathbf{a_T} = \dfrac{6t + 12t^3}{\sqrt{1+t^2+t^4}};$ $\qquad$ normal component: $\quad \mathbf{a_N} = 6\sqrt{\dfrac{1+4t^2+t^4}{1+t^2+t^4}}$

51. By Exercise 50

$$\kappa = \frac{\left|\left(e^{a\theta}\right)^2 + 2\left(ae^{a\theta}\right)^2 - \left(e^{a\theta}\right)\left(a^2 e^{a\theta}\right)\right|}{\left[\left(e^{a\theta}\right)^2 + \left(ae^{a\theta}\right)^2\right]^{3/2}} = \frac{e^{-a\theta}}{\sqrt{1+a^2}}.$$

53. By Exercise 50,

$$\kappa = \frac{\left|a^2(1-\cos\theta)^2 + 2a^2\sin^2\theta - a^2(1-\cos\theta)(\cos\theta)\right|}{\left[a^2(1-\cos\theta)^2 + a^2\sin^2\theta\right]^{3/2}} = \frac{3a^2(1-\cos\theta)}{[2a^2(1-\cos\theta)]^{3/2}} = \frac{3ar}{[2ar]^{3/2}} = \frac{3}{2\sqrt{2ar}}.$$

PROJECT 14.5A

1. The system of equations generated by the specified conditions is:

$a + b + c + d = 3$ $\qquad\qquad$ $27a + 9b + 3c + d = 7$

$6a + 2b = 0$ $\qquad\qquad$ $27\alpha + 9\beta + 3\gamma + \delta = 7$

$729\alpha + 81\beta + 9\gamma + \delta = -2$ $\qquad\qquad$ $54\alpha + 2\beta = 0$

$27a + 6b + c = 27\alpha + 6\beta + \gamma$ $\qquad\qquad$ $18a + 2b = 18\alpha + 2\beta$

$a \cong -0.1094$ $\qquad$ $b \cong 0.3281$ $\qquad$ $c \cong 2.1094$ $\qquad$ $d \cong 0.6719$

$\alpha \cong 0.0365$ $\qquad$ $\beta \cong -0.9844$ $\qquad$ $\gamma \cong 6.0469$ $\qquad$ $\delta \cong -3.2656$

3. (a), (b) The system of equations generated by the specified conditions (and the derivative conditions of Problem 1) is:

$27\alpha + 9b + 3c + d = 10$ $\qquad\qquad$ $64\alpha + 16b + 4c + d = 15$

$18a + 2b = 0$ $\qquad\qquad$ $64\alpha + 64\beta + 4\gamma + \delta = 15$

$216\alpha + 36\beta + 6\gamma + \delta = 35$ $\qquad\qquad$ $36\alpha + 2\beta = 0$

$48a + 8b + c = 48\alpha + 8\beta + \gamma$ $\qquad\qquad$ $24\alpha + 2b = 24\alpha + 2\beta$

(c) $a = b = 0$, $c = 5$, $d = -5$; $\alpha = 1.25$, $\beta = -15$, $\gamma = 65$, $\delta = -85$

PROJECT 14.5B

1. $\dfrac{d\mathbf{T}}{dt} = \dfrac{d\mathbf{T}}{ds}\dfrac{ds}{dt} \Longrightarrow \dfrac{d\mathbf{T}}{ds} = \dfrac{d\mathbf{T}/dt}{ds/dt} = \dfrac{\mathbf{T}'(t)}{\|\mathbf{T}'(t)\|}\dfrac{\|\mathbf{T}/(t)\|}{ds/dt} = \kappa\,\mathbf{N}.$

3. $\dfrac{d\mathbf{N}}{ds} = \dfrac{d}{ds}(\mathbf{B}\times\mathbf{T}) = \left(\mathbf{B}\times\dfrac{d\mathbf{T}}{ds}\right) + \left(\dfrac{d\mathbf{B}}{ds}\times\mathbf{T}\right) = (\mathbf{B}\times\kappa\mathbf{N}) + \tau(\mathbf{N}\times\mathbf{T})$

$$= -\kappa(\mathbf{N}\times\mathbf{B}) - \tau(\mathbf{T}\times\mathbf{N}) = -\kappa\mathbf{T} - \tau\mathbf{B}$$

SECTION 14.6

1. (a) $\mathbf{r}'(t) = \dfrac{a\omega}{2}\left(e^{\omega t} - e^{-\omega t}\right)\mathbf{i} + \dfrac{b\omega}{2}\left(e^{\omega t} + e^{-\omega t}\right)\mathbf{j},$ $\quad \mathbf{r}'(0) = b\omega\mathbf{j}$

(b) $\mathbf{r}''(t) = \dfrac{a\omega^2}{2}\left(e^{\omega t} + e^{-\omega t}\right)\mathbf{i} + \dfrac{b\omega^2}{2}\left(e^{\omega t} - e^{-\omega t}\right)\mathbf{j} = \omega^2\mathbf{r}(t)$

(c) The torque τ is $\mathbf{0}$: $\quad \tau(t) = \mathbf{r}(t)\times m\mathbf{a}(t) = \mathbf{r}(t)\times m\omega^2\mathbf{r}(t) = \mathbf{0}.$

The angular momentum $\mathbf{L}(t)$ is constant since $\mathbf{L}'(t) = \tau(t) = \mathbf{0}.$

3. We begin with the force equation $\mathbf{F}(t) = \alpha\mathbf{k}$. In general, $\mathbf{F}(t) = m\,\mathbf{a}(t)$, so that here

$$\mathbf{a}(t) = \frac{\alpha}{m}\,\mathbf{k}.$$

Integration gives

$$\mathbf{v}(t) = C_1\mathbf{i} + C_2\mathbf{j} + \left(\frac{\alpha}{m}\,t + C_3\right)\mathbf{k}.$$

Since $\mathbf{v}(0) = 2\,\mathbf{j}$, we can conclude that $C_1 = 0$, $C_2 = 2$, $C_3 = 0$. Thus

$$\mathbf{v}(t) = 2\,\mathbf{j} + \frac{\alpha}{m}\,t\mathbf{k}.$$

Another integration gives

$$\mathbf{r}(t) = D_1\mathbf{i} + (2t + D_2)\,\mathbf{j} + \left(\frac{\alpha}{2m}\,t^2 + D_3\right)\mathbf{k}.$$

Since $\mathbf{r}(0) = y_0\,\mathbf{j} + z_0\,\mathbf{k}$, we have $D_1 = 0$, $D_2 = y_0$, $D_3 = z_0$, and therefore

$$\mathbf{r}(t) = (2t + y_0)\,\mathbf{j} + \left(\frac{\alpha}{2m}\,t^2 + z_0\right)\mathbf{k}.$$

The conditions of the problem require that t be restricted to nonnegative values.

To obtain an equation for the path in Cartesian coordinates, we write out the components

$$x(t) = 0, \quad y(t) = 2t + y_0, \quad z(t) = \frac{\alpha}{2m}\,t^2 + z_0. \qquad (t \geq 0)$$

From the second equation we have

$$t = \tfrac{1}{2}\,[y(t) - y_0]. \qquad (y(t) \geq y_0)$$

Substituting this into the third equation, we get

$$z(t) = \frac{\alpha}{8m}\,[y(t) - y_0]^2 + z_0. \qquad (y(t) \geq y_0)$$

Eliminating t altogether, we have

$$z = \frac{\alpha}{8m}\,(y - y_0)^2 + z_0. \qquad (y \geq y_0)$$

Since $x = 0$, the path of the object is a parabolic arc in the yz-plane.

Answers to (a) through (d):

(a) velocity: $\mathbf{v}(t) = 2\,\mathbf{j} + \dfrac{\alpha}{m}\,t\,\mathbf{k}.$
\qquad (b) speed: $v(t) = \dfrac{1}{m}\,\sqrt{4m^2 + \alpha^2 t^2}.$

(c) momentum: $\mathbf{p}(t) = 2m\,\mathbf{j} + \alpha\,t\,\mathbf{k}.$

(d) path in vector form: $\mathbf{r}(t) = (2t + y_0)\,\mathbf{j} + \left(\dfrac{\alpha}{2m}\,t^2 + z_0\right)\mathbf{k}, \quad t \geq 0.$

path in Cartesian coordinates: $z = \dfrac{\alpha}{8m}\,(y - y_0)^2 + z_0, \quad y \geq y_0, \quad x = 0.$

5. $\mathbf{F}(t) = m\,\mathbf{a}(t) = m\,\mathbf{r}''(t) = 2m\mathbf{k}$

7. From $\mathbf{F}(t) = m\,\mathbf{a}(t)$ we obtain

$$\mathbf{a}(t) = \pi^2[a\cos\pi t\,\mathbf{i} + b\sin\pi t\,\mathbf{j}].$$

By direct calculation using $\mathbf{v}(0) = -\pi b\mathbf{j} + \mathbf{k}$ and $\mathbf{r}(0) = b\mathbf{j}$ we obtain

$$\mathbf{v}(t) = a\pi \sin \pi t \,\mathbf{i} - b\pi \cos \pi t \,\mathbf{j} + \mathbf{k}$$

$$\mathbf{r}(t) = a(1 - \cos \pi t)\,\mathbf{i} + b(1 - \sin \pi t)\,\mathbf{j} + t\mathbf{k}.$$

(a) $\mathbf{v}(1) = b\pi \mathbf{j} + \mathbf{k}$

(b) $\|\mathbf{v}(1)\| = \sqrt{\pi^2 b^2 + 1}$

(c) $\mathbf{a}(1) = -\pi^2 a\mathbf{i}$

(d) $m\,\mathbf{v}(1) = m\,(\pi b\mathbf{j} + \mathbf{k})$

(e) $\mathbf{L}(1) = \mathbf{r}(1) \times m\,\mathbf{v}(1) = [2a\mathbf{i} + b\mathbf{j} + \mathbf{k}] \times [m\,(b\pi\mathbf{j} + \mathbf{k})]$
$\qquad = m\,[b(1-\pi)\mathbf{i} - 2a\mathbf{j} + 2ab\pi\mathbf{k}]$

(f) $\tau(1) = \mathbf{r}(1) \times \mathbf{F}(1) = [2a\mathbf{i} + b\mathbf{j} + \mathbf{k}] \times [-m\pi^2 a\mathbf{i}] = -m\pi^2 a\,[\mathbf{j} - b\mathbf{k}]$

9. We have $m\mathbf{v} = m\mathbf{v}_1 + m\mathbf{v}_2$ and $\frac{1}{2}mv^2 = \frac{1}{2}m{v_1}^2 + \frac{1}{2}m{v_2}^2.$

Therefore $\mathbf{v} = \mathbf{v}_1 + \mathbf{v}_2$ and $v^2 = {v_1}^2 + {v_2}^2.$

Since $v^2 = \mathbf{v} \cdot \mathbf{v} = (\mathbf{v}_1 + \mathbf{v}_2) \cdot (\mathbf{v}_1 + \mathbf{v}_2) = {v_1}^2 + {v_2}^2 + 2(\mathbf{v}_1 \cdot \mathbf{v}_2),$

we have $\mathbf{v}_1 \cdot \mathbf{v}_2 = 0$ and $\mathbf{v}_1 \perp \mathbf{v}_2.$

11. $\mathbf{r}''(t) = \mathbf{a},$ $\mathbf{r}'(t) = \mathbf{v}(0) + t\mathbf{a},$ $\mathbf{r}(t) = \mathbf{r}(0) + t\,\mathbf{v}(0) + \frac{1}{2}\,t^2\,\mathbf{a}.$

If neither $\mathbf{v}(0)$ nor $\mathbf{a}$ is zero, the displacement $\mathbf{r}(t) - \mathbf{r}(0)$ is a linear combination of $\mathbf{v}(0)$ and $\mathbf{a}$ and thus remains on the plane determined by these vectors. The equation of this plane can be written

$$[\mathbf{a} \times \mathbf{v}(0)] \cdot [\mathbf{r} - \mathbf{r}(0)] = 0.$$

(If either $\mathbf{v}(0)$ or $\mathbf{a}$ is zero, the motion is restricted to a straight line; if both of these vectors are zero, the particle remains at its initial position $\mathbf{r}(0)$.)

13. $\mathbf{r}(t) = \mathbf{i} + t\,\mathbf{j} + \left(\dfrac{qE_0}{2m}\right) t^2 \mathbf{k}$
          15. $\mathbf{r}(t) = \left(1 + \dfrac{t^3}{6m}\right)\mathbf{i} + \dfrac{t^4}{12m}\,\mathbf{j} + t\,\mathbf{k}$

17. $\dfrac{d}{dt}\left(\dfrac{1}{2}mv^2\right) = mv\dfrac{dv}{dt} = m\left(\mathbf{v} \cdot \dfrac{d\mathbf{v}}{dt}\right) = m\dfrac{d\mathbf{v}}{dt} \cdot \mathbf{v} = \mathbf{F} \cdot \dfrac{d\mathbf{r}}{dt}$

$$= 4r^2\left(\mathbf{r} \cdot \dfrac{d\mathbf{r}}{dt}\right) = 4r^2\left(r\dfrac{dr}{dt}\right) = 4r^3\dfrac{dr}{dt} = \dfrac{d}{dt}\left(r^4\right).$$

Therefore $d/dt\left(\frac{1}{2}mv^2 - r^4\right) = 0$ and $\frac{1}{2}mv^2 - r^4$ is a constant E.   Evaluating E from $t = 0,$ we find that $E = 2m.$

Thus $\frac{1}{2}mv^2 - r^4 = 2m$ and $v = \sqrt{4 + (2/m)\,r^4}\,.$

SECTION 14.7

1. On Earth: year of length $T,$ average distance from sun $d.$

On Venus: year of length $\alpha\,T,$ average distance from sun $0.72d.$

Therefore

$$\dfrac{(\alpha\,T)^2}{T^2} = \dfrac{(0.72d)^3}{d^3}.$$

This gives $\alpha^2 = (0.72)^3 \cong 0.372$ and $\alpha \cong 0.615.$ Answer: about 61.5% of an Earth year.

3.
$$\left(\frac{dx}{dt}\right)^2 + \left(\frac{dy}{dt}\right)^2 = \left[\frac{d}{dt}(r\cos\theta)\right]^2 + \left[\frac{d}{dt}(r\sin\theta)\right]^2$$

$$= \left[r(-\sin\theta)\frac{d\theta}{dt} + \frac{dr}{dt}\cos\theta\right]^2 + \left[r\cos\theta\frac{d\theta}{dt} + \frac{dr}{dt}\sin\theta\right]^2$$

$$= r^2\sin^2\theta\left(\frac{d\theta}{dt}\right)^2 + \left(\frac{dr}{dt}\right)^2\cos^2\theta + r^2\cos^2\theta\left(\frac{d\theta}{dt}\right)^2 + \left(\frac{dr}{dt}\right)^2\sin^2\theta$$

$$= \left(\frac{dr}{dt}\right)^2 + r^2\left(\frac{d\theta}{dt}\right)^2$$

5.   Substitute
$$r = \frac{a}{1+e\cos\theta}, \quad \left(\frac{dr}{d\theta}\right)^2 = \left[\frac{-a}{(1+e\cos\theta)^2}\cdot(-e\sin\theta)\right]^2 = \frac{(ae\sin\theta)^2}{(1+e\cos\theta)^4}$$

into the right side of the equation and you will see that, with a and e^2 as given, the expression reduces to E.

REVIEW EXERCISES

1. $\mathbf{f}'(t) = 6t\,\mathbf{i} - 15t^2\,\mathbf{j}, \qquad \mathbf{f}''(t) = 6\,\mathbf{i} - 30t\,\mathbf{j}$

3. $\mathbf{f}'(t) = (e^t\cos t - e^t\sin t)\,\mathbf{i} + 2\sin 2t\,\mathbf{j}, \qquad \mathbf{f}''(t) = -2e^t\sin t\,\mathbf{i} + 4\cos 2t\,\mathbf{j}$

5. $\displaystyle\int_0^2 \left[2t\,\mathbf{i} + (t^2-1)\,\mathbf{j}\right]\,dt = \left[t^2\,\mathbf{i} + \left(\frac{1}{3}t^3 - t\right)\mathbf{j}\right]_0^2 = 4\,\mathbf{i} + \frac{2}{3}\,\mathbf{j}$

7.

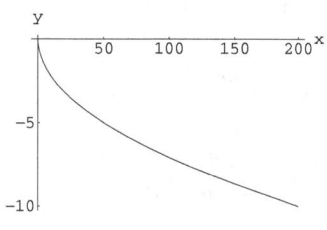

9.

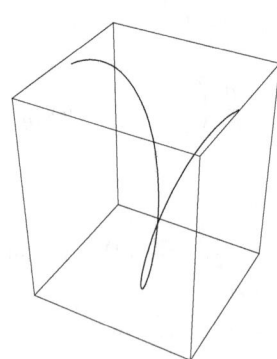

11.   (a) $\mathbf{r}(t) = 2\cos\left(t+\frac{\pi}{2}\right)\mathbf{i} + 4\sin\left(t+\frac{\pi}{2}\right)\mathbf{j}$ (b) $\mathbf{r}(t) = -2\cos 2t\,\mathbf{i} + 4\sin 2t\,\mathbf{j}$

13. $\mathbf{f}(t) = \frac{1}{3}t^3\,\mathbf{i} + \left(\frac{1}{2}e^{2t} + t\right)\mathbf{j} + \frac{1}{3}(2t+1)^{3/2}\mathbf{k} + \mathbf{C}.$

$\mathbf{f}(0) = \mathbf{i} - 3\,\mathbf{j} + 3\mathbf{k} \Longrightarrow \mathbf{C} = \mathbf{i} - \frac{7}{2}\mathbf{j} + \frac{8}{3}\mathbf{k}; \quad \mathbf{f}(t) = \left(\frac{1}{3}t^3 + 1\right)\mathbf{i} + \left(\frac{1}{2}e^{2t} + t - \frac{7}{2}\right)\mathbf{j} + \left(\frac{1}{3}(2t+1)^{3/2} + \frac{8}{3}\right)\mathbf{k}$

15. $\mathbf{f}'(t) = (6\,\mathbf{i} + 12t^3\,\mathbf{j}) + (8t\,\mathbf{i} - 12\,\mathbf{k}) = (6+8t)\,\mathbf{i} + 12t^3\,\mathbf{j} - 12\mathbf{k}$

17. $\mathbf{f}(t) = (t^2 + 2t^3)\,\mathbf{i} - \left(2t^2 + \dfrac{1}{t^2}\right)\mathbf{j} + (t^4 - t)\,\mathbf{k}, \quad \mathbf{f}'(t) = (2t + 6t^2)\,\mathbf{i} - \left(4t - \dfrac{2}{t^3}\right)\mathbf{j} + (4t^3 - 1)\,\mathbf{k}$

19. $\mathbf{r}'(t) = 2\mathbf{r}(t) \Longrightarrow \mathbf{r}(t) = \mathbf{r}_0 e^{2t}$

$\mathbf{r}(0) = (1,2,1) \Longrightarrow \mathbf{r}_0 = (1,2,1) \quad \text{and} \quad \mathbf{r}(t) = (e^{2t}, 2e^{2t}, e^{2t})$

21. The tip of $\mathbf{r}(t)$ is $P(1,1,1)$ when $t = 0$.

$\mathbf{r}'(t) = (2t + 2)\,\mathbf{i} + 3\,\mathbf{j} + (3t^2 + 1)\,\mathbf{k}, \quad \mathbf{r}'(0) = 2\,\mathbf{i} + 3\,\mathbf{j} + \mathbf{k}$

Scalar parametric equations for the tangent line are: $x = 1 + 2t, \quad y = 1 + 3t, \quad z = 1 + t$.

23. $\mathbf{r}_1(t) = (2,1,1)$ at $t = 1; \quad \mathbf{r}_2(u) = (2,1,1)$ at $u = -1$. Therefore the curves intersect at the point $(2,1,1)$.

$\mathbf{r}_1'(t) = (2\,\mathbf{i} + 2t\,\mathbf{j} + \mathbf{k}, \quad \mathbf{r}_1(1) = 2\,\mathbf{i} + 2\,\mathbf{j} + \mathbf{k}; \quad \mathbf{r}_2(u) = -\mathbf{i} - 2u\,\mathbf{j} + 2u\,\mathbf{k}, \quad \mathbf{r}_2'(-1) = -\mathbf{i} + 2\,\mathbf{j} - 2\,\mathbf{k}.$

Since $\mathbf{r}_1'(1) \cdot \mathbf{r}_2'(-1) = 0,$ the angle of intersection is $\pi/2$ radians

25. $\mathbf{r}(t) = t\,\mathbf{i} + e^{2t}\,\mathbf{j}, \quad \mathbf{r}(0) = \mathbf{j};$

$\mathbf{r}'(t) = \mathbf{i} + 2e^{2t}\,\mathbf{j}, \quad \mathbf{r}'(0) = \mathbf{i} + 2\,\mathbf{j}$

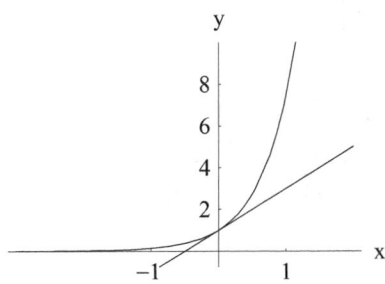

27. $\mathbf{r}'(t) = t\cos t\,\mathbf{i} + t\sin t\,\mathbf{j} + \sqrt{3}t\mathbf{k}; \quad \|\mathbf{r}'\| = \dfrac{ds}{dt} = 2t$

unit tangent vector: $\mathbf{T} = \dfrac{\mathbf{r}'(t)}{\|\mathbf{r}'(t)\|} = \dfrac{1}{2}(\cos t\,\mathbf{i} + \sin t\,\mathbf{j} + \sqrt{3}\,\mathbf{k}).$

$\mathbf{T}'(t) = -\dfrac{1}{2}\sin t\,\mathbf{i} + \dfrac{1}{2}\cos t\,\mathbf{j}; \quad \|\mathbf{T}'(t)\| = \dfrac{1}{2}.$

principal normal vector: $\mathbf{N} = \dfrac{\mathbf{T}'(t)}{\|\mathbf{T}'(t)\|} = -\sin t\,\mathbf{i} + \cos t\,\mathbf{j}$

29. $\mathbf{r}'(t) = 2\,\mathbf{i} + \dfrac{1}{t}\,\mathbf{j} - 2t\mathbf{k}; \quad \|\mathbf{r}'(t)\| = \dfrac{1 + 2t^2}{t}.$

$\mathbf{T}(t) = \dfrac{\mathbf{r}'(t)}{\|\mathbf{r}'(t)\|} = \dfrac{2t}{2t^2 + 1}\,\mathbf{i} + \dfrac{1}{2t^2 + 1}\,\mathbf{j} - \dfrac{2t^2}{2t^2 + 1}\mathbf{k}; \quad \mathbf{T}(1) = \dfrac{2}{3}\,\mathbf{i} + \dfrac{1}{3}\,\mathbf{j} - \dfrac{2}{3}\,\mathbf{k}$

$\mathbf{T}'(t) = \dfrac{2 - 4t^2}{(2t^2 + 1)^2}\,\mathbf{i} - \dfrac{4t}{(2t^2 + 1)^2}\,\mathbf{j} - \dfrac{4t}{(2t^2 + 1)^2}\mathbf{k}; \quad \mathbf{T}'(1) = -\dfrac{2}{9}\,\mathbf{i} - \dfrac{4}{9}\,\mathbf{j} - \dfrac{4}{9}\,\mathbf{k}, \quad \|\mathbf{T}'(1)\| = \dfrac{2}{3}$

$\mathbf{N}(1) = \dfrac{\mathbf{T}'(1)}{\|\mathbf{T}'(1)\|} = -\dfrac{1}{3}\,\mathbf{i} - \dfrac{2}{3}\,\mathbf{j} - \dfrac{2}{3}\,\mathbf{k}$

A normal vector for the osculating plane is: $(2\,\mathbf{i} + \mathbf{j} - 2\,\mathbf{k}) \times (\mathbf{i} + 2\,\mathbf{j} + 2\,\mathbf{k}) = 6\,\mathbf{i} - 6\,\mathbf{j} + 3\,\mathbf{k}.$

Since $\mathbf{r}(1) = 2\mathbf{i} - \mathbf{k}$, an equation for the osculating plane is

$$6(x - 2) - 6y + 3(z + 1) = 0 \quad \text{or} \quad 2x - 2y + z = 3.$$

31. $\mathbf{r}'(t) = 2\mathbf{i} + t^{1/2}\mathbf{j}; \quad L = \int_0^5 \|\mathbf{r}'(t)\| dt = \int_0^5 \sqrt{4 + t}\, dt = \dfrac{38}{3}$

33. $\mathbf{r}'(t) = \cosh t\,\mathbf{i} + \sinh t\,\mathbf{j} + \mathbf{k}; \quad \|\mathbf{r}'(t)\| = \sqrt{\cosh^2 t + \sinh^2 t + 1} = \sqrt{2}\,\cosh t;$

$L = \displaystyle\int_0^1 \sqrt{2}\,\cosh t\, dt = \left[\sqrt{2}\,\sinh t\right]_0^1 = \sqrt{2}\,\sinh 1.$

35. (a) $\mathbf{r}'(t) = -\sin t\,\mathbf{i} + \cos t\,\mathbf{j} + t^{1/2}\,\mathbf{k}; \quad \|\mathbf{r}'(t)\| = \sqrt{1 + t}.$

$s = \displaystyle\int_0^t \|\mathbf{r}'(u)\|\, du = \int_0^t \sqrt{1 + t}\, dt = \left[\frac{2}{3}(1 + u)^{3/2}\right]_0^t = \frac{2}{3}(1 + t)^{3/2} - \frac{2}{3}$

(b) $t = \left(\dfrac{3}{2}s + 1\right)^{2/3} - 1 = \phi(s); \quad \mathbf{R}(s) = \cos \phi(s)\,\mathbf{i} + \sin \phi(s)\,\mathbf{j} + \dfrac{2}{3}[\phi(s)]^{3/2}\,\mathbf{k}$

(c) $\mathbf{R}'(s) = \left[-\sin \phi(s)\,\mathbf{i} + \cos \phi(s)\,\mathbf{j} + \phi(s)^{1/2}\,\mathbf{k}\right]\phi'(s)$

$\|\mathbf{R}'(s)\| = \phi'(s)\sqrt{1 + \phi(s)} = \dfrac{2}{3}\left[\dfrac{3}{2}s + 1\right]^{-1/3} \left(\dfrac{3}{2}\right)\sqrt{\left(\dfrac{3}{2}s + 1\right)^{2/3}} = 1$

37. $\mathbf{r}''(t) = -\cos t\,\mathbf{i} - \sin t\,\mathbf{j}$ and $\mathbf{r}'(0) = \mathbf{k} \implies \mathbf{r}'(t) = -\sin t\,\mathbf{i} + (\cos t - 1)\mathbf{j} + \mathbf{k}.$

Thus: velocity $\mathbf{v} = -\sin t\,\mathbf{i} + (\cos t - 1)\mathbf{j} + \mathbf{k}$ and speed $\|\mathbf{v}\| = \sqrt{3 - 2\cos t}.$

$\mathbf{r}'(t) = -\sin t\,\mathbf{i} + (\cos t - 1)\mathbf{j} + \mathbf{k}$ and $\mathbf{r}(0) = \mathbf{i} \implies \mathbf{r}(t) = \cos t\,\mathbf{i} + (\sin t - t)\mathbf{j} + t\,\mathbf{k}.$

39. $y' = \dfrac{3}{2}x^{1/2}, \quad y'' = \dfrac{3}{4}x^{-1/2};$

$\kappa = \dfrac{|y''|}{[1 + (y')^2]^{3/2}} = \dfrac{\frac{3}{4}x^{-1/2}}{[1 + \frac{9}{4}x]^{3/2}} = \dfrac{6}{\sqrt{x}(4 + 9x)^{3/2}}$

41. $x(t) = 2e^{-t}, \; y(t) = e^{-2t} \implies x'(t) = -2e^{-t}, \; y'(t) = -2e^{-2t} \implies x''(t) = 2e^{-t}, \; y''(t) = 4e^{-2t}$

$\kappa = \dfrac{|(-2e^{-t})(4e^{-2t}) - (-2e^{-2t})(2e^{-t})|}{[4e^{-2t} + 4e^{-4t}]^{3/2}} = \dfrac{1}{2(1 + e^{-2t})^{3/2}} = \dfrac{e^{3t}}{2(e^{2t} + 1)^{3/2}}$

43. $\mathbf{r}'(t) = -3\sin 3t\,\mathbf{i} - 4\mathbf{j} + 3\cos 3t\,\mathbf{k}, \quad \dfrac{ds}{dt} = |\mathbf{r}'(t)| = 5$

$\mathbf{T}(t) = -\dfrac{3}{5}\sin 3t\,\mathbf{i} - \dfrac{4}{5}\mathbf{j} + \dfrac{3}{5}\cos 3t\,\mathbf{k}, \quad \mathbf{T}'(t) = -\dfrac{9}{5}\cos 3t\,\mathbf{i} - \dfrac{9}{5}\sin 3t\,\mathbf{k}; \quad \|\mathbf{T}'(t)\| = 9/5$

$\kappa = \dfrac{\|\mathbf{T}'(t)\|}{ds/dt} = \dfrac{9}{25}$

45. $y' = \sinh(x/a), \quad y'' = \dfrac{1}{a}\cosh(x/a); \quad \kappa = \dfrac{|y''|}{[1 + (y')^2]^{\frac{3}{2}}} = \dfrac{1}{a\cosh^2(x/a)} = \dfrac{a}{y^2}$

47. $\mathbf{r}'(t) = -\dfrac{4}{5}\sin t\,\mathbf{i} + \dfrac{3}{5}\sin t\,\mathbf{j} + \cos t\,\mathbf{k};\quad \dfrac{ds}{dt} = \|\mathbf{r}'(t)\| = 1$

$\mathbf{T} = \mathbf{r}'(t) = -\dfrac{4}{5}\sin t\,\mathbf{i} + \dfrac{3}{5}\sin t\,\mathbf{j} + \cos t\,\mathbf{k},\quad \mathbf{T}'(t) = -\dfrac{4}{5}\cos t\,\mathbf{i} + \dfrac{3}{5}\cos t\,\mathbf{j} - \sin t\,\mathbf{k};\quad \|\mathbf{T}'(t)\| = 1$

$\kappa = 1;\qquad \mathbf{a_T} = \dfrac{d^2 s}{dt^2} = 0,\quad \mathbf{a_N} = \kappa\left(\dfrac{ds}{dt}\right)^2 = 1$

CHAPTER 15

SECTION 15.1

1. dom (f) = the first and third quadrants, including the axes;   range $(f) = [0, \infty)$

3. dom (f) = the set of all points (x, y) except those on the line $y = -x$;   range $(f) = (-\infty, 0) \cup (0, \infty)$

5. dom (f) = the entire plane;   range $(f) = (-1, 1)$ since
    $$\frac{e^x - e^y}{e^x + e^y} = \frac{e^x + e^y - 2e^y}{e^x + e^y} = 1 - \frac{2}{e^{x-y} + 1}$$
    and the last quotient takes on all values between 0 and 2.

7. dom (f) = the first and third quadrants, excluding the axes;   range $(f) = (-\infty, \infty)$

9. dom (f) = the set of all points (x, y) with $x^2 < y$ —in other words, the set of all points of the plane
    above the parabola $y = x^2$;   range $(f) = (0, \infty)$

11. dom (f) = the set of all points (x, y) with $-3 \le x \le 3$, $-2 \le y \le 2$ (a rectangle);
     range $(f) = [-2, 3]$

13. dom (f) = the set of all points (x, y, z) not on the plane $x + y + z = 0$;   range $(f) = \{-1, 1\}$

15. dom (f) = the set of all points (x, y, z) with $|y| < |x|$;   range $(f) = (-\infty, 0]$

17. dom (f) = the set of all points (x, y) with $x^2 + y^2 < 9$ —in other words, the set of all points of the
     plane inside the circle $x^2 + y^2 = 9$;   range $(f) = [2/3, \infty)$

19. dom (f) = the set of all points (x, y, z) with $x + 2y + 3z > 0$ — in other words, the set of all points in
     space that lie on the same side of the plane $x + 2y + 3z = 0$ as the point $(1, 1, 1)$;   range $(f) = (-\infty, \infty)$

21. dom (f) = all of space;   range $(f) = (0, 1]$

23. dom $(f) = \{x : x \ge 0\}$;   range $(f) = [0, \infty)$
     dom $(g) = \{(x, y) : x \ge 0, \ y \ \text{real}\}$;   range $(g) = [0, \infty)$
     dom $(h) = \{(x, y, z) : x \ge 0, \ y, z \ \text{real}\}$;   range $(h) = [0, \infty)$

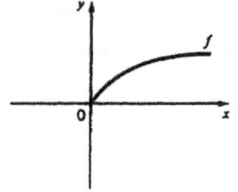 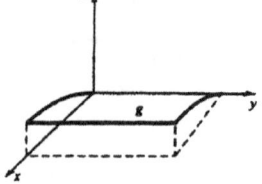

25. $\displaystyle\lim_{h\to 0}\frac{f(x+h,y)-f(x,y)}{h}=\lim_{h\to 0}\frac{2(x+h)^2-y-(2x^2-y)}{h}=\lim_{h\to 0}\frac{4xh+2h^2}{h}=4x$

$\displaystyle\lim_{h\to 0}\frac{f(x,y+h)-f(x,y)}{h}=\lim_{h\to 0}\frac{2x^2-(y+h)-(2x^2-y)}{h}=-1$

27. $\displaystyle\lim_{h\to 0}\frac{f(x+h,y)-f(x,y)}{h}=\lim_{h\to 0}\frac{3(x+h)-(x+h)y+2y^2-(3x-xy+2y^2)}{h}=\lim_{h\to 0}\frac{3h-hy}{h}=3-y$

$\displaystyle\lim_{h\to 0}\frac{f(x,y+h)-f(x,y)}{h}=\lim_{h\to 0}\frac{3x-x(y+h)+2(y+h)^2-(3x-xy+2y^2)}{h}$

$\displaystyle\qquad\qquad =\lim_{h\to 0}\frac{-xh+4yh+2h^2}{h}=-x+4y$

29. $\displaystyle\lim_{h\to 0}\frac{f(x+h,y)-f(x,y)}{h}=\lim_{h\to 0}\frac{\cos[(x+h)y]-\cos[xy]}{h}$

$\displaystyle\qquad\qquad =\lim_{h\to 0}\frac{\cos[xy]\cos[hy]-\sin[xy]\sin[hy]-\cos[xy]}{h}$

$\displaystyle\qquad\qquad =\cos[xy]\left(\lim_{h\to 0}\frac{\cos[hy]-1}{h}\right)-\sin[xy]\lim_{h\to 0}\frac{\sin hy}{h}$

$\displaystyle\qquad\qquad =y\,\cos[xy]\left(\lim_{h\to 0}\frac{\cos[hy]-1}{hy}\right)-y\,\sin[xy]\lim_{h\to 0}\frac{\sin hy}{hy}$

$\displaystyle\qquad\qquad =-y\,\sin[xy]$

and

$\displaystyle\lim_{h\to 0}\frac{f(x,y+h)-f(x,y)}{h}=\lim_{h\to 0}\frac{\cos[x(y+h)]-\cos[xy]}{h}$

$\displaystyle\qquad\qquad =\lim_{h\to 0}\frac{\cos[xy]\cos[hx]-\sin[xy]\sin[hx]-\cos[xy]}{h}$

$\displaystyle\qquad\qquad =\cos[xy]\left(\lim_{h\to 0}\frac{\cos[hx]-1}{h}\right)-\sin[xy]\lim_{h\to 0}\frac{\sin hx}{h}$

$\displaystyle\qquad\qquad =x\,\cos[xy]\left(\lim_{h\to 0}\frac{\cos[hx]-1}{hx}\right)-x\,\sin[xy]\lim_{h\to 0}\frac{\sin hx}{hx}$

$\displaystyle\qquad\qquad =-x\,\sin[xy]$

31. (a) $f(x,y)=Ay$ (b) $f(x,y)=\pi x^2 y$ (b) $f(x,y)=|2\mathbf{i}\times(x\mathbf{i}+y\mathbf{j})|=2|y|$

33. Surface area: $S=2lw+2lh+2hw=20\implies w=\dfrac{20-2lh}{2l+2h}=\dfrac{10-lh}{l+h}$

Volume: $V=lwh=\dfrac{lh(10-lh)}{l+h}$

35. $V=\pi r^2 h+\dfrac{4}{3}\pi r^3$

SECTION 15.2

1. an elliptic cone

3. a parabolic cylinder

5. a hyperboloid of one sheet

7. a sphere

9. an elliptic paraboloid

11. a hyperbolic paraboloid

13.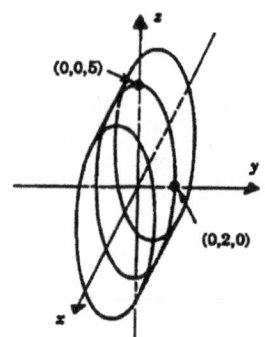

15.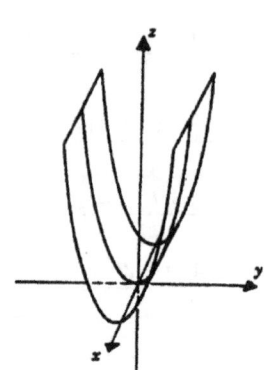

17.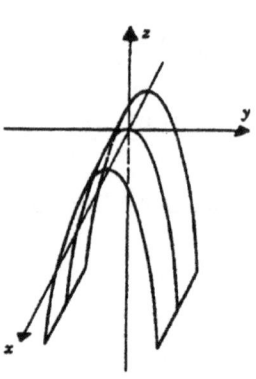

19.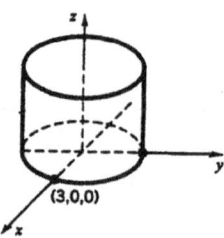

21.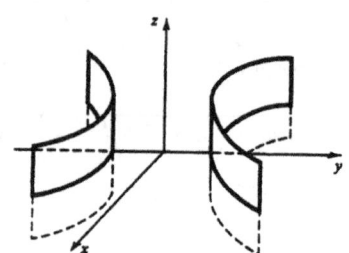

23.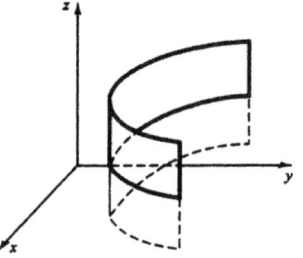

25. elliptic paraboloid
xy-trace: the origin
xz-trace: the parabola $x^2 = 4z$
yz-trace: the parabola $y^2 = 9z$
surface has the form of Figure 15.2.5

27. elliptic cone
xy-trace: the origin
xz-trace: the lines $x = \pm 2z$
yz-trace: the lines $y = \pm 3z$
surface has the form of Figure 15.2.4

29. hyperboloid of two sheets
xy-trace: none
xz-trace: the hyperbola $4z^2 - x^2 = 4$
yz-trace: the hyperbola $9z^2 - y^2 = 9$
surface has the form of Figure 15.2.3

31. hyperboloid of two sheets
xy-trace: the hyperbola $9x^2 - 4y^2 = 36$
xz-trace: the hyperbola $x^2 - 4z^2 = 4$
yz-trace: none
see Figure 15.2.3

33. elliptic paraboloid
xy-trace: the parabola $x^2 = 9y$
xz-trace: the origin
yz-trace: the parabola $z^2 = 4y$
surface has the form of Figure 15.2.5

35. hyperboloid of two sheets
xy-trace: the hyperbola $9y^2 - 4x^2 = 36$
xz-trace: none
yz-trace: the hyperbola $y^2 - 4z^2 = 4$
see Figure 15.2.3

37. paraboloid of revolution
xy-trace: the origin
xz-trace: the parabola $x^2 = 4z$
yz-trace: the parabola $y^2 = 4z$
surface has the form of Figure 15.2.5

39. (a) an elliptic paraboloid (vertex down if A and B are both positive, vertex up if A and B are both negative)

(b) a hyperbolic paraboloid

(c) the xy-plane if A and B are both zero; otherwise a parabolic cylinder

41. $x^2 + y^2 - 4z = 0$ (paraboloid of revolution)

43. (a) a circle

(b) (i) $\sqrt{x^2 + y^2} = -3z$ (ii) $\sqrt{x^2 + z^2} = \frac{1}{3}y$

45. $x + 2y + 3\left(\dfrac{x + y - 6}{2}\right) = 6$ or $5x + 7y = 30$, a line

47. $\left.\begin{array}{l} x^2 + y^2 + (z - 1)^2 = \frac{3}{2} \\ x^2 + y^2 - z^2 = 1 \end{array}\right\}$ $(z^2 + 1) + (z - 1)^2 = \dfrac{3}{2}$; $(2z - 1)^2 = 0$, $z = \dfrac{1}{2}$ so that $x^2 + y^2 = \dfrac{5}{4}$

49. $x^2 + y^2 + \left(x^2 + 3y^2\right) = 4$ or $x^2 + 2y^2 = 2$, an ellipse

51. $x^2 + y^2 = (2 - y)^2$ or $x^2 = -4(y - 1)$, a parabola

53. (a) Set $x = a\cos u\,\cos v$, $y = b\cos u\,\sin v$, $z = c\sin u$. Then: $\dfrac{x^2}{a^2} + \dfrac{y^2}{b^2} + \dfrac{z^2}{c^2} = 1$.

(b)

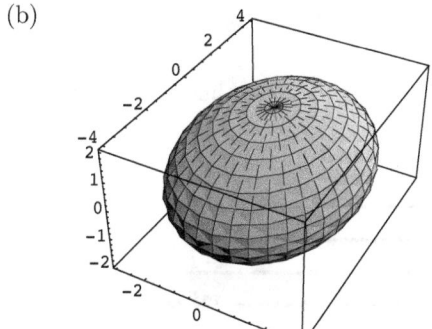

55. (a) Set $x = av\cos u$,
  $y = bv\sin u$,
  $z = cv$

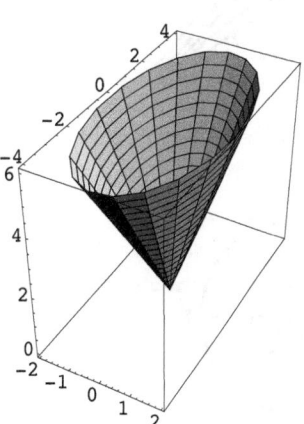

SECTION 15.3

1. lines of slope 1: $y = x - c$

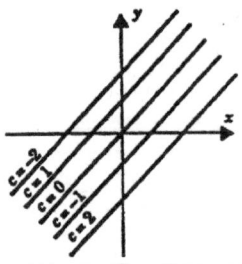

3. parabolas: $y = x^2 - c$

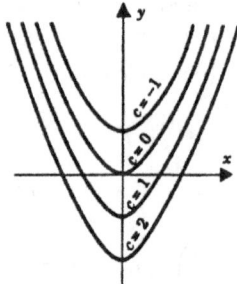

5. the y-axis and the lines $y = \left(\dfrac{1-c}{c}\right) x$ with the origin omitted throughout

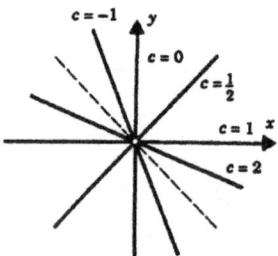

7. the cubics $y = x^3 - c$

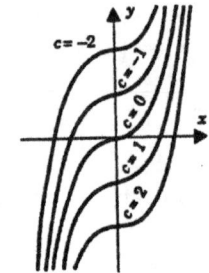

9. the lines $y = \pm x$ and the hyperbolas $x^2 - y^2 = c$

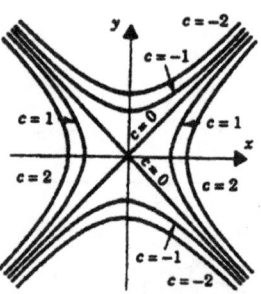

11. pairs of horizontal lines $y = \pm\sqrt{c}$ and the x-axis

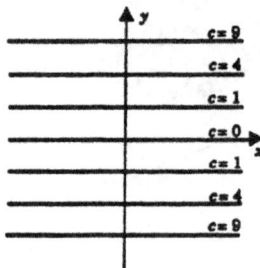

13. the circles $x^2 + y^2 = e^c, c$ real

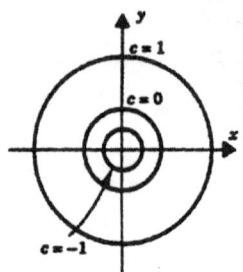

15. the curves $y = e^{cx^2}$ with the point $(0, 1)$ omitted

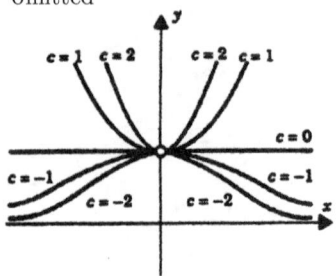

17. the coordinate axes and pairs of lines
$$y = \pm\frac{\sqrt{1-c}}{\sqrt{c}}\, x, \quad \text{with the origin omitted}$$
throughout

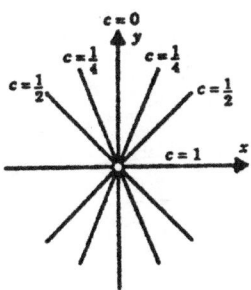

19. $x + 2y + 3z = 0,$ plane through the origin

21. $z = \sqrt{x^2 + y^2},$ the upper nappe of the circular cone $z^2 = x^2 + y^2$ (Figure 15.2.4)

23. the elliptic paraboloid $\dfrac{x^2}{18} + \dfrac{y^2}{8} = z$ (Figure 15.2.5)

25. (i) hyperboloid of two sheets   (Figure 15.2.3)

(ii) circular cone   (Figure 15.2.4)

(iii) hyperboloid of one sheet   (Figure 15.2.2)

27. The level curves of f are: $1 - 4x^2 - y^2 = c$. Substituting $P(0,1)$ into this equation, we have
$$1 - 4(0)^2 - (1)^2 = c \quad \Longrightarrow \quad c = 0$$
The level curve that contains P is: $1 - 4x^2 - y^2 = 0,$ or $4x^2 + y^2 = 1.$

29. The level curves of f are: $y^2 \arctan x = c$. Substituting $P(1,2)$ into this equation, we have
$$4 \arctan 1 = c \quad \Longrightarrow \quad c = \pi$$
The level curve that contains P is: $y^2 \tan^{-1} x = \pi.$

31. The level surfaces of f are: $x^2 + 2y^2 - 2xyz = c$. Substituting $P(-1,2,1)$ into this equation, we have
$$(-1)^2 + 2(2)^2 - 2(-1)(2)(1) = c \quad \Longrightarrow \quad c = 13$$
The level surface that contains P is: $x^2 + 2y^2 - 2xyz = 13.$

33. (a)

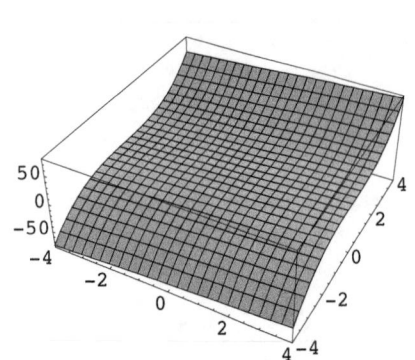

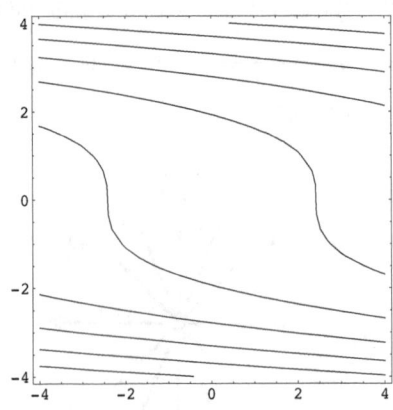

(b)

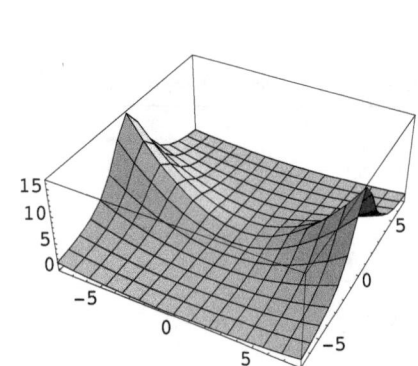

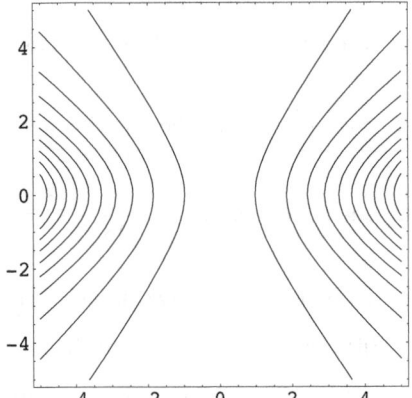

35. (a) $\dfrac{3x + 2y + 1}{4x^2 + 9} = \dfrac{3}{5}$ (b) $x^2 + 2y^2 - z^2 = 21$

37. $\dfrac{GmM}{x^2 + y^2 + z^2} = c \implies x^2 + y^2 + z^2 = \dfrac{GmM}{c}$; the surfaces of constant gravitational force are concentric spheres.

39. (a) $T(x, y, z) = \dfrac{k}{\sqrt{x^2 + y^2 + z^2}}$, where k is a constant.

(b) $\dfrac{k}{\sqrt{x^2 + y^2 + z^2}} = c \implies x^2 + y^2 + z^2 = \dfrac{k^2}{c^2}$; the level surfaces are concentric spheres.

(c) $T(1, 2, 1) = \dfrac{k}{\sqrt{1^2 + 2^2 + 1^2}} = 50 \implies k = 50\sqrt{6} \implies T(x, y, z) = \dfrac{50\sqrt{6}}{\sqrt{x^2 + y^2 + z^2}}$

41. $f(x, y) = y^2 - y^3$; F   **43.** $f(x, y) = \cos\sqrt{x^2 + y^2}$; A   **45.** $f(x, y) = xye^{-(x^2+y^2)/2}$; E

PROJECT 15.3

1.

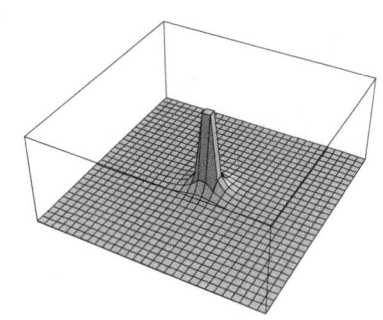

3.

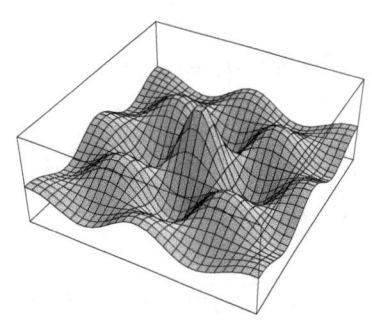

SECTION 15.4

1. $\dfrac{\partial f}{\partial x} = 6x - y, \quad \dfrac{\partial f}{\partial y} = 1 - x$

3. $\dfrac{\partial \rho}{\partial \phi} = \cos\phi\cos\theta, \quad \dfrac{\partial \rho}{\partial \theta} = -\sin\phi\sin\theta$

5. $\dfrac{\partial f}{\partial x} = e^{x-y} + e^{y-x}, \quad \dfrac{\partial f}{\partial y} = -e^{x-y} - e^{y-x}$

7. $\dfrac{\partial g}{\partial x} = \dfrac{(AD - BC)y}{(Cx + Dy)^2}, \quad \dfrac{\partial g}{\partial y} = \dfrac{(BC - AD)x}{(Cx + Dy)^2}$

9. $\dfrac{\partial u}{\partial x} = y + z, \quad \dfrac{\partial u}{\partial y} = x + z, \quad \dfrac{\partial u}{\partial z} = x + y$

11. $\dfrac{\partial f}{\partial x} = z\cos(x - y), \quad \dfrac{\partial f}{\partial y} = -z\cos(x - y), \quad \dfrac{\partial f}{\partial z} = \sin(x - y)$

13. $\dfrac{\partial \rho}{\partial \theta} = e^{\theta+\phi}\left[\cos(\theta - \phi) - \sin(\theta - \phi)\right], \quad \dfrac{\partial \rho}{\partial \phi} = e^{\theta+\phi}\left[\cos(\theta - \phi) + \sin(\theta - \phi)\right]$

15. $\dfrac{\partial f}{\partial x} = 2xy\sec xy + x^2 y(\sec xy)(\tan xy)y = 2xy\sec xy + x^2 y^2 \sec xy \tan xy$

$\dfrac{\partial f}{\partial y} = x^2 \sec xy + x^2 y(\sec xy)(\tan xy)x = x^2 \sec xy + x^3 y \sec xy \tan xy$

17. $\dfrac{\partial h}{\partial x} = \dfrac{x^2 + y^2 - x(2x)}{\left(x^2 + y^2\right)^2} = \dfrac{y^2 - x^2}{\left(x^2 + y^2\right)^2} \qquad \dfrac{\partial h}{\partial y} = \dfrac{-2xy}{\left(x^2 + y^2\right)^2}$

19. $\dfrac{\partial f}{\partial x} = \dfrac{(y\cos x)\sin y - (x\sin y)(-y\sin x)}{(y\cos x)^2} = \dfrac{\sin y(\cos x + x\sin x)}{y\cos^2 x}$

$\dfrac{\partial f}{\partial y} = \dfrac{(y\cos x)(x\cos y) - (x\sin y)\cos x}{(y\cos x)^2} = \dfrac{x(y\cos y - \sin y)}{y^2 \cos x}$

21. $\dfrac{\partial h}{\partial x} = 2f(x)f'(x)g(y), \quad \dfrac{\partial h}{\partial y} = [f(x)]^2 g'(y)$

23. $\dfrac{\partial f}{\partial x} = (y^2 \ln z)z^{xy^2}, \quad \dfrac{\partial f}{\partial y} = (2xy\ln z)z^{xy^2}, \quad \dfrac{\partial f}{\partial z} = xy^2 z^{xy^2 - 1}$

25. $\dfrac{\partial h}{\partial r} = 2re^{2t}\cos(\theta - t)$ $\qquad\qquad\qquad\qquad \dfrac{\partial h}{\partial \theta} = -r^2 e^{2t}\sin(\theta - t)$

$\dfrac{\partial h}{\partial t} = 2r^2 e^{2t}\cos(\theta - t) + r^2 e^{2t}\sin(\theta - t) = r^2 e^{2t}[2\cos(\theta - t) + \sin(\theta - t)]$

27. $\dfrac{\partial f}{\partial x} = z\,\dfrac{1}{1 + (y/x)^2}\left(\dfrac{-y}{x^2}\right) = -\dfrac{yz}{x^2 + y^2}$ $\qquad \dfrac{\partial f}{\partial y} = z\,\dfrac{1}{1 + (y/x)^2}\left(\dfrac{1}{x}\right) = \dfrac{xz}{x^2 + y^2}$

$\dfrac{\partial f}{\partial x} = \arctan(y/x)$

29. $f_x(x, y) = e^x \ln y, \quad f_x(0, e) = 1; \quad f_y(x, y) = \dfrac{1}{y}\,e^x, \quad f_y(0, e) = e^{-1}$

31. $f_x(x, y) = \dfrac{y}{(x + y)^2}, \quad f_x(1, 2) = \dfrac{2}{9}; \quad f_y(x, y) = \dfrac{-x}{(x + y)^2}, \quad f_y(1, 2) = -\dfrac{1}{9}$

33. $f_x(x, y) = \lim\limits_{h \to 0} \dfrac{(x + h)^2 y - x^2 y}{h} = \lim\limits_{h \to 0} y\left(\dfrac{2xh + h^2}{h}\right) = y\lim\limits_{h \to 0}(2x + h) = 2xy$

$f_x(x, y) = \lim\limits_{h \to 0} \dfrac{x^2(y + h) - x^2 y}{h} = \lim\limits_{h \to 0} \dfrac{x^2 h}{h} = \lim\limits_{h \to 0} x^2 = x^2$

35. $f_x(x, y) = \lim\limits_{h \to 0} \dfrac{\ln\left(y(x + h)^2\right) - \ln x^2 y}{h} = \lim\limits_{h \to 0} \dfrac{\ln y + 2\ln(x + h) - 2\ln x - \ln y}{h}$

$\qquad = 2\lim\limits_{h \to 0} \dfrac{\ln(x + h) - \ln x}{h} = 2\dfrac{d}{dx}(\ln x) = \dfrac{2}{x}$

$f_y(x, y) = \lim\limits_{h \to 0} \dfrac{\ln\left(x^2(y + h)\right) - \ln x^2 y}{h} = \lim\limits_{h \to 0} \dfrac{\ln x^2 + \ln(y + h) - \ln x^2 - \ln y}{h}$

$\qquad\qquad\qquad = \lim\limits_{h \to 0} \dfrac{\ln(y + h) - \ln y}{h} = \dfrac{d}{dy}(\ln y) = \dfrac{1}{y}$

37. $f_y(x, y) = \lim\limits_{h \to 0} \dfrac{1}{h}\left\{\dfrac{1}{(x + h) - y} - \dfrac{1}{x - y}\right\} = \lim\limits_{h \to 0} \dfrac{1}{h}\left\{\dfrac{-h}{(x + h - y)(x - y)}\right\}$

$\qquad = \lim\limits_{h \to 0} \dfrac{-1}{(x + h - y)(x - y)} = \dfrac{-1}{(x - y)^2}$

$f_y(x, y) = \lim\limits_{h \to 0} \dfrac{1}{h}\left\{\dfrac{1}{x - (y + h)} - \dfrac{1}{x - y}\right\} = \lim\limits_{h \to 0} \dfrac{1}{h}\left\{\dfrac{h}{(x - y - h)(x - y)}\right\}$

$\qquad = \lim\limits_{h \to 0} \dfrac{1}{(x - y - h)(x - y)} = \dfrac{1}{(x - y)^2}$

39. $f_x(x, y, z) = \lim\limits_{h \to 0} \dfrac{(x + h)y^2 z - xy^2 z}{h} = \lim\limits_{h \to 0} y^2 z = y^2 z$

$f_y(x, y, z) = \lim\limits_{h \to 0} \dfrac{x(y + h)^2 z - xy^2 z}{h} = \lim\limits_{h \to 0} \dfrac{xz(2yh + h^2)}{h}$

$\qquad\qquad = \lim\limits_{h \to 0} xz(2y + h) = 2xyz$

$f_z(x, y, z) = \lim\limits_{h \to 0} \dfrac{xy^2(z + h) - xy^2 z}{h} = \lim\limits_{h \to 0} xy^2 = xy^2$

41. (b) The slope of the tangent line to C at the point $P(x_0, y_0, f(x_0, y_0))$ is $f_y(x_0, y_0)$

Thus, equations for the tangent line are:

$$x = x_0, \quad z - z_0 = f_y(x_0, y_0)(y - y_0)$$

43. Let $z = f(x, y) = x^2 + y^2$. Then $f(2, 1) = 5$, $\quad f_y(x, y) = 2y$ and $\quad f_y(2, 1) = 2$;

equations for the tangent line are: $\quad x = 2, \quad z - 5 = 2(y - 1)$

45. Let $z = f(x, y) = \dfrac{x^2}{y^2 - 3}$. Then $f(3, 2) = 9$, $\quad f_x(x, y) = \dfrac{2x}{y^2 - 3}$ and $\quad f_x(3, 2) = 6$;

equations for the tangent line are: $\quad y = 2, \quad z - 9 = 6(x - 3)$

47. (a) $m_x = -6$; $\quad$ tangent line: $\quad y = 2, \quad z = -6x + 13$

(b) $m_y = 18$; $\quad$ tangent line: $\quad x = 1, \quad z = 18y - 29$

49. $u_x(x, y) = 2x = v_y(x, y)$; $\quad u_y(x, y) = -2y = -v_x(x, y)$

51. $u_x(x, y) = \dfrac{1}{2} \dfrac{1}{x^2 + y^2} 2x = \dfrac{x}{x^2 + y^2}$; $\quad v_y(x, y) = \dfrac{1}{1 + (y/x)^2} \left(\dfrac{1}{x}\right) = \dfrac{x}{x^2 + y^2}$

Thus, $u_x(x, y) = v_y(x, y)$.

$u_y(x, y) = \dfrac{1}{2} \dfrac{1}{x^2 + y^2} 2y = \dfrac{y}{x^2 + y^2}$; $\quad v_x(x, y) = \dfrac{1}{1 + (y/x)^2} \left(\dfrac{-y}{x^2}\right) = \dfrac{-y}{x^2 + y^2}$

Thus, $u_y(x, y) = -v_x(x, y)$.

53. (a) f depends only on y. $\qquad\qquad\qquad$ (b) f depends only on x.

55. (a) $\dfrac{75\sqrt{3}}{2}$ in.^{2}

(b) $\dfrac{\partial A}{\partial b} = \dfrac{1}{2} c \sin\theta$; $\quad$ at time t_0, $\quad \dfrac{\partial A}{\partial b} = \dfrac{15\sqrt{3}}{4}$

(c) $\dfrac{\partial A}{\partial \theta} = \dfrac{1}{2} bc \cos\theta$; $\quad$ at time t_0, $\quad \dfrac{\partial A}{\partial \theta} = \dfrac{75}{2}$

(d) with $\quad h = \dfrac{\pi}{180}$, $\quad A(b, c, \theta + h) - A(b, c, \theta) \cong h \dfrac{\partial A}{\partial \theta} = \dfrac{\pi}{180} \dfrac{75}{2} = \dfrac{5\pi}{24}$ in.^{2}

(e) $0 = \dfrac{1}{2} \sin\theta \left(b \dfrac{\partial c}{\partial b} + c\right)$; $\quad$ at time t_0, $\quad \dfrac{\partial c}{\partial b} = \dfrac{-c}{b} = -\dfrac{3}{2}$

57. (a) y_0-section: $\mathbf{r}(x) = x\mathbf{i} + y_0\mathbf{j} + f(x, y_0)\mathbf{k}$

$\quad$ tangent line: $\mathbf{R}(t) = [x_0\mathbf{i} + y_0\mathbf{j} + f(x_0, y_0)\mathbf{k}] + t\left[\mathbf{i} + \dfrac{\partial f}{\partial x}(x_0, y_0)\mathbf{k}\right]$

(b) x_0-section: $\mathbf{r}(y) = x_0\mathbf{i} + y\mathbf{j} + f(x_0, y)\mathbf{k}$

$\quad$ tangent line: $\mathbf{R}(t) = [x_0\mathbf{i} + y_0\mathbf{j} + f(x_0, y_0)\mathbf{k}] + t\left[\mathbf{j} + \dfrac{\partial f}{\partial y}(x_0, y_0)\mathbf{k}\right]$

(c) For (x, y, z) in the plane

$$[(x - x_0)\mathbf{i} + (y - y_0)\mathbf{j} + (z - f(x_0, y_0))\mathbf{k}] \cdot \left[\left(\mathbf{i} + \frac{\partial f}{\partial x}(x_0, y_0)\mathbf{k}\right) \times \left(\mathbf{j} + \frac{\partial f}{\partial y}(x_0, y_0)\mathbf{k}\right)\right] = 0.$$

From this it follows that

$$z - f(x_0, y_0) = (x - x_0)\frac{\partial f}{\partial x}(x_0, y_0) + (y - y_0)\frac{\partial f}{\partial y}(x_0, y_0).$$

59. (a) Set $u = ax + by$. Then

$$b\frac{\partial w}{\partial x} - a\frac{\partial w}{\partial y} = b(a\,g'(u)) - a(b\,g'(u)) = 0.$$

(b) Set $u = x^m y^n$. Then

$$nx\frac{\partial w}{\partial x} - my\frac{\partial w}{\partial y} = nx\left[mx^{m-1}y^n g'(u)\right] - my\left[nx^m y^{n-1} g'(u)\right] = 0.$$

61. $V\dfrac{\partial P}{\partial V} = V\left(-\dfrac{kT}{V^2}\right) = -k\dfrac{T}{V} = -P; \qquad V\dfrac{\partial P}{\partial V} + T\dfrac{\partial P}{\partial T} = -k\dfrac{T}{V} + T\left(\dfrac{k}{V}\right) = 0$

SECTION 15.5

1. interior $= \{(x, y) : 2 < x < 4, \quad 1 < y < 3\}$ (the inside of the rectangle), boundary $=$ the union of the four boundary line segments; set is closed.

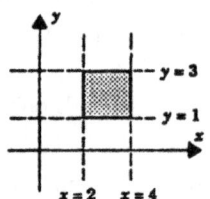

3. interior $=$ the entire set (region between the two concentric circles), boundary $=$ the two circles, one of radius 1, the other of radius 2; set is open.

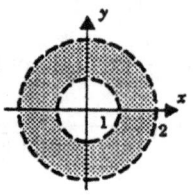

5. interior $= \{(x, y) : 1 < x^2 < 4\} =$
$\{(x, y) : -2 < x < -1\} \cup \{(x, y) : 1 < x < 2\}$
(two vertical strips without the boundary lines),
boundary $= \{(x, y) : x = -2, x = -1, x = 1,$
or $x = 2\}$ (four vertical lines); set is neither open
nor closed.

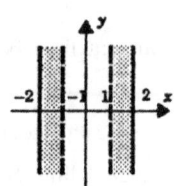

7. interior = region below the parabola $y = x^2$,
boundary = the parabola $y = x^2$;   the set is closed.

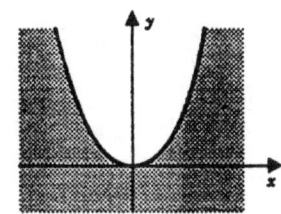

9. interior = $\{(x, y, z) : x^2 + y^2 < 1, 0 < z \leq 4\}$
(the inside of the cylinder), boundary = the total
surface of the cylinder (the curved part, the top, and
the bottom);   the set is closed.

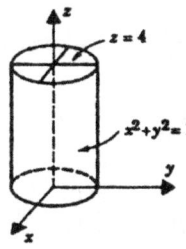

11. (a) ϕ (b) S (c)   closed

13. interior = $\{x : 1 < x < 3\}$, boundary = $\{1, 3\}$;   set is closed.

15. interior = the entire set, boundary = $\{1\}$;   set is open.

17. interior = $\{x : |x| > 1\}$, boundary = $\{1, -1\}$;   set is neither open nor closed.

19. interior = ϕ, boundary = $\{$the entire set$\} \cup \{0\}$;   the set is neither open nor closed.

SECTION 15.6

1. $\dfrac{\partial^2 f}{\partial x^2} = 2A, \quad \dfrac{\partial^2 f}{\partial y^2} = 2C, \quad \dfrac{\partial^2 f}{\partial y \partial x} = \dfrac{\partial^2 f}{\partial x \partial y} = 2B$

3. $\dfrac{\partial^2 f}{\partial x^2} = Cy^2 e^{xy}, \quad \dfrac{\partial^2 f}{\partial y^2} = Cx^2 e^{xy}, \quad \dfrac{\partial^2 f}{\partial y \partial x} = \dfrac{\partial^2 f}{\partial x \partial y} = Ce^{xy}(xy + 1)$

5. $\dfrac{\partial^2 f}{\partial x^2} = 2, \quad \dfrac{\partial^2 f}{\partial y^2} = 4(x + 3y^2 + z^3), \quad \dfrac{\partial^2 f}{\partial z^2} = 6z(2x + 2y^2 + 5z^3)$

$\dfrac{\partial^2 f}{\partial x \partial y} = \dfrac{\partial^2 f}{\partial y \partial x} = 4y, \quad \dfrac{\partial^2 f}{\partial z \partial x} = \dfrac{\partial^2 f}{\partial x \partial z} = 6z^2, \quad \dfrac{\partial^2 f}{\partial z \partial y} = \dfrac{\partial^2 f}{\partial y \partial z} = 12yz^2$

7. $\dfrac{\partial^2 f}{\partial x^2} = \dfrac{1}{(x + y)^2} - \dfrac{1}{x^2}, \quad \dfrac{\partial^2 f}{\partial y^2} = \dfrac{1}{(x + y)^2}, \quad \dfrac{\partial^2 f}{\partial y \partial x} = \dfrac{\partial^2 f}{\partial x \partial y} = \dfrac{1}{(x + y)^2}$

9. $\dfrac{\partial^2 f}{\partial x^2} = 2(y + z), \quad \dfrac{\partial^2 f}{\partial y^2} = 2(x + z), \quad \dfrac{\partial^2 f}{\partial z^2} = 2(x + y)$

all the second mixed partials are $2(x + y + z)$

11. $\dfrac{\partial^2 f}{\partial x^2} = y(y-1)x^{y-2}$, $\quad \dfrac{\partial^2 f}{\partial y^2} = (\ln x)^2 x^y$, $\quad \dfrac{\partial^2 f}{\partial y \partial x} = \dfrac{\partial^2 f}{\partial x \partial y} = x^{y-1}(1 + y \ln x)$

13. $\dfrac{\partial^2 f}{\partial x^2} = ye^x$, $\quad \dfrac{\partial^2 f}{\partial y^2} = xe^y$, $\quad \dfrac{\partial^2 f}{\partial y \partial x} = \dfrac{\partial^2 f}{\partial x \partial y} = e^y + e^x$

15. $\dfrac{\partial^2 f}{\partial x^2} = \dfrac{y^2 - x^2}{(x^2 + y^2)^2}$, $\quad \dfrac{\partial^2 f}{\partial y^2} = \dfrac{x^2 - y^2}{(x^2 + y^2)^2}$, $\quad \dfrac{\partial^2 f}{\partial y \partial x} = \dfrac{\partial^2 f}{\partial x \partial y} = -\dfrac{2xy}{(x^2 + y^2)^2}$

17. $\dfrac{\partial^2 f}{\partial x^2} = -2\,y^2 \cos 2xy$, $\quad \dfrac{\partial^2 f}{\partial y^2} = -2\,x^2 \cos 2xy$, $\quad \dfrac{\partial^2 f}{\partial y \partial x} = \dfrac{\partial^2 f}{\partial x \partial y} = -[\sin 2xy + 2xy \cos 2xy]$

19. $\dfrac{\partial^2 f}{\partial x^2} = 0$, $\quad \dfrac{\partial^2 f}{\partial y^2} = xz \sin y$, $\quad \dfrac{\partial^2 f}{\partial z^2} = -xy \sin z$,

$\dfrac{\partial^2 f}{\partial y \partial x} = \dfrac{\partial^2 f}{\partial x \partial y} = \sin z - z \cos y$, $\quad \dfrac{\partial^2 f}{\partial x \partial z} = \dfrac{\partial^2 f}{\partial z \partial x} = y \cos z - \sin y$, $\quad \dfrac{\partial^2 f}{\partial y \partial z} = \dfrac{\partial^2 f}{\partial z \partial y} = x \cos z - x \cos y$

21. $x^2 \dfrac{\partial^2 u}{\partial x^2} + 2xy \dfrac{\partial^2 u}{\partial x \partial y} + y^2 \dfrac{\partial^2 u}{\partial y^2} = x^2 \left(\dfrac{-2y^2}{(x+y)^3} \right) + 2xy \left(\dfrac{2xy}{(x+y)^3} \right) + y^2 \left(\dfrac{-2x^2}{(x+y)^3} \right) = 0$

23.   (a) no, since $\quad \dfrac{\partial^2 f}{\partial y \partial x} \neq \dfrac{\partial^2 f}{\partial x \partial y}$ (b)   no, since $\quad \dfrac{\partial^2 f}{\partial y \partial x} \neq \dfrac{\partial^2 f}{\partial x \partial y}$ for $x \neq y$

25. $\dfrac{\partial^3 f}{\partial x^2 \partial y} = \dfrac{\partial}{\partial x} \left(\dfrac{\partial^2 f}{\partial x \partial y} \right) = \dfrac{\partial}{\partial x} \left(\dfrac{\partial^2 f}{\partial y \partial x} \right) = \dfrac{\partial^2}{\partial x \partial y} \left(\dfrac{\partial f}{\partial x} \right) = \dfrac{\partial^2}{\partial y \partial x} \left(\dfrac{\partial f}{\partial x} \right) = \dfrac{\partial}{\partial y} \left(\dfrac{\partial^2 f}{\partial x^2} \right) = \dfrac{\partial^3 f}{\partial y \partial x^2}$

⌞ by def.   ⌞ (15.6.5)   ⌞ by def.   ⌞ (15.6.5)   ⌞ by def.   ⌞ by def.

27.   (a) $\lim\limits_{x \to 0} \dfrac{(x)(0)}{x^2 + 0} = \lim\limits_{x \to 0} 0 = 0$ (b) $\lim\limits_{y \to 0} \dfrac{(0)(y)}{0 + y^2} = \lim\limits_{y \to 0} 0 = 0$

(c) $\lim\limits_{x \to 0} \dfrac{(x)(mx)}{x^2 + (mx)^2} = \lim\limits_{x \to 0} \dfrac{m}{1 + m^2} = \dfrac{m}{1 + m^2}$

(d) $\lim\limits_{\theta \to 0^+} \dfrac{(\theta \cos \theta)(\theta \sin \theta)}{(\theta \cos \theta)^2 + (\theta \sin \theta)^2} = \lim\limits_{\theta \to 0^+} \cos \theta \sin \theta = 0$

(e) By L'Hospital's rule $\quad \lim\limits_{x \to 0} \dfrac{f(x)}{x} = \lim\limits_{x \to 0} f'(x) = f'(0)$.   Thus

$$\lim_{x \to 0} \dfrac{xf(x)}{x^2 + [f(x)]^2} = \lim_{x \to 0} \dfrac{f(x)/x}{1 + [f(x)/x]^2} = \dfrac{f'(0)}{1 + [f'(0)]^2}.$$

(f) $\lim\limits_{\theta \to (\pi/3)^-} = \dfrac{(\cos \theta \sin 3\theta)(\sin \theta \sin 3\theta)}{(\cos \theta \sin 3\theta)^2 + (\sin \theta \sin 3\theta)^2} = \lim\limits_{\theta \to (\pi/3)^-} \cos \theta \sin \theta = \dfrac{1}{4}\sqrt{3}$

(g) $\lim\limits_{t \to \infty} \dfrac{(1/t)(\sin t)/t}{1/t^2 + (\sin^2 t)/t^2} = \lim\limits_{t \to \infty} \dfrac{\sin t}{1 + \sin^2 t}$;   does not exist

29. (a) $\dfrac{\partial g}{\partial x}(0,0) = \lim\limits_{h\to 0} \dfrac{g(h,0) - g(0,0)}{h} = \lim\limits_{h\to 0} 0 = 0,$

$\dfrac{\partial g}{\partial y}(0,0) = \lim\limits_{h\to 0} \dfrac{g(0,h) - g(0,0)}{h} = \lim\limits_{h\to 0} 0 = 0$

(b) as (x,y) tends to $(0,0)$ along the x-axis, $\ g(x,y) = g(x,0) = 0\ $ tends to 0;

as (x,y) tends to $(0,0)$ along the line $y = x$, $\ g(x,y) = g(x,x) = \frac{1}{2}\ $ tends to $\frac{1}{2}$

31. For $y \neq 0$, $\qquad \dfrac{\partial f}{\partial x}(0,y) = \lim\limits_{h\to 0} \dfrac{f(h,y) - f(0,y)}{h} = \lim\limits_{h\to 0} \dfrac{y(y^2 - h^2)}{h^2 + y^2} = y.$

Since $\qquad \dfrac{\partial f}{\partial x}(0,0) = \lim\limits_{h\to 0} \dfrac{f(h,0) - f(0,0)}{h} = \lim\limits_{h\to 0} 0 = 0,$

we have $\qquad \dfrac{\partial f}{\partial x}(0,y) = y$ for all y.

For $x \neq 0$, $\qquad \dfrac{\partial f}{\partial y}(x,0) = \lim\limits_{h\to 0} \dfrac{f(x,h) - f(x,0)}{h} = \lim\limits_{h\to 0} \dfrac{x(h^2 - x^2)}{x^2 + h^2} = -x.$

Since $\qquad \dfrac{\partial f}{\partial y}(0,0) = \lim\limits_{h\to 0} \dfrac{f(0,h) - f(0,0)}{h} = \lim\limits_{h\to 0} 0 = 0,$

we have $\qquad \dfrac{\partial f}{\partial y}(x,0) = -x\quad$ for all x.

Therefore $\qquad \dfrac{\partial^2 f}{\partial y \partial x}(0,y) = 1$ for all $y\quad$ and $\quad \dfrac{\partial^2 f}{\partial x \partial y}(x,0) = -1$ for all x.

In particular $\qquad \dfrac{\partial^2 f}{\partial y \partial x}(0,0) = 1\quad$ while $\quad \dfrac{\partial^2 f}{\partial x \partial y}(0,0) = -1.$

33. Since $f_{xy}(x,y) = 0$, $\ f_x(x,y)\ $ must be a function of x alone, and $f_y(x,y)\ $ must be a function of y alone. Then f must be of the form

$f(x,y) = g(x) + h(y).$

35.

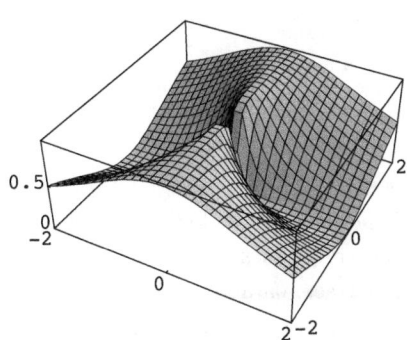

PROJECT 15.6

1. (a) $\dfrac{\partial u}{\partial x} = \dfrac{x^2y^2 + 2xy^3}{(x+y)^2}$, $\dfrac{\partial u}{\partial y} = \dfrac{x^2y^2 + 2x^3y}{(x+y)^2}$

$$x\frac{\partial u}{\partial x} + y\frac{\partial u}{\partial y} = \frac{3x^2y^2(x+y)}{(x+y)^2} = 3u$$

(b) $\dfrac{\partial u}{\partial x} = 2xy + z^2$, $\dfrac{\partial u}{\partial y} = 2yz + x^2$, $\dfrac{\partial u}{\partial z} = 2xz + y^2$

$$\frac{\partial u}{\partial x} + \frac{\partial u}{\partial y} + \frac{\partial u}{\partial z} = 2xy + z^2 + 2yz + x^2 + 2xz + y^2 = (x+y+z)^2$$

3. (i) $\dfrac{\partial^2 f}{\partial t^2} = \dfrac{\partial^2 f}{\partial x^2} = 0 \implies \dfrac{\partial^2 f}{\partial t^2} - c^2\dfrac{\partial^2 f}{\partial x^2} = 0$

(ii) $\dfrac{\partial^2 f}{\partial t^2} = -5c^2\sin(x+ct)\cos(2x+2ct) - 4c^2\cos(x+ct)\sin(2x+2ct)$

$\dfrac{\partial^2 f}{\partial x^2} = -5\sin(x+ct)\cos(2x+2ct) - 4\cos(x+ct)\sin(2x+2ct)$

It now follows that $\dfrac{\partial^2 f}{\partial t^2} - c^2\dfrac{\partial^2 f}{\partial x^2} = 0$

(iii) $\dfrac{\partial^2 f}{\partial t^2} = -\dfrac{c^2}{(x+ct)^2}$, $\dfrac{\partial^2 f}{\partial x^2} = -\dfrac{1}{(x+ct)^2} \implies \dfrac{\partial^2 f}{\partial t^2} - c^2\dfrac{\partial^2 f}{\partial x^2} = 0$

(iv) $\dfrac{\partial^2 f}{\partial t^2} = c^2k^2\left(Ae^{kx} + Be^{-kx}\right)\left(Ce^{ckt} + De^{-ckt}\right)$, $\dfrac{\partial^2 f}{\partial x^2} = k^2\left(Ae^{kx} + Be^{-kx}\right)\left(Ce^{ckt} + De^{-ckt}\right)$

It now follows that $\dfrac{\partial^2 f}{\partial t^2} - c^2\dfrac{\partial^2 f}{\partial x^2} = 0$

REVIEW EXERCISES

1. domain $\{(x,y) : y > x^2\}$,   range $(0, \infty)$

3. domain $\{(x,y,x) : z \geq x^2 + y^2\}$,   dange $[0, +\infty)$

5. (a) $f(x,y) = \dfrac{1}{3}\pi x^2 y$;

(b) $f(x,y) = \dfrac{1}{2}yx^2$;

(c) $\theta = \arccos\dfrac{x + 2y}{\sqrt{5}\sqrt{x^2 + y^2}}$

7. ellipsoid
xy−trace: ellipse $4x^2 + 9y^2 = 36$
xz−trace: ellipse $4x^2 + 36z^2 = 36$
yz−trace: ellipse $9y^2 + 36z^2 = 36$

9. hyperbolic paraboloid
xy−trace: lines $x = \pm y$
xz−trace: parabola $z = -x^2$
yz−trace: parabola $z = y^2$

11. cone
xy–trace: lines $x = \pm y$
xz–trace: lines $x = \pm z$
yz–trace: $(0,0)$

13.

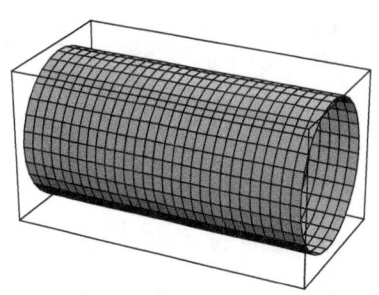

15.

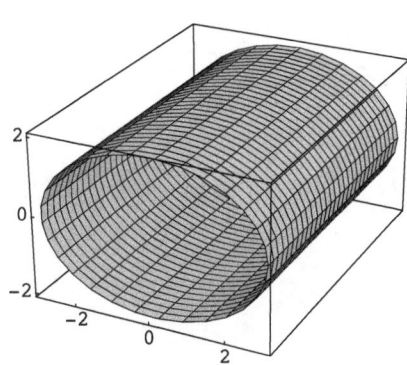

17. $c = 0, \Longrightarrow 0 = 2x^2 + 3y^2 \Longrightarrow (0,0)$
$c = 6, \Longrightarrow 6 = 2x^2 + 3y^2$, ellipse
$c = 12, \Longrightarrow 6 = 2x^2 + 3y^2$, ellipse

19. $c = -4, \Longrightarrow x = -4y^2$, parabola
$c = -1, \Longrightarrow x = -y^2$, parabola
$c = 1, \Longrightarrow x = y^2$, parabola
$c = 4, \Longrightarrow x = 4y^2$, parabola
the origin is omitted

21. $c = 6$, $2x + y + 3z = 6$, plane

23. (a) $f(0,0) = 1$, level curve: $f(x,y) = 1$
(b) $f(\ln 2, 1) = 4$, level curve: $f(x,y) = 4$
(c) $f(1,-1) = 2e$, level curve: $f(x,y) = 2e$

25.
$$f_x = \lim_{h \to 0} \frac{f(x+h,y) - f(x,y)}{h} = \lim_{h \to 0} \frac{(x+h)^2 + 2(x+h)y - x^2 - 2xy}{h}$$
$$= \lim_{h \to 0}(2x + h + 2y) = 2x + 2y$$
$$f_y = \lim_{h \to 0} \frac{f(x,y+h) - f(x,y)}{h} = \lim_{h \to 0} \frac{x^2 + 2x(y+h) - x^2 - 2xy}{h} = 2x$$

27. $f_x = 2xy - 2y^3;$ $f_y = x^2 - 6xy^2$

29. $\dfrac{\partial z}{\partial x} = 2x \sin(xy^2) + x^2y^2 \cos(xy^2);$ $\dfrac{\partial z}{\partial y} = 2x^3 y \cos(xy^2).$

31. $h_x = -e^{-x} \cos(2x - y) - 2e^{-x} \sin(2x - y)$ $h_y = e^{-x} \sin(2x - y)$

33. $f_x = \dfrac{2y^2 + 2yz}{(x + y + z)^2};$ $f_y = \dfrac{2x^2 + 2xz}{(x + y + z)^2};$ $f_z = \dfrac{-2xy}{(x + y + z)^2}$

35. $\dfrac{\partial g}{\partial x} = \dfrac{x}{x^2 + y^2 + z^2}$; $\dfrac{\partial g}{\partial y} = \dfrac{y}{x^2 + y^2 + z^2}$; $\dfrac{\partial g}{\partial z} = \dfrac{z}{x^2 + y^2 + z^2}$.

37. $f_x = 3x^2 y^2 - 4y^3 + 2$, $f_y = 2x^3 y - 12xy^2 - 1$;

$f_{xx} = 6xy^2$, $f_{yy} = 2x^3 - 24xy$, $f_{yx} = f_{xy} = 6x^2 y - 12y^2$

39. $g_x = y \sin xy + xy^2 \cos xy$, $g_{xx} = 2y^2 \cos xy - xy^3 \sin xy$;

$g_y = x \sin xy + x^2 y \cos xy$, $g_{yy} = 2x^2 \cos xy - yx^3 \sin xy$,

$g_{xy} = g_{yx} = \sin xy + 3xy \cos xy - x^2 y^2 \sin xy$

41. $f_x = 2xe^{2y} \cos(2z + 1)$, $f_y = 2x^2 e^{2y} \cos(2z + 1)$, $f_z = -2x^2 e^{2y} \sin(2z + 1)$;

$f_{xx} = 2e^{2y} \cos(2z + 1)$, $f_{yy} = 4x^2 e^{2y} \cos(2z + 1)$, $f_{zz} = -4x^2 e^{2y} \cos(2z + 1)$;

$f_{xy} = f_{yx} = 4xe^{2y} \cos(2z + 1)$, $f_{xz} = f_{zx} = -4xe^{2y} \sin(2z + 1)$, $f_{yz} = f_{zy} = -4x^2 e^{2y} \sin(2z + 1)$

43. $\dfrac{\partial z}{\partial x} = 4x + 6\big|_{(1,2,8)} = 10$

$x = 1 + t$, $y = 2$, $z = 8 + 10t$

45. (a) $z_y(2, 1) = \dfrac{-6y}{2\sqrt{20 - 2x^2 - 3y^2}}(2, 1) = -1$; the equation for l_1 is:

$x = 2$; $y = 1 - t$; $z = 3 + t$

(b) $z_x(2, 1) = \dfrac{-4x}{2\sqrt{20 - 2x^2 - 3y^2}}(2, 1) = -\dfrac{4}{3}$; the equation for l_2 is:

$x = 2 - \dfrac{3}{4}t$; $y = 1$; $z = 3 + t$

(c) The normal vector for this plane is: $-\mathbf{i} - \dfrac{3}{4}\mathbf{j} - \dfrac{3}{4}\mathbf{k}$ or $4\mathbf{i} + 3\mathbf{j} + 3\mathbf{k}$;

an equation for the plane is: $4(x - 2) + 3(y - 1) + 3(z - 3) = 0$.

47. Open.

interior: $\{(x, y) : 0 < x^2 + y^2 < 4\}$

boundary: $\{(0, 0)\} \cup \{(x, y) : x^2 + y^2 = 4\}$

49. Closed.

interior: $\{(x, y, z) : 0 < x < 2,\ 0 < y,\ 0 < z,\ y^2 + z^2 < 4\}$

boundary: the quarter disks $x = 0,\ y^2 + z^2 \leq 4$; $x = 2,\ y^2 + z^2 \leq 4$;

the squares $z = 0,\ 0 \leq x, y \leq 2$; $y = 0, 0 \leq x,\ z \leq 2$;   and

the cylindrical surface $y^2 + z^2 = 4,\ 0 \leq x \leq 2,\ y,\ z \geq 0$

51. (a) $f_x = yg'(xy)$, $f_y = xg'(xy)$; $xf_x - yf_y = xyg' - xyg' = 0$

(b) $f_{xx} = y^2 g''(xy)$, $f_{yy} = x^2 g''(xy)$; $x^2 f_{xx} - y^2 f_{yy} = x^2 y^2 g'' - x^2 y^2 g'' = 0$

53. No. $\dfrac{\partial^2 f}{\partial y \partial x} = x^2 e^{xy} \neq y^2 e^{xy} = \dfrac{\partial^2 f}{\partial x \partial y}$

CHAPTER 16

SECTION 16.1

1. $\nabla f = (6x - y)\,\mathbf{i} + (1 - x)\,\mathbf{j}$ **3.** $\nabla f = e^{xy}[\,(xy + 1)\,\mathbf{i} + x^2\mathbf{j}]$

5. $\nabla f = \left[2y^2\sin(x^2 + 1) + 4x^2y^2\cos(x^2 + 1)\right]\mathbf{i} + 4xy\sin(x^2 + 1)\,\mathbf{j}$

7. $\nabla f = (e^{x-y} + e^{y-x})\,\mathbf{i} + (-e^{x-y} - e^{y-x})\,\mathbf{j} = (e^{x-y} + e^{y-x})(\mathbf{i} - \mathbf{j})$

9. $\nabla f = (z^2 + 2xy)\,\mathbf{i} + (x^2 + 2yz)\,\mathbf{j} + (y^2 + 2zx)\,\mathbf{k}$

11. $\nabla f = e^{-z}(2xy\,\mathbf{i} + x^2\,\mathbf{j} - x^2y\,\mathbf{k})$

13. $\nabla f = e^{x+2y}\cos\left(z^2 + 1\right)\,\mathbf{i} + 2e^{x+2y}\cos\left(z^2 + 1\right)\,\mathbf{j} - 2ze^{x+2y}\sin\left(z^2 + 1\right)\,\mathbf{k}$

15. $\nabla f = \left[2y\cos(2xy) + \dfrac{2}{x}\right]\mathbf{i} + 2x\cos(2xy)\,\mathbf{j} + \dfrac{1}{z}\,\mathbf{k}$

17. $\nabla f = (4x - 3y)\,\mathbf{i} + (8y - 3x)\,\mathbf{j}$;   at $(2, 3)$, $\nabla f = -\mathbf{i} + 18\mathbf{j}$

19. $\nabla f = \dfrac{2x}{x^2 + y^2}\,\mathbf{i} + \dfrac{2y}{x^2 + y^2}\,\mathbf{j}$;   at $(2, 1)$, $\nabla f = \dfrac{4}{5}\,\mathbf{i} + \dfrac{2}{5}\,\mathbf{j}$

21. $\nabla f = (\sin xy + xy\cos xy)\,\mathbf{i} + x^2\cos xy\,\mathbf{j}$;   at $(1, \pi/2)$, $\nabla f = \mathbf{i}$

23. $\nabla f = -e^{-x}\sin(z + 2y)\,\mathbf{i} + 2e^{-x}\cos(z + 2y)\,\mathbf{j} + e^{-x}\cos(z + 2y)\,\mathbf{k}$;

at $(0, \pi/4, \pi/4)$, $\nabla f = -\frac{1}{2}\sqrt{2}\,(\mathbf{i} + 2\mathbf{j} + \mathbf{k})$

25. $\nabla f = \mathbf{i} - \dfrac{y}{\sqrt{y^2 + z^2}}\,\mathbf{j} - \dfrac{z}{\sqrt{y^2 + z^2}}\,\mathbf{k}$;   at $(2, -3, 4)$, $\nabla f = \mathbf{i} + \dfrac{3}{5}\,\mathbf{j} - \dfrac{4}{5}\,\mathbf{k}$

27. (a) $\nabla f(0, 2) = 4\,\mathbf{i}$ (b) $\nabla f\left(\frac{1}{4}\pi, \frac{1}{6}\pi\right) = \left(-1 - \dfrac{-1 + \sqrt{3}}{2\sqrt{2}}\right)\mathbf{i} + \left(-\dfrac{1}{2} - \dfrac{-1 + \sqrt{3}}{\sqrt{2}}\right)\mathbf{j}$

(c) $\nabla f(1, e) = (1 - 2e)\,\mathbf{i} - 2\,\mathbf{j}$

29. For the function $f(x, y) = 3x^2 - xy + y$, we have

$$f(\mathbf{x} + \mathbf{h}) - f(\mathbf{x}) = f(x + h_1, y + h_2) - f(x, y)$$

$$= 3(x + h_1)^2 - (x + h_1)(y + h_2) + (y + h_2) - \left[3x^2 - xy + y\right]$$

$$= [(6x - y)\,\mathbf{i} + (1 - x)\,\mathbf{j}]\,\cdot\,(h_1\,\mathbf{i} + h_2\,\mathbf{j}) + 3h_1^2 - h_1 h_2$$

$$= [(6x - y)\,\mathbf{i} + (1 - x)\,\mathbf{j}]\,\cdot\,\mathbf{h} + 3h_1^2 - h_1 h_2$$

The remainder $g(\mathbf{h}) = 3h_1^2 - h_1 h_2 = (3h_1\,\mathbf{i} - h_1\,\mathbf{j}) \cdot (h_1\,\mathbf{i} + h_2\,\mathbf{j}),$ and

$$\frac{|g(\mathbf{h})|}{\|\mathbf{h}\|} = \frac{\|3h_1\,\mathbf{i} - h_1\,\mathbf{j}\| \cdot \|\mathbf{h}\| \cdot \cos\theta}{\|\mathbf{h}\|} \leq \|3h_1\,\mathbf{i} - h_1\,\mathbf{j}\|$$

Since $\|3h_1\,\mathbf{i} - h_1\,\mathbf{j}\| \to 0$ as $\mathbf{h} \to \mathbf{0}$ it follows that

$$\nabla f = (6x - y)\,\mathbf{i} + (1 - x)\,\mathbf{j}$$

31.   For the function $f(x, y, z) = x^2 y + y^2 z + z^2 x,$ we have

$$f(\mathbf{x} + \mathbf{h}) - f(\mathbf{x}) = f(x + h_1, y + h_2, z + h_3) - f(x, y, z)$$

$$= (x + h_1)^2\,(y + h_2) + (y + h_2)^2\,(z + h_3) + (z + h_3)^2\,(x + h_1) - \left(x^2 y + y^2 z + z^2 x\right)$$

$$= \left(2xy + z^2\right) h_1 + \left(2yz + x^2\right) h_2 + \left(2xz + y^2\right) h_3 + \left(2xh_2 + yh_1 + h_1 h_2\right) h_1 +$$

$$\quad \left(2yh_3 + zh_2 + h_2 h_3\right) h_2 + \left(2zh_1 + xh_3 + h_1 h_3\right) h_3$$

$$= \left[\left(2xy + z^2\right)\mathbf{i} + \left(2yz + x^2\right)\mathbf{j} + \left(2xz + y^2\right)\mathbf{k}\right] \cdot \mathbf{h} + g(\mathbf{h}) \cdot \mathbf{h},$$

where $g(\mathbf{h}) = (2xh_2 + yh_1 + h_1 h_2)\,\mathbf{i} + (2yh_3 + zh_2 + h_2 h_3)\,\mathbf{j} + (2zh_1 + xh_3 + h_1 h_3)\,\mathbf{k}$

Since $\dfrac{|g(\mathbf{h})|}{\|\mathbf{h}\|} \to 0$ as $\mathbf{h} \to \mathbf{0}$ it follows that

$$\nabla f = \left(2xy + z^2\right)\mathbf{i} + \left(2yz + x^2\right)\mathbf{j} + \left(2xz + y^2\right)\mathbf{k}$$

33. $\nabla f = \mathbf{F}(x, y) = 2xy\,\mathbf{i} + \left(1 + x^2\right)\mathbf{j} \;\Rightarrow\; \dfrac{\partial f}{\partial x} = 2xy \;\Rightarrow\; f(x, y) = x^2 y + g(y)$ for some function g.

Now, $\dfrac{\partial f}{\partial y} = x^2 + g'(y) = 1 + x^2 \;\Rightarrow\; g'(y) = 1 \;\Rightarrow\; g(y) = y + C, \;\; C$ a constant.

Thus, $f(x, y) = x^2 y + y + C$

35. $\nabla f = \mathbf{F}(x, y) = (x + \sin y)\,\mathbf{i} + (x \cos y - 2y)\,\mathbf{j} \;\Rightarrow\; \dfrac{\partial f}{\partial x} = x + \sin y \;\Rightarrow\; f(x, y) = \tfrac{1}{2}\,x^2 + x \sin y + g(y)$

for some function g.

Now, $\dfrac{\partial f}{\partial y} = x \cos y + g'(y) = x \cos y - 2y \;\Rightarrow\; g'(y) = -2y \;\Rightarrow\; g(y) = -y^2 + C, \;\; C$ a constant.

Thus, $f(x, y) = \tfrac{1}{2}\,x^2 + x \sin y - y^2 + C.$

37.   With $r = (x^2 + y^2 + z^2)^{1/2}$ we have

$$\frac{\partial r}{\partial x} = \frac{x}{r}, \quad \frac{\partial r}{\partial y} = \frac{y}{r}, \quad \frac{\partial r}{\partial z} = \frac{z}{r}.$$

(a)

$$\nabla(\ln r) = \frac{\partial}{\partial x}\,(\ln r)\,\mathbf{i} + \frac{\partial}{\partial y}\,(\ln r)\,\mathbf{j} + \frac{\partial}{\partial z}\,(\ln r)\mathbf{k}$$

$$= \frac{1}{r}\,\frac{\partial r}{\partial x}\,\mathbf{i} + \frac{1}{r}\,\frac{\partial r}{\partial y}\,\mathbf{j} + \frac{1}{r}\,\frac{\partial r}{\partial z}\,\mathbf{k}$$

$$= \frac{x}{r^2}\,\mathbf{i} + \frac{y}{r^2}\,\mathbf{j} + \frac{z}{r^2}\,\mathbf{k} = \frac{\mathbf{r}}{r^2}$$

(b)
$$\boldsymbol{\nabla}(\sin r) = \frac{\partial}{\partial x}(\sin r)\mathbf{i} + \frac{\partial}{\partial y}(\sin r)\mathbf{j} + \frac{\partial}{\partial z}(\sin r)\mathbf{k}$$

$$= \cos r \frac{\partial r}{\partial x}\mathbf{i} + \cos r \frac{\partial r}{\partial y}\mathbf{j} + \cos r \frac{\partial r}{\partial z}\mathbf{k}$$

$$= (\cos r)\frac{x}{r}\mathbf{i} + (\cos r)\frac{y}{r}\mathbf{j} + (\cos r)\frac{z}{r}\mathbf{k}$$

$$= \left(\frac{\cos r}{r}\right)\mathbf{r}$$

(c) $\boldsymbol{\nabla}e^r = \left(\dfrac{e^r}{r}\right)\mathbf{r}$ [same method as in (a) and (b)]

39. (a) $\boldsymbol{\nabla}f = 2x\,\mathbf{i} + 2y\,\mathbf{j} = \mathbf{0} \implies x = y = 0;\quad \boldsymbol{\nabla}f = \mathbf{0}$ at $(0,0)$.

(b)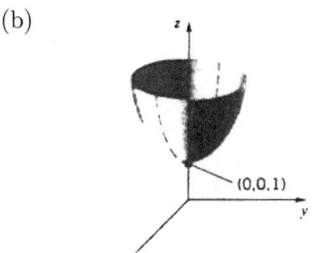

(c) f has an absolute minimum at $(0,0)$

(0,0,1)

41. (a) Let $\mathbf{c} = c_1\mathbf{i} + c_2\mathbf{j} + c_3\mathbf{k}$. First, we take $\mathbf{h} = h\mathbf{i}$. Since $\mathbf{c}\cdot\mathbf{h}$ is $o(\mathbf{h})$,

$$0 = \lim_{\mathbf{h}\to\mathbf{0}} \frac{\mathbf{c}\cdot\mathbf{h}}{\|\mathbf{h}\|} = \lim_{h\to 0}\frac{c_1 h}{h} = c_1.$$

Similarly, $c_2 = 0$ and $c_3 = 0$.

(b) $(\mathbf{y}-\mathbf{z})\cdot\mathbf{h} = [f(\mathbf{x}+\mathbf{h}) - f(\mathbf{x}) - \mathbf{z}\cdot\mathbf{h}] + [\mathbf{y}\cdot\mathbf{h} - f(\mathbf{x}+\mathbf{h}) + f(\mathbf{x})] = o(\mathbf{h}) + o(\mathbf{h}) = o(\mathbf{h})$,
so that, by part (a), $\mathbf{y} - \mathbf{z} = \mathbf{0}$.

43. (a) In Section 15.6 we showed that f was not continuous at $(0,0)$. It is therefore not differentiable at $(0,0)$.

(b) For $(x,y) \neq (0,0)$, $\dfrac{\partial f}{\partial x} = \dfrac{2y(y^2 - x^2)}{(x^2 + y^2)^2}$. As (x,y) tends to $(0,0)$ along the positive y-axis,
$$\frac{\partial f}{\partial x} = \frac{2y^3}{y^4} = \frac{2}{y}\quad\text{tends to }\infty.$$

SECTION 16.2

1. $\boldsymbol{\nabla}f = 2x\mathbf{i} + 6y\mathbf{j},\quad \boldsymbol{\nabla}f(1,1) = 2\mathbf{i} + 6\mathbf{j},\quad \mathbf{u} = \tfrac{1}{2}\sqrt{2}\,(\mathbf{i} - \mathbf{j}),\qquad f'_{\mathbf{u}}(1,1) = \boldsymbol{\nabla}f(1,1)\cdot\mathbf{u} = -2\sqrt{2}$

3. $\boldsymbol{\nabla}f = (e^y - ye^x)\,\mathbf{i} + (xe^y - e^x)\,\mathbf{j},\quad \boldsymbol{\nabla}f(1,0) = \mathbf{i} + (1-e)\mathbf{j},\quad \mathbf{u} = \dfrac{1}{5}(3\mathbf{i} + 4\mathbf{j}),$

$f'_{\mathbf{u}}(1,0) = \boldsymbol{\nabla}f(1,0)\cdot\mathbf{u} = \dfrac{1}{5}(7 - 4e)$

5. $\nabla f = \dfrac{(a-b)y}{(x+y)^2}\,\mathbf{i} + \dfrac{(b-a)x}{(x+y)^2}\,\mathbf{j}, \quad \nabla f(1,1) = \dfrac{a-b}{4}(\mathbf{i}-\mathbf{j}), \quad \mathbf{u} = \dfrac{1}{2}\sqrt{2}\,(\mathbf{i}-\mathbf{j}),$

$f_{\mathbf{u}}'(1,1) = \nabla f(1,1)\cdot\mathbf{u} = \dfrac{1}{4}\sqrt{2}\,(a-b)$

7. $\nabla f = \dfrac{2x}{x^2+y^2}\,\mathbf{i} + \dfrac{2y}{x^2+y^2}\,\mathbf{j}, \quad \nabla f(0,1) = 2\,\mathbf{j}, \quad \mathbf{u} = \dfrac{1}{\sqrt{65}}(8\,\mathbf{i}+\mathbf{j}),$

$f_{\mathbf{u}}'(0,1) = \nabla f(0,1)\cdot\mathbf{u} = \dfrac{2}{\sqrt{65}}$

9. $\nabla f = (y+z)\mathbf{i} + (x+z)\mathbf{j} + (y+x)\mathbf{k}, \quad \nabla f(1,-1,1) = 2\mathbf{j}, \quad \mathbf{u} = \tfrac{1}{6}\sqrt{6}\,(\mathbf{i}+2\mathbf{j}+\mathbf{k}),$

$f_{\mathbf{u}}'(1,-1,1) = \nabla f(1,-1,1)\cdot\mathbf{u} = \dfrac{2}{3}\sqrt{6}$

11. $\nabla f = 2\left(x+y^2+z^3\right)\left(\mathbf{i}+2y\mathbf{j}+3z^2\mathbf{k}\right), \quad \nabla f(1,-1,1) = 6(\mathbf{i}-2\mathbf{j}+3\mathbf{k}), \quad \mathbf{u} = \tfrac{1}{2}\sqrt{2}\,(\mathbf{i}+\mathbf{j}),$

$f_{\mathbf{u}}'(1,-1,1) = \nabla f(1,-1,1)\cdot\mathbf{u} = -3\sqrt{2}$

13. $\nabla f = \tan^{-1}(y+z)\,\mathbf{i} + \dfrac{x}{1+(y+z)^2}\,\mathbf{j} + \dfrac{x}{1+(y+z)^2}\,\mathbf{k}, \quad \nabla f(1,0,1) = \dfrac{\pi}{4}\mathbf{i} + \dfrac{1}{2}\mathbf{j} + \dfrac{1}{2}\mathbf{k},$

$\mathbf{u} = \dfrac{1}{\sqrt{3}}(\mathbf{i}+\mathbf{j}-\mathbf{k}), \qquad f_{\mathbf{u}}'(1,0,1) = \nabla f(1,0,1)\cdot\mathbf{u} = \dfrac{\pi}{4\sqrt{3}} = \dfrac{\sqrt{3}}{12}\,\pi$

15. $\nabla f = \dfrac{x}{x^2+y^2}\,\mathbf{i} + \dfrac{y}{x^2+y^2}\,\mathbf{j}, \quad \mathbf{u} = \dfrac{1}{\sqrt{x^2+y^2}}\,(-x\mathbf{i}-y\mathbf{j}), \quad f_{\mathbf{u}}'(x,y) = \nabla f\cdot\mathbf{u} = -\dfrac{1}{\sqrt{x^2+y^2}}$

17. $\nabla f = (2Ax+2By)\,\mathbf{i} + (2Bx+2Cy)\,\mathbf{j}, \quad \nabla f(a,b) = (2aA+2bB)\mathbf{i} + (2aB+2bC)\,\mathbf{j}$

(a) $\mathbf{u} = \tfrac{1}{2}\sqrt{2}\,(-\mathbf{i}+\mathbf{j}), \quad f_{\mathbf{u}}'(a,b) = \nabla f(a,b)\cdot\mathbf{u} = \sqrt{2}\,[a(B-A)+b(C-B)]$

(b) $\mathbf{u} = \tfrac{1}{2}\sqrt{2}\,(\mathbf{i}-\mathbf{j}), \quad f_{\mathbf{u}}'(a,b) = \nabla f(a,b)\cdot\mathbf{u} = \sqrt{2}\,[a(A-B)+b(B-C)]$

19. $\nabla f = e^{y^2-z^2}(\mathbf{i}+2xy\mathbf{j}-2xz\mathbf{k}), \quad \nabla f(1,2,-2) = \mathbf{i}+4\mathbf{j}+4\mathbf{k}, \quad \mathbf{r}'(t) = \mathbf{i}-2\sin(t-1)\mathbf{j}-2e^{t-1}\mathbf{k},$

at $(1,2,-2)$ $t=1$, $\mathbf{r}'(1) = \mathbf{i}-2\mathbf{k}, \quad \mathbf{u} = \tfrac{1}{5}\sqrt{5}\,(\mathbf{i}-2\mathbf{k}), \quad f_{\mathbf{u}}'(1,2,-2) = \nabla f(1,2,-2)\cdot\mathbf{u} = -\dfrac{7}{5}\sqrt{5}$

21. $\nabla f = (2x+2yz)\,\mathbf{i} + \left(2xz-z^2\right)\mathbf{j} + (2xy-2yz)\,\mathbf{k}, \quad \nabla f(1,1,2) = 6\,\mathbf{i}-2\,\mathbf{k}$

The vectors $\mathbf{v} = \pm(2\mathbf{i}+\mathbf{j}-3\mathbf{k})$ are direction vectors for the given line; $\mathbf{u} = \pm\left(\dfrac{1}{\sqrt{14}}[2\mathbf{i}+\mathbf{j}-3\mathbf{k}]\right)$

are corresponding unit vectors; $\quad f_{\mathbf{u}}'(1,1,2) = \nabla f(1,1,2)\cdot(\pm\mathbf{u}) = \pm\dfrac{18}{\sqrt{14}}$

23. $\nabla f = 2y^2e^{2x}\,\mathbf{i} + 2ye^{2x}\,\mathbf{j}, \quad \nabla f(0,1) = 2\,\mathbf{i}+2\,\mathbf{j}, \quad \|\nabla f\| = 2\sqrt{2}, \quad \dfrac{\nabla f}{\|\nabla f\|} = \dfrac{1}{\sqrt{2}}(\mathbf{i}+\mathbf{j})$

f increases most rapidly in the direction $\mathbf{u} = \dfrac{1}{\sqrt{2}}(\mathbf{i}+\mathbf{j})$; the rate of change is $2\sqrt{2}$.

f decreases most rapidly in the direction $\mathbf{v} = -\dfrac{1}{\sqrt{2}}(\mathbf{i}+\mathbf{j})$; the rate of change is $-2\sqrt{2}$.

25. $\nabla f = \dfrac{x}{\sqrt{x^2 + y^2 + z^2}}\,\mathbf{i} + \dfrac{y}{\sqrt{x^2 + y^2 + z^2}}\,\mathbf{j} + \dfrac{z}{\sqrt{x^2 + y^2 + z^2}}\,\mathbf{k},$

$\nabla f(1, -2, 1) = \dfrac{1}{\sqrt{6}}(\mathbf{i} - 2\mathbf{j} + \mathbf{k}), \quad \|\nabla f\| = 1$

f increases most rapidly in the direction $\mathbf{u} = \dfrac{1}{\sqrt{6}}(\mathbf{i} - 2\mathbf{j} + \mathbf{k});$ the rate of change is 1.

f decreases most rapidly in the direction $\mathbf{v} = -\dfrac{1}{\sqrt{6}}(\mathbf{i} - 2\mathbf{j} + \mathbf{k});$ the rate of change is -1.

27. $\nabla f = f'(x_0)\,\mathbf{i}.$ If $f'(x_0) \neq 0,$ the gradient points in the direction in which f increases: to the right if $f'(x_0) > 0,$ to the left if $f'(x_0) < 0.$

29. (a) $\displaystyle\lim_{h \to 0}\dfrac{f(h, 0) - f(0, 0)}{h} = \lim_{h \to 0}\dfrac{\sqrt{h^2}}{h} = \lim_{h \to 0}\dfrac{|h|}{h}$ does not exist

   (b) no; by Theorem 16.2.5 f cannot be differentiable at $(0, 0)$

31. $\nabla\lambda(x, y) = -\tfrac{8}{3}x\mathbf{i} - 6y\mathbf{j}$

   (a) $\nabla\lambda(1, -1) = -\dfrac{8}{3}\mathbf{i} = 6\mathbf{j}, \quad \mathbf{u} = \dfrac{-\nabla\lambda(1, -1)}{\|\nabla\lambda(1, -1)\|} = \dfrac{\tfrac{8}{3}\mathbf{i} - 6\mathbf{j}}{\tfrac{2}{3}\sqrt{97}}, \quad \lambda'_{\mathbf{u}}(1, -1) = \nabla\lambda(1, -1) \cdot \mathbf{u} = -\dfrac{2}{3}\sqrt{97}$

   (b) $\mathbf{u} = \mathbf{i}, \quad \lambda'_{\mathbf{u}}(1, 2) = \nabla\lambda(1, 2) \cdot \mathbf{u} = \left(-\tfrac{8}{3}\mathbf{i} - 12\mathbf{j}\right) \cdot \mathbf{i} = -\tfrac{8}{3}$

   (c) $\mathbf{u} = \tfrac{1}{2}\sqrt{2}\,(\mathbf{i} + \mathbf{j}), \quad \lambda'_{\mathbf{u}}(2, 2) = \nabla\lambda(2, 2) \cdot \mathbf{u} = \left(-\tfrac{16}{3}\mathbf{i} - 12\mathbf{j}\right) \cdot \left[\tfrac{1}{2}\sqrt{2}\,(\mathbf{i} + \mathbf{j})\right] = -\tfrac{26}{3}\sqrt{2}$

33. (a) The projection of the path onto the xy-plane is the curve

$$C: \ \mathbf{r}(t) = x(t)\mathbf{i} + y(t)\mathbf{j}$$

which begins at $(1, 1)$ and at each point has its tangent vector in the direction of $-\nabla f.$ Since

$$\nabla f = 2x\mathbf{i} + 6y\mathbf{j},$$

we have the initial-value problems

$$x'(t) = -2x(t), \quad x(0) = 1 \qquad \text{and} \qquad y'(t) = -6y(t), \quad y(0) = 1.$$

From Theorem 7.6.1 we find that

$$x(t) = e^{-2t} \qquad \text{and} \qquad y(t) = e^{-6t}.$$

Eliminating the parameter $t,$ we find that C is the curve $y = x^3$ from $(1, 1)$ to $(0, 0).$

   (b) Here

$$x'(t) = -2x(t), \quad x(0) = 1 \qquad \text{and} \qquad y'(t) = -6y(t), \quad y(0) = -2$$

so that

$$x(t) = e^{-2t} \qquad \text{and} \qquad y(t) = -2e^{-6t}.$$

Eliminating the parameter $t,$ we find that the projection of the path onto the xy-plane is the curve $y = -2x^3$ from $(1, -2)$ to $(0, 0).$

35. The projection of the path onto the xy-plane is the curve

$$C: \ \mathbf{r}(t) = x(t)\mathbf{i} + y(t)\mathbf{j}$$

which begins at (a, b) and at each point has its tangent vector in the direction of $-\nabla f = -\left(2a^2 x\mathbf{i} + 2b^2 y\mathbf{j}\right)$. We can satisfy these conditions by setting

$$x'(t) = -2a^2 x(t), \quad x(0) = a^2 \qquad \text{and} \qquad y'(t) = -2b^2 y(t), \quad y(0) = b$$

so that

$$x(t) = ae^{-2a^2 t} \quad \text{and} \qquad y(t) = be^{-2b^2 t}.$$

Since

$$\left[\frac{x}{a}\right]^{b^2} = \left(e^{-2a^2 t}\right)^{b^2} = \left[\frac{y}{b}\right]^{a^2},$$

C is the curve $(b)^{a^2} x^{b^2} = (a)^{b^2} y^{a^2}$ from (a, b) to $(0, 0)$.

37. We want the curve

$$C: \ \mathbf{r}(t) = x(t)\mathbf{i} + y(t)\mathbf{j}$$

which begins at $(\pi/4, 0)$ and at each point has its tangent vector in the direction of

$$\nabla T = -\sqrt{2}\,e^{-y}\sin x\,\mathbf{i} - \sqrt{2}\,e^{-y}\cos x\,\mathbf{j}.$$

From

$$x'(t) = -\sqrt{2}\,e^{-y}\sin x \qquad \text{and} \qquad y'(t) = -\sqrt{2}\,e^{-y}\cos x$$

we obtain

$$\frac{dy}{dx} = \frac{y'(t)}{x'(t)} = \cot x$$

so that

$$y = \ln|\sin x| + C.$$

Since $y = 0$ when $x = \pi/4$, we get $C = \ln\sqrt{2}$ and $y = \ln|\sqrt{2}\sin x|$. As $\nabla T(\pi/4, 0) = -\mathbf{i} - \mathbf{j}$, the curve $y = \ln|\sqrt{2}\,\sin x|$ is followed in the direction of decreasing x.

39. (a)

$$\lim_{h \to 0} \frac{f\left(2+h, (2+h)^2\right) - f(2, 4)}{h} = \lim_{h \to 0} \frac{3(2+h)^2 + (2+h)^2 - 16}{h}$$

$$= \lim_{h \to 0} 4\left[\frac{4h + h^2}{h}\right] = \lim_{h \to 0} 4(4+h) = 16$$

(b)

$$\lim_{h \to 0} \frac{f\left(\dfrac{h+8}{4}, 4+h\right) - f(2, 4)}{h} = \lim_{h \to 0} \frac{3\left(\dfrac{h+8}{4}\right)^2 + (4+h) - 16}{h}$$

$$= \lim_{h \to 0} \frac{\frac{3}{16}h^2 + 3h + 12 + 4 + h - 16}{h}$$

$$= \lim_{h \to 0} \left(\tfrac{3}{16}h + 4\right) = 4$$

(c) The limits computed in (a) and (b) are not directional derivatives. In (a) and (b) we have, in essence, computed $\nabla f(2,4) \cdot \mathbf{r}_0$ taking $\mathbf{r}_0 = \mathbf{i} + 4\mathbf{j}$ in (a) and $\mathbf{r}_0 = \frac{1}{4}\mathbf{i} + \mathbf{j}$ in (b). In neither case is $\mathbf{r}_0$ a unit vector.

41. (a) $\mathbf{u} = \cos\theta\,\mathbf{i} + \sin\theta\,\mathbf{j}, \qquad \nabla f(x,y) = \dfrac{\partial f}{\partial x}\,\mathbf{i} + \dfrac{\partial f}{\partial y}\,\mathbf{j};$

$$f'_{\mathbf{u}}(x,y) = \nabla f \cdot \mathbf{u} = \left(\frac{\partial f}{\partial x}\,\mathbf{i} + \frac{\partial f}{\partial y}\,\mathbf{j}\right) \cdot (\cos\theta\,\mathbf{i} + \sin\theta\,\mathbf{j}) = \frac{\partial f}{\partial x}\,\cos\theta + \frac{\partial f}{\partial y}\,\sin\theta$$

(b) $\nabla f = \left(3x^2 + 2y - y^2\right)\mathbf{i} + (2x - 2xy)\,\mathbf{j}, \qquad \nabla f(-1,2) = 3\,\mathbf{i} + 2\,\mathbf{j}$

$$f'_{\mathbf{u}}(-1,2) = 3\,\cos(2\pi/3) + 2\,\sin(2\pi/3) = \frac{2\sqrt{3} - 3}{2}$$

43. $\nabla(fg) = \dfrac{\partial(fg)}{\partial x}\,\mathbf{i} + \dfrac{\partial(fg)}{\partial y}\,\mathbf{j} + \dfrac{\partial(fg)}{\partial z}\,\mathbf{k} = \left(f\dfrac{\partial g}{\partial x} + g\dfrac{\partial f}{\partial x}\right)\mathbf{i} + \left(f\dfrac{\partial g}{\partial y} + g\dfrac{\partial f}{\partial y}\right)\mathbf{j} + \left(f\dfrac{\partial g}{\partial z} + g\dfrac{\partial f}{\partial z}\right)\mathbf{k}$

$$= f\left(\frac{\partial g}{\partial x}\,\mathbf{i} + \frac{\partial g}{\partial y}\,\mathbf{j} + \frac{\partial g}{\partial z}\,\mathbf{k}\right) + g\left(\frac{\partial f}{\partial x}\,\mathbf{i} + \frac{\partial f}{\partial y}\,\mathbf{j} + \frac{\partial f}{\partial z}\,\mathbf{k}\right)$$

$$= f\,\nabla g + g\,\nabla f$$

45. $\nabla f^n = \dfrac{\partial f^n}{\partial x}\,\mathbf{i} + \dfrac{\partial f^n}{\partial y}\,\mathbf{j} + \dfrac{\partial f^n}{\partial z}\,\mathbf{k} = nf^{n-1}\dfrac{\partial f}{\partial x}\,\mathbf{i} + nf^{n-1}\dfrac{\partial f}{\partial y}\,\mathbf{j} + nf^{n-1}\dfrac{\partial f}{\partial z}\,\mathbf{k} = nf^{n-1}\,\nabla f$

SECTION 16.3

1. $f(\mathbf{b}) = f(1,3) = -2; \quad f(\mathbf{a}) = f(0,1) = 0; \quad f(\mathbf{b}) - f(\mathbf{a}) = -2$

$\nabla f = \left(3x^2 - y\right)\mathbf{i} - x\,\mathbf{j}; \quad \mathbf{b} - \mathbf{a} = \mathbf{i} + 2\,\mathbf{j} \quad \text{and} \quad \nabla f \cdot (\mathbf{b} - \mathbf{a}) = 3x^2 - y - 2x$

The line segment joining $\mathbf{a}$ and $\mathbf{b}$ is parametrized by

$$x = t, \quad y = 1 + 2t, \quad 0 \le t \le 1$$

Thus, we need to solve the equation

$$3t^2 - (1 + 2t) - 2t = -2, \quad \text{which is the same as} \quad 3t^2 - 4t + 1 = 0, \ \ 0 \le t \le 1$$

The solutions are: $t = \frac{1}{3}, t = 1$. Thus, $\mathbf{c} = (\frac{1}{3}, \frac{5}{3})$ satisfies the equation.
Note that the endpoint $\mathbf{b}$ also satisfies the equation.

3. (a) $f(x,y,z) = a_1 x + a_2 y + a_3 z + C$ \qquad (b) $f(x,y,z) = g(x,y,z) + a_1 x + a_2 y + a_3 z + C$

5. (a) U is not connected

    (b) (i) $g(\mathbf{x}) = f(\mathbf{x}) - 1$ \qquad (ii) $g(\mathbf{x}) = -f(\mathbf{x})$

7. $\nabla f = 2xy\,\mathbf{i} + x^2\,\mathbf{j};$

$\nabla f(\mathbf{r}(t)) \cdot \mathbf{r}'(t) = \left(2\mathbf{i} + e^{2t}\mathbf{j}\right) \cdot \left(e^t\mathbf{i} - e^{-t}\mathbf{j}\right) = e^t$

9. $\nabla f = \dfrac{-2x}{1 + (y^2 - x^2)^2}\,\mathbf{i} + \dfrac{2y}{1 + (y^2 - x^2)^2}\,\mathbf{j}, \quad \nabla f(\mathbf{r}(t)) = \dfrac{-2\sin t}{1 + \cos^2 2t}\,\mathbf{i} + \dfrac{2\cos t}{1 + \cos^2 2t}\,\mathbf{j}$

$\nabla f(\mathbf{r}(t)) \cdot \mathbf{r}'(t) = \left(\dfrac{-2\sin t}{1 + \cos^2 2t}\,\mathbf{i} + \dfrac{2\cos t}{1 + \cos^2 2t}\,\mathbf{j}\right) \cdot (\cos t\,\mathbf{i} - \sin t\,\mathbf{j}) = \dfrac{-4\sin t\cos t}{1 + \cos^2 2t} = \dfrac{-2\sin 2t}{1 + \cos^2 2t}$

11. $\nabla f = (e^y - ye^{-x})\,\mathbf{i} + (xe^y + e^{-x})\,\mathbf{j}; \quad \nabla f(\mathbf{r}(t)) = (t^t - \ln t)\,\mathbf{i} + \left(t^t \ln t + \dfrac{1}{t}\right)\mathbf{j}$

$\nabla f(\mathbf{r}(t)) \cdot \mathbf{r}'(t) = \left((t^t - \ln t)\,\mathbf{i} + \left(t^t \ln t + \dfrac{1}{t}\right)\mathbf{j}\right) \cdot \left(\dfrac{1}{t}\,\mathbf{i} + [1 + \ln t]\,\mathbf{j}\right) = t^t\left(\dfrac{1}{t} + \ln t + [\ln t]^2\right) + \dfrac{1}{t}$

13. $\nabla f = y\,\mathbf{i} + (x - z)\,\mathbf{j} - y\,\mathbf{k};$

$\nabla f(\mathbf{r}(t)) \cdot \mathbf{r}'(t) = \left(t^2\,\mathbf{i} + \left(t - t^3\right)\mathbf{j} - t^2\,\mathbf{k}\right) \cdot \left(\mathbf{i} + 2t\,\mathbf{j} + 3t^2\,\mathbf{k}\right) = 3t^2 - 5t^4$

15. $\nabla f = 2x\,\mathbf{i} + 2y\,\mathbf{j} + \mathbf{k};$

$\nabla f(\mathbf{r}(t)) \cdot \mathbf{r}'(t) = (2a\cos \omega t\,\mathbf{i} + 2b\sin \omega t\,\mathbf{j} + \mathbf{k}) \cdot (-a\omega \sin \omega t\,\mathbf{i} + b\omega \cos \omega t\,\mathbf{j} + b\omega\,\mathbf{k})$

$= 2\omega\left(b^2 - a^2\right)\sin \omega t \cos \omega t + b\omega$

17. $\dfrac{du}{dt} = \dfrac{\partial u}{\partial x}\dfrac{dx}{dt} + \dfrac{\partial u}{\partial y}\dfrac{dy}{dt} = (2x - 3y)(-\sin t) + (4y - 3x)(\cos t)$

$= 2\cos t \sin t + 3\sin^2 t - 3\cos^2 t = \sin 2t - 3\cos 2t$

19. $\dfrac{du}{dt} = \dfrac{\partial u}{\partial x}\dfrac{dx}{dt} + \dfrac{\partial u}{\partial y}\dfrac{dy}{dt}$

$= (e^x \sin y + e^y \cos x)\left(\tfrac{1}{2}\right) + (e^x \cos y + e^y \sin x)(2)$

$= e^{t/2}\left(\tfrac{1}{2}\sin 2t + 2\cos 2t\right) + e^{2t}\left(\tfrac{1}{2}\cos \tfrac{1}{2}t + 2\sin \tfrac{1}{2}t\right)$

21. $\dfrac{du}{dt} = \dfrac{\partial u}{\partial x}\dfrac{dx}{dt} + \dfrac{\partial u}{\partial y}\dfrac{dy}{dt} = (e^x \sin y)(2t) + (e^x \cos y)(\pi)$

$= e^{t^2}[2t\,\sin(\pi t) + \pi\,\cos(\pi t)]$

23. $\dfrac{du}{dt} = \dfrac{\partial u}{\partial x}\dfrac{dx}{dt} + \dfrac{\partial u}{\partial y}\dfrac{dy}{dt} + \dfrac{\partial u}{\partial z}\dfrac{dz}{dt}$

$= (y + z)(2t) + (x + z)(1 - 2t) + (y + x)(2t - 2)$

$= (1 - t)(2t) + (2t^2 - 2t + 1)(1 - 2t) + t(2t - 2)$

$= 1 - 4t + 6t^2 - 4t^3$

25. $V = \dfrac{1}{3}\pi r^2 h, \quad \dfrac{dV}{dt} = \dfrac{\partial V}{\partial r}\dfrac{dr}{dt} + \dfrac{\partial V}{\partial h}\dfrac{dh}{dt} = \left(\dfrac{2}{3}\pi rh\right)\dfrac{dr}{dt} + \left(\dfrac{1}{3}\pi r^2\right)\dfrac{dh}{dt}.$

At the given instant,

$$\dfrac{dV}{dt} = \dfrac{2}{3}\pi(280)(3) + \dfrac{1}{3}\pi(196)(-2) = \dfrac{1288}{3}\pi.$$

The volume is increasing at the rate of $\dfrac{1288}{3}\pi$ in.^{3}/ sec .

27. $A = \frac{1}{2} xy \sin\theta;$ $\dfrac{dA}{dt} = \dfrac{\partial A}{\partial x}\dfrac{dx}{dt} + \dfrac{\partial A}{\partial y}\dfrac{dy}{dt} + \dfrac{\partial A}{\partial \theta}\dfrac{d\theta}{dt} = \frac{1}{2}\left[(y\sin\theta)\dfrac{dx}{dt} + (x\sin\theta)\dfrac{dy}{dt} + (xy\cos\theta)\dfrac{d\theta}{dt}\right].$

At the given instant

$$\frac{dA}{dt} = \frac{1}{2}\left[(2\sin 1)(0.25) + (1.5\sin 1)(0.25) - (2(1.5)\cos 1)(0.1)\right] \cong 0.2871\,\text{ft}^2/s \cong 41.34\,\text{in}^2/s$$

29. $\dfrac{\partial u}{\partial s} = \dfrac{\partial u}{\partial x}\dfrac{\partial x}{\partial s} + \dfrac{\partial u}{\partial y}\dfrac{\partial y}{\partial s} = (2x - y)(\cos t) + (-x)(t\cos s)$

$\qquad = 2s\cos^2 t - t\sin s\cos t - st\cos s\cos t$

$\dfrac{\partial u}{\partial t} = \dfrac{\partial u}{\partial x}\dfrac{\partial x}{\partial t} + \dfrac{\partial u}{\partial y}\dfrac{\partial y}{\partial t} = (2x - y)(-s\sin t) + (-x)(\sin s)$

$\qquad = -2s^2\cos t\sin t + st\sin s\sin t - s\cos t\sin s$

31. $\dfrac{\partial u}{\partial s} = \dfrac{\partial u}{\partial x}\dfrac{\partial x}{\partial s} + \dfrac{\partial u}{\partial y}\dfrac{\partial y}{\partial s} = (2x\tan y)(2st) + \left(x^2\sec^2 y\right)(1)$

$\qquad = 4s^3t^2\tan\left(s + t^2\right) + s^4t^2\sec^2\left(s + t^2\right)$

$\dfrac{\partial u}{\partial t} = \dfrac{\partial u}{\partial x}\dfrac{\partial x}{\partial t} + \dfrac{\partial u}{\partial y}\dfrac{\partial y}{\partial t} = (2x\tan y)\left(s^2\right) + \left(x^2\sec^2 y\right)(2t)$

$\qquad = 2s^4t\tan\left(s + t^2\right) + 2s^4t^3\sec^2\left(s + t^2\right)$

33. $\dfrac{\partial u}{\partial s} = \dfrac{\partial u}{\partial x}\dfrac{\partial x}{\partial s} + \dfrac{\partial u}{\partial y}\dfrac{\partial y}{\partial s} + \dfrac{\partial u}{\partial z}\dfrac{\partial z}{\partial s}$

$\qquad = (2x - y)(\cos t) + (-x)(-\cos(t - s)) + 2z(t\cos s)$

$\qquad = 2s\cos^2 t - \sin(t - s)\cos t + s\cos t\cos(t - s) + 2t^2\sin s\cos s$

$\dfrac{\partial u}{\partial t} = \dfrac{\partial u}{\partial x}\dfrac{\partial x}{\partial t} + \dfrac{\partial u}{\partial y}\dfrac{\partial y}{\partial t} + \dfrac{\partial u}{\partial z}\dfrac{\partial z}{\partial t}$

$\qquad = (2x - y)(-s\sin t) + (-x)(\cos(t - s)) + 2z(\sin s)$

$\qquad = -2s^2\cos t\sin t + s\sin(t - s)\sin t - s\cos t\cos(t - s) + 2t\sin^2 s$

35. $\dfrac{d}{dt}[f(\mathbf{r}(t))] = \left[\boldsymbol{\nabla} f(\mathbf{r}(t)) \cdot \dfrac{\mathbf{r}'(t)}{\|\mathbf{r}'(t)\|}\right]\|\mathbf{r}'(t)\|$

$\qquad = f'_{\mathbf{u}(t)}(\mathbf{r}(t))\,\|\mathbf{r}'(t)\|$ where $\mathbf{u}(t) = \dfrac{\mathbf{r}'(t)}{\|\mathbf{r}'(t)\|}$

37. (a) $(\cos r)\dfrac{\mathbf{r}}{r}$ $\qquad\qquad\qquad\qquad\qquad$ (b) $(r\cos r + \sin r)\dfrac{\mathbf{r}}{r}$

39. (a) $(r\cos r - \sin r)\dfrac{\mathbf{r}}{r^3}$ $\qquad\qquad\qquad$ (b) $\left(\dfrac{\sin r - r\cos r}{\sin^2 r}\right)\dfrac{\mathbf{r}}{r}$

41. (a)

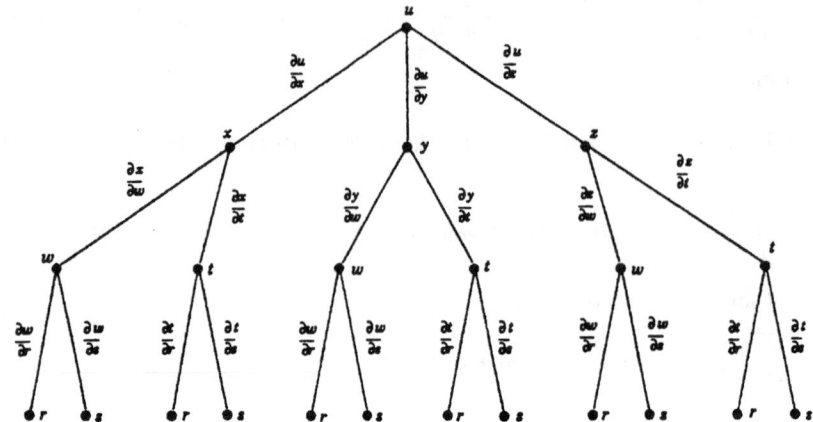

(b) $\dfrac{\partial u}{\partial r} = \dfrac{\partial u}{\partial x}\left(\dfrac{\partial x}{\partial w}\dfrac{\partial w}{\partial r} + \dfrac{\partial x}{\partial t}\dfrac{\partial t}{\partial r}\right) + \dfrac{\partial u}{\partial y}\left(\dfrac{\partial y}{\partial w}\dfrac{\partial w}{\partial r} + \dfrac{\partial y}{\partial t}\dfrac{\partial t}{\partial r}\right) + \dfrac{\partial u}{\partial z}\left(\dfrac{\partial z}{\partial w}\dfrac{\partial w}{\partial r} + \dfrac{\partial z}{\partial t}\dfrac{\partial t}{\partial r}\right).$

    To obtain $\partial u/\partial s$, replace each r by s.

43. $\dfrac{du}{dt} = \dfrac{\partial u}{\partial x}\dfrac{dx}{dt} + \dfrac{\partial u}{\partial y}\dfrac{dy}{dt}$

$\dfrac{d^2u}{dt^2} = \dfrac{\partial u}{\partial x}\dfrac{d^2x}{dt^2} + \dfrac{dx}{dt}\left[\dfrac{\partial^2 u}{\partial x^2}\dfrac{dx}{dt} + \dfrac{\partial^2 u}{\partial y\partial x}\dfrac{dy}{dt}\right] + \dfrac{\partial u}{\partial y}\dfrac{d^2y}{dt^2} + \dfrac{dy}{dt}\left[\dfrac{\partial^2 u}{\partial x\,\partial y}\dfrac{dx}{dt} + \dfrac{\partial^2 u}{\partial y^2}\dfrac{dy}{dt}\right]$

and the result follows.

45. (a) $\dfrac{\partial u}{\partial r} = \dfrac{\partial u}{\partial x}\dfrac{\partial x}{\partial r} + \dfrac{\partial u}{\partial y}\dfrac{\partial y}{\partial r} = \dfrac{\partial u}{\partial x}\cos\theta + \dfrac{\partial u}{\partial y}\sin\theta$

       $\dfrac{\partial u}{\partial\theta} = \dfrac{\partial u}{\partial x}\dfrac{\partial x}{\partial\theta} + \dfrac{\partial u}{\partial y}\dfrac{\partial y}{\partial\theta} = \dfrac{\partial u}{\partial x}(-r\sin\theta) + \dfrac{\partial u}{\partial y}(r\cos\theta)$

  (b) $\left(\dfrac{\partial u}{\partial r}\right)^2 = \left(\dfrac{\partial u}{\partial x}\right)^2\cos^2\theta + 2\dfrac{\partial u}{\partial x}\dfrac{\partial u}{\partial y}\cos\theta\sin\theta + \left(\dfrac{\partial u}{\partial y}\right)^2\sin^2\theta,$

      $\dfrac{1}{r^2}\left(\dfrac{\partial u}{\partial\theta}\right)^2 = \left(\dfrac{\partial u}{\partial x}\right)^2\sin^2\theta - 2\dfrac{\partial u}{\partial x}\dfrac{\partial u}{\partial y}\cos\theta\sin\theta + \left(\dfrac{\partial u}{\partial y}\right)^2\cos^2\theta,$

      $\left(\dfrac{\partial u}{\partial r}\right)^2 + \dfrac{1}{r^2}\left(\dfrac{\partial u}{\partial\theta}\right)^2 = \left(\dfrac{\partial u}{\partial x}\right)^2(\cos^2\theta + \sin^2\theta) + \left(\dfrac{\partial u}{\partial y}\right)^2(\sin^2\theta + \cos^2\theta) = \left(\dfrac{\partial u}{\partial x}\right)^2 + \left(\dfrac{\partial u}{\partial y}\right)^2$

47. Solve the equations in Exercise 45 (a) for $\dfrac{\partial u}{\partial x}$ and $\dfrac{\partial u}{\partial y}$:

$$\dfrac{\partial u}{\partial x} = \dfrac{\partial u}{\partial r}\cos\theta - \dfrac{1}{r}\dfrac{\partial u}{\partial\theta}\sin\theta, \qquad \dfrac{\partial u}{\partial y} = \dfrac{\partial u}{\partial r}\sin\theta + \dfrac{1}{r}\dfrac{\partial u}{\partial\theta}\cos\theta$$

Then $\nabla u = \dfrac{\partial u}{\partial x}\mathbf{i} + \dfrac{\partial u}{\partial y}\mathbf{j} = \dfrac{\partial u}{\partial r}(\cos\theta\,\mathbf{i} + \sin\theta\,\mathbf{j}) + \dfrac{1}{r}\dfrac{\partial u}{\partial\theta}(-\sin\theta\,\mathbf{i} + \cos\theta\,\mathbf{j})$

49. $u(x, y) = x^2 - xy + y^2 = r^2 - r^2 \cos\theta \sin\theta = r^2 \left(1 - \frac{1}{2}\sin 2\theta\right)$

$$\frac{\partial u}{\partial r} = r(2 - \sin 2\theta), \quad \frac{\partial u}{\partial \theta} = -r^2 \cos 2\theta$$

$$\boldsymbol{\nabla} u = \frac{\partial u}{\partial r}\,\mathbf{e_r} + \frac{1}{r}\frac{\partial u}{\partial \theta}\,\mathbf{e_\theta} = r(2 - \sin 2\theta)\mathbf{e_r} - r\,\cos 2\theta\,\mathbf{e_\theta}$$

51. From Exercise 45 (a),

$$\frac{\partial^2 u}{\partial r^2} = \frac{\partial^2 u}{\partial x^2}\cos^2\theta + 2\frac{\partial^2 u}{\partial y\,\partial x}\sin\theta\,\cos\theta + \frac{\partial^2 u}{\partial y^2}\sin^2\theta$$

$$\frac{\partial^2 u}{\partial \theta^2} = \frac{\partial^2 u}{\partial x^2}\,r^2\sin^2\theta - 2\frac{\partial^2 u}{\partial y\,\partial x}\,r^2\sin\theta\,\cos\theta + \frac{\partial^2 u}{\partial y^2}\,r^2\cos^2\theta - r\left(\frac{\partial u}{\partial x}\cos\theta + \frac{\partial u}{\partial y}\sin\theta\right).$$

The term in parentheses is $\dfrac{\partial u}{\partial r}$. Now divide the second equation by r^2 and add the two equations. The result follows.

53. Set $u = xe^y + ye^x - 2x^2 y$. Then

$$\frac{\partial u}{\partial x} = e^y + ye^x - 4xy, \quad \frac{\partial u}{\partial y} = xe^y + e^x - 2x^2$$

$$\frac{dy}{dx} = -\frac{\partial u/\partial x}{\partial u/\partial y} = -\frac{e^y + ye^x - 4xy}{xe^y + e^x - 2x^2}.$$

55. Set $u = x\cos xy + y\cos x - 2$. Then

$$\frac{\partial u}{\partial x} = \cos xy - xy\,\sin xy - y\,\sin x, \quad \frac{\partial u}{\partial y} = -x^2\sin xy + \cos x$$

$$\frac{dy}{dx} = -\frac{\partial u/\partial x}{\partial u/\partial y} = \frac{\cos xy - xy\,\sin xy - y\,\sin x}{x^2\sin xy - \cos x}.$$

57. Set $u = \cos xyz + \ln\left(x^2 + y^2 + z^2\right)$. Then

$$\frac{\partial u}{\partial x} = -yz\,\sin xyz + \frac{2x}{x^2 + y^2 + z^2}, \quad \frac{\partial u}{\partial y} = -xz\,\sin xyz + \frac{2y}{x^2 + y^2 + z^2}, \quad \text{and}$$

$$\frac{\partial u}{\partial z} = -xy\,\sin xyz + \frac{2z}{x^2 + y^2 + z^2}.$$

$$\frac{\partial z}{\partial x} = -\frac{\partial u/\partial x}{\partial u/\partial z} = -\frac{2x - yz\left(x^2 + y^2 + z^2\right)\sin xyz}{2z - xy\left(x^2 + y^2 + z^2\right)\sin xyz},$$

$$\frac{\partial z}{\partial y} = -\frac{\partial u/\partial y}{\partial u/\partial z} = -\frac{2y - xz\left(x^2 + y^2 + z^2\right)\sin xyz}{2z - xy\left(x^2 + y^2 + z^2\right)\sin xyz}.$$

59. $\dfrac{\partial \mathbf{u}}{\partial s} = \dfrac{\partial \mathbf{u}}{\partial x}\dfrac{\partial x}{\partial s} + \dfrac{\partial \mathbf{u}}{\partial y}\dfrac{\partial y}{\partial s}, \quad \dfrac{\partial \mathbf{u}}{\partial t} = \dfrac{\partial \mathbf{u}}{\partial x}\dfrac{\partial x}{\partial t} + \dfrac{\partial \mathbf{u}}{\partial y}\dfrac{\partial y}{\partial t}$

SECTION 16.4

1. Set $f(x, y) = x^2 + xy + y^2$.   Then,

$$\nabla f = (2x + y)\mathbf{i} + (x + 2y)\mathbf{j}, \quad \nabla f(-1, -1) = -3\mathbf{i} - 3\mathbf{j}.$$

normal vector $\mathbf{i} + \mathbf{j}$; tangent vector $\mathbf{i} - \mathbf{j}$

tangent line $x + y + 2 = 0$; normal line $x - y = 0$

3. Set $f(x, y) = \left(x^2 + y^2\right)^2 - 9\left(x^2 - y^2\right)$.   Then,

$$\nabla f = [4x(x^2 + y^2) - 18x]\mathbf{i} + \left[4y\left(x^2 + y^2\right) + 18y\right]\mathbf{j}, \quad \nabla f\left(\sqrt{2}, 1\right) = -6\sqrt{2}\,\mathbf{i} + 30\mathbf{j}.$$

normal vector $\sqrt{2}\,\mathbf{i} - 5\,\mathbf{j}$; tangent vector $5\mathbf{i} + \sqrt{2}\,\mathbf{j}$

tangent line $\sqrt{2}x - 5y + 3 = 0$; normal line $5x + \sqrt{2}\,y - 6\sqrt{2} = 0$

5. Set $f(x, y) = xy^2 - 2x^2 + y + 5x$.   Then,

$$\nabla f = \left(y^2 - 4x + 5\right)\mathbf{i} + (2xy + 1)\,\mathbf{j}, \quad \nabla f(4, 2) = -7\mathbf{i} + 17\mathbf{j}.$$

normal vector $7\mathbf{i} - 17\mathbf{j}$; tangent vector $17\mathbf{i} + 7\mathbf{j}$

tangent line $7x - 17y + 6 = 0$; normal line $17x + 7y - 82 = 0$

7. Set $f(x, y) = 2x^3 - x^2y^2 - 3x + y$.   Then,

$$\nabla f = \left(6x^2 - 2xy^2 - 3\right)\mathbf{i} + \left(-2x^2y + 1\right)\mathbf{j}, \quad \nabla f(1, -2) = -5\mathbf{i} + 5\mathbf{j}.$$

normal vector $\mathbf{i} - \mathbf{j}$; tangent vector $\mathbf{i} + \mathbf{j}$

tangent line $x - y - 3 = 0$; normal line $x + y + 1 = 0$

9. Set $f(x, y) = x^2y + a^2y$.   By (15.4.4)

$$m = -\frac{\partial f/\partial x}{\partial f/\partial y} = -\frac{2xy}{x^2 + a^2}.$$

At $(0, a)$ the slope is 0.

11. Set $f(x, y, z) = x^3 + y^3 - 3xyz$.   Then,

$$\nabla f = \left(3x^2 - 3yz\right)\mathbf{i} + \left(3y^2 - 3xz\right)\mathbf{j} - 3xy\mathbf{k}, \quad \nabla f\left(1, 2, \tfrac{3}{2}\right) = -6\mathbf{i} + \tfrac{15}{2}\mathbf{j} - 6\mathbf{k};$$

tangent plane at $\left(1, 2, \tfrac{3}{2}\right)$: $-6(x - 1) + \tfrac{15}{2}(y - 2) - 6\left(z - \tfrac{3}{2}\right) = 0$, which reduces to $4x - 5y + 4z = 0$.

Normal: $x = 1 + 4t, \quad y = 2 - 5t, \quad z = \tfrac{3}{2} + 4t$

13. Set $z = g(x, y) = axy$.   Then, $\nabla g = ay\mathbf{i} + ax\mathbf{j}, \quad \nabla g\left(1, \dfrac{1}{a}\right) = \mathbf{i} + a\mathbf{j}.$

tangent plane at $\left(1, \dfrac{1}{a}, 1\right)$: $z - 1 = 1(x - 1) + a\left(y - \dfrac{1}{a}\right)$, which reduces to $x + ay - z - 1 = 0$

Normal: $x = 1 + t, \quad y = \tfrac{1}{a} + at, \quad z = 1 - t$

15. Set $z = g(x,y) = \sin x + \sin y + \sin(x+y)$.   Then,

$$\nabla g = [\cos x + \cos(x+y)]\,\mathbf{i} + [\cos y + \cos(x+y)]\,\mathbf{j}, \quad \nabla g(0,0) = 2\mathbf{i} + 2\mathbf{j};$$

tangent plane at $(0,0,0)$: $\ z - 0 = 2(x-0) + 2(y-0), \quad 2x + 2y - z = 0.$
Normal: $\ x = 2t, \quad y = 2t, \quad z = -t$

17. Set $f(x,y,z) = b^2 c^2 x^2 - a^2 c^2 y^2 - a^2 b^2 z^2$.   Then,

$$\nabla f(x_0, y_0, z_0) = 2b^2 c^2 x_0 \mathbf{i} - 2a^2 c^2 y_0 \mathbf{j} - 2a^2 b^2 z_0 \mathbf{k};$$

tangent plane at (x_0, y_0, z_0):

$$2b^2 c^2 x_0 (x - x_0) - 2a^2 c^2 y_0 (y - y_0) - 2a^2 b^2 z_0 (z - z_0) = 0,$$

which can be rewritten as follows:

$$b^2 c^2 x_0 x - a^2 c^2 y_0 y - a^2 b^2 z_0 z = b^2 c^2 x_0{}^2 - a^2 c^2 y_0{}^2 - a^2 b^2 z_0{}^2$$
$$= f(x_0, y_0, z_0) = a^2 b^2 c^2.$$

Normal: $\ x = x_0 + 2b^2 c^2 x_0 t, \quad y = y_0 - 2a^2 c^2 y_0 t, \quad z = z_0 - 2a^2 b^2 z_0 t$

19. Set $z = g(x,y) = xy + a^3 x^{-1} + b^3 y^{-1}$.

$$\nabla g = \left(y - a^3 x^{-2}\right)\mathbf{i} + \left(x - b^3 y^{-2}\right)\mathbf{j}, \quad \nabla g = \mathbf{0} \implies y = a^3 x^{-2} \ \text{and} \ x = b^3 y^{-2}.$$

Thus,

$$y = a^3 b^{-6} y^4, \quad y^3 = b^6 a^{-3}, \quad y = b^2/a, \quad x = b^3 y^{-2} = a^2/b \ \text{and} \ g\left(a^2/b,\, b^2/a\right) = 3ab.$$

The tangent plane is horizontal at $\left(a^2/b,\, b^2/a,\, 3ab\right)$.

21. Set $z = g(x,y) = xy$.   Then, $\nabla g = y\mathbf{i} + x\mathbf{j}$.

$$\nabla g = \mathbf{0} \implies x = y = 0.$$

The tangent plane is horizontal at $(0,0,0)$.

23. Set $z = g(x,y) = 2x^2 + 2xy - y^2 - 5x + 3y - 2$.   Then,

$$\nabla g = (4x + 2y - 5)\,\mathbf{i} + (2x - 2y + 3)\,\mathbf{j}.$$

$$\nabla g = \mathbf{0} \implies 4x + 2y - 5 = 0 = 2x - 2y + 3 \implies x = \tfrac{1}{3}, \quad y = \tfrac{11}{6}.$$

The tangent plane is horizontal at $\left(\tfrac{1}{3}, \tfrac{11}{6}, -\tfrac{1}{12}\right)$.

25. $\dfrac{x - x_0}{(\partial f/\partial x)(x_0, y_0, z_0)} = \dfrac{y - y_0}{(\partial f/\partial y)(x_0, y_0, z_0)} = \dfrac{z - z_0}{(\partial f/\partial z)(x_0, y_0, z_0)}$

27. Since the tangent planes meet at right angles, the normals ∇F and ∇G meet at right angles:

$$\frac{\partial F}{\partial x}\frac{\partial G}{\partial x} + \frac{\partial F}{\partial y}\frac{\partial G}{\partial y} + \frac{\partial F}{\partial z}\frac{\partial G}{\partial z} = 0.$$

29. The tangent plane at an arbitrary point (x_0, y_0, z_0) has equation

$$y_0 z_0 (x - x_0) + x_0 z_0 (y - y_0) + x_0 y_0 (z - z_0) = 0,$$

which simplifies to

$$y_0 z_0 x + x_0 z_0 y + x_0 y_0 z = 3 x_0 y_0 z_0 \quad \text{and thus to} \quad \frac{x}{3x_0} + \frac{y}{3y_0} + \frac{z}{3z_0} = 1.$$

The volume of the pyramid is

$$V = \frac{1}{3} Bh = \frac{1}{3} \left[\frac{(3x_0)(3y_0)}{2} \right] (3z_0) = \frac{9}{2} x_0 y_0 z_0 = \frac{9}{2} a^3.$$

31. The point $(2, 3, -2)$ is the tip of $\mathbf{r}(1)$.

Since $\mathbf{r}'(t) = 2\mathbf{i} - \dfrac{3}{t^2}\mathbf{j} - 4t\mathbf{k}$, we have $\mathbf{r}'(1) = 2\mathbf{i} - 3\mathbf{j} - 4\mathbf{k}$.

Now set $f(x, y, z) = x^2 + y^2 + 3z^2 - 25$. The function has gradient $2x\mathbf{i} + 2y\mathbf{j} + 6z\mathbf{k}$.

At the point $(2, 3, -2)$,

$$\nabla f = 2(2\mathbf{i} + 3\mathbf{j} - 6\mathbf{k}).$$

The angle θ between $\mathbf{r}'(1)$ and the gradient gives

$$\cos\theta = \frac{(2\mathbf{i} - 3\mathbf{j} - 4\mathbf{k})}{\sqrt{29}} \cdot \frac{(2\mathbf{i} + 3\mathbf{j} - 6\mathbf{k})}{7} = \frac{19}{7\sqrt{29}} \cong 0.504.$$

Therefore $\theta \cong 1.043$ radians. The angle between the curve and the plane is

$$\frac{\pi}{2} - \theta \cong 1.571 - 1.043 \cong 0.528 \text{ radians.}$$

33. Set $f(x, y, z) = x^2 y^2 + 2x + z^3$. Then,

$$\nabla f = (2xy^2 + 2)\,\mathbf{i} + 2x^2 y\mathbf{j} + 3z^2\mathbf{k}, \quad \nabla f(2, 1, 2) = 6\mathbf{i} + 8\mathbf{j} + 12\mathbf{k}.$$

The plane tangent to $f(x, y, z) = 16$ at $(2, 1, 2)$ has equation

$$6(x - 2) + 8(y - 1) + 12(z - 2) = 0, \quad \text{or} \quad 3x + 4y + 6z = 22.$$

Next, set $g(x, y, z) = 3x^2 + y^2 - 2z$. Then,

$$\nabla g = 6x\mathbf{i} + 2y\mathbf{j} - 2\mathbf{k}, \quad \nabla g(2, 1, 2) = 12\mathbf{i} + 2\mathbf{j} - 2\mathbf{k}.$$

The plane tangent to $g(x, y, z) = 9$ at $(2, 1, 2)$ is

$$12(x - 2) + 2(y - 1) - 2(z - 2) = 0, \quad \text{or} \quad 6x + y - z = 11.$$

35. A normal vector to the sphere at $(1, 1, 2)$ is

$$2x\mathbf{i} + (2y - 4)\,\mathbf{j} + (2z - 2)\mathbf{k} = 2\mathbf{i} - 2\mathbf{j} + 2\mathbf{k}.$$

A normal vector to the paraboloid at $(1, 1, 2)$ is

$$6x\mathbf{i} + 4y\mathbf{j} - 2\mathbf{k} = 6\mathbf{i} + 4\mathbf{j} - 2\mathbf{k}.$$

Since

$$(2\mathbf{i} - 2\mathbf{j} + 2\mathbf{k}) \cdot (6\mathbf{i} + 4\mathbf{j} - 2\mathbf{k}) = 0,$$

the surfaces intersect at right angles.

37. (a) $3x + 4y + 6 = 0$ since plane p is vertical.

(b) $y = -\frac{1}{4}(3x + 6) = -\frac{1}{4}[3(4t - 2) + 6] = -3t$

$z = x^2 + 3y^2 + 2 = (4t - 2)^2 + 3(-3t)^2 + 2 = 43t^2 - 16t + 6$

$\mathbf{r}(t) = (4t - 2)\mathbf{i} - 3t\mathbf{j} + (43t^2 - 16t + 6)\mathbf{k}$

(c) From part (b) the tip of $\mathbf{r}(1)$ is $(2, -3, 33)$.   We take

$\mathbf{r}'(1) = 4\mathbf{i} - 3\mathbf{j} + 70\mathbf{j}$ as $\mathbf{d}$ to write

$$\mathbf{R}(s) = (2\mathbf{i} - 3\mathbf{j} + 33\mathbf{k}) + s(4\mathbf{i} - 3\mathbf{j} + 70\mathbf{k}).$$

(d) Set $g(x, y) = x^2 + 3y^2 + 2$.   Then,

$$\nabla g = 2x\mathbf{i} + 6y\mathbf{j} \quad \text{and} \quad \nabla g(2, -3) = 4\mathbf{i} - 18\mathbf{j}.$$

An equation for the plane tangent to $z = g(x, y)$ at $(2, -3, 33)$ is

$$z - 33 = 4(x - 2) - 18(y + 3) \quad \text{which reduces to} \quad 4x - 18y - z = 29.$$

(e) Substituting t for x in the equations for p and p_1, we obtain

$$3t + 4y + 6 = 0 \quad \text{and} \quad 4t - 18y - z = 29.$$

From the first equation

$$y = -\tfrac{3}{4}(t + 2)$$

and then from the second equation

$$z = 4t - 18\left[-\tfrac{3}{4}(t + 2)\right] - 29 = \tfrac{35}{2}t - 2.$$

Thus,

$(*) \qquad \mathbf{r}(t) = t\mathbf{i} - \left(\tfrac{3}{4}t + \tfrac{3}{2}\right)\mathbf{j} + \left(\tfrac{35}{2}t - 2\right)\mathbf{k}.$

Lines l and l' are the same. To see this, consider how l and l' are formed; to assure yourself, replace t in $(*)$ by $4s + 2$ to obtain $\mathbf{R}(s)$ found in part (c).

39. (a) normal vector: $2\mathbf{i} + 2\mathbf{j} + 4\mathbf{k}$;   normal line: $x = 1 + 2t, \; y = 2 + 2t, \; z = 2 + 4t$

(b) tangent plane: $2(x - 1) + 2(y - 2) + 4(z - 2) = 0 \quad \text{or} \quad x + y + 2z - 7 = 0$

(c)

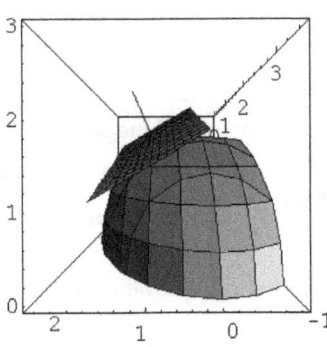

41. (a)

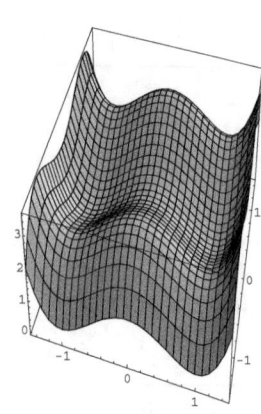

(b)

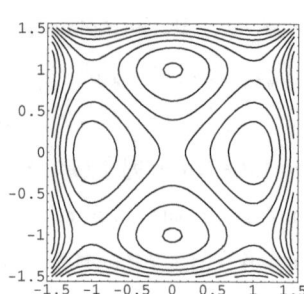

(c) $\nabla f = \left(4x^3 - 4x\right)\mathbf{i} - \left(4y^3 - 4y\right)\mathbf{j}$;

$\nabla f = 0 :$ $4x^3 - 4x = 0 \implies x = 0, \pm 1;$ $4y^3 - 4y = 0 \implies y = 0, \pm 1$

$\nabla f = 0$ at $(0,0)$, $(\pm 1, 0)$, $(0, \pm 1)$, $(\pm 1, \pm 1)$

SECTION 16.5

1. $\nabla f = (2 - 2x)\mathbf{i} - 2y\,\mathbf{j} = \mathbf{0}$ only at $(1,0)$.

The difference

$$f(1 + h, k) - f(1,0) = \left[2(1+h) - (1+h)^2 - k^2\right] - 1 = -h^2 - k^2 \leq 0$$

for all small h and k; there is a local maximum of 1 at $(1,0)$.

3. $\nabla f = (2x + y + 3)\mathbf{i} + (x + 2y)\mathbf{j} = \mathbf{0}$ only at $(-2, 1)$.

The difference

$$f(-2 + h,\, 1 + k) - f(-2, 1)$$
$$= \left[(-2+h)^2 + (-2+h)(1+k) + (1+k)^2 + 3(-2+h) + 1\right] - (-2) = h^2 + hk + k^2$$

is nonnegative for all small h and k. To see this, note that

$$h^2 + hk + k^2 \geq h^2 - 2|h||k| + k^2 = (|h| - |k|)^2 \geq 0;$$

there is a local minimum of -2 at $(-2, 1)$.

5. $\nabla f = (2x + y - 6)\mathbf{i} + (x + 2y)\mathbf{j} = \mathbf{0}$ only at $(4, -2)$.

$f_{xx} = 2, \quad f_{xy} = 1, \quad f_{yy} = 2.$

At $(4, -2)$, $D = 3 > 0$ and $A = 2 > 0$ so we have a local min; the value is -10.

7. $\nabla f = (3x^2 - 6y)\mathbf{i} + (3y^2 - 6x)\mathbf{j} = \mathbf{0}$ at $(2, 2)$ and $(0, 0)$.

$f_{xx} = 6x, \quad f_{xy} = -6, \quad f_{yy} = 6y, \quad D = 36xy - 36.$

At $(2, 2)$, $D = 108 > 0$ and $A = 12 > 0$ so we have a local min; the value is -8.

At $(0, 0)$, $D = -36 < 0$ so we have a saddle point.

9. $\nabla f = (3x^2 - 6y + 6)\,\mathbf{i} + (2y - 6x + 3)\,\mathbf{j} = \mathbf{0}$ at $(5, \frac{27}{2})$ and $(1, \frac{3}{2})$.

$f_{xx} = 6x, \quad f_{xy} = -6, \quad f_{yy} = 2, \quad D = 12x - 36.$

At $(5, \frac{27}{2})$, $D = 24 > 0$ and $A = 30 > 0$ so we have a local min; the value is $-\frac{117}{4}$.

At $(1, \frac{3}{2})$, $D = -24 < 0$ so we have a saddle point.

11. $\nabla f = \sin y\,\mathbf{i} + x \cos y\,\mathbf{j} = \mathbf{0}$ at $(0, n\pi)$ for all integral n.

$f_{xx} = 0, \quad f_{xy} = \cos y, \quad f_{yy} = -x \sin y.$

Since $D = -\cos^2 n\pi = -1 < 0$, each stationary point is a saddle point.

13. $\nabla f = (2xy + 1 + y^2)\,\mathbf{i} + (x^2 + 2xy + 1)\,\mathbf{j} = \mathbf{0}$ at $(1, -1)$ and $(-1, 1)$.

$f_{xx} = 2y, \quad f_{xy} = 2x + 2y, \quad f_{yy} = 2x, \quad D = 4xy - 4(x + y)^2.$

At both $(1, -1)$ and $(-1, 1)$ we have saddle points since $D = -4 < 0$.

15. $\nabla f = (y - x^{-2})\,\mathbf{i} + (x - 8y^{-2})\,\mathbf{j} = \mathbf{0}$ only at $(\frac{1}{2}, 4)$.

$f_{xx} = 2x^{-3}, \quad f_{xy} = 1, \quad f_{yy} = 16y^{-3}, \quad D = 32x^{-3}y^{-3} - 1.$

At $(\frac{1}{2}, 4)$, $D = 3 > 0$ and $A = 16 > 0$ so we have a local min; the value is 6.

17. $\nabla f = (y - x^{-2})\,\mathbf{i} + (x - y^{-2})\,\mathbf{j} = \mathbf{0}$ only at $(1, 1)$.

$f_{xx} = 2x^{-3}, \quad f_{xy} = 1, \quad f_{yy} = 2y^{-3}, \quad D = 4x^{-3}y^{-3} - 1.$

At $(1, 1)$, $D = 3 > 0$ and $A = 2 > 0$ so we have a local min; the value is 3.

19. $\nabla f = \dfrac{2(x^2 - y^2 - 1)}{(x^2 + y^2 + 1)^2}\,\mathbf{i} + \dfrac{4xy}{(x^2 + y^2 + 1)^2}\,\mathbf{j} = \mathbf{0}$ at $(1, 0)$ and $(-1, 0)$.

$f_{xx} = \dfrac{-4x^3 + 12xy^2 + 12x}{(x^2 + y^2 + 1)^3}, \quad f_{xy} = \dfrac{4y^3 + 4y - 12x^2y}{(x^2 + y^2 + 1)^3}, \quad f_{yy} = \dfrac{4x^3 + 4x - 12xy^2}{(x^2 + y^2 + 1)^3}.$

point	A	B	C	D	result
$(1, 0)$	1	0	1	1	loc. min.
$(-1, 0)$	-1	0	-1	1	loc. max.

$f(1, 0) = -1; \quad f(-1, 0) = 1$

21. $\nabla f = (4x^3 - 4x)\,\mathbf{i} + 2y\,\mathbf{j} = \mathbf{0}$ at $(0, 0)$, $(1, 0)$, and $(-1, 0)$.

$f_{xx} = 12x^2 - 4, \quad f_{xy} = 0, \quad f_{yy} = 2.$

point	A	B	C	D	result
$(0, 0)$	-4	0	2	-8	saddle
$(1, 0)$	8	0	2	16	loc. min.
$(-1, 0)$	8	0	2	16	loc. min.

$f(\pm 1, 0) = -3.$

23. $\nabla f = \cos x \, \sin y \, \mathbf{i} + \sin x \, \cos y \, \mathbf{j} = \mathbf{0}$ at $\left(\frac{1}{2}\pi, \frac{1}{2}\pi\right)$, $\left(\frac{1}{2}\pi, \frac{3}{2}\pi\right)$, (π, π), $\left(\frac{3}{2}\pi, \frac{1}{2}\pi\right)$, $\left(\frac{3}{2}\pi, \frac{3}{2}\pi\right)$.

$f_{xx} = -\sin x \, \sin y, \quad f_{xy} = \cos x \, \cos y, \quad f_{yy} = -\sin x \, \sin y$

point	A	B	C	D	result
$\left(\frac{1}{2}\pi, \frac{1}{2}\pi\right)$	-1	0	-1	1	loc. max.
$\left(\frac{1}{2}\pi, \frac{3}{2}\pi\right)$	1	0	1	1	loc. min.
(π, π)	0	1	0	-1	saddle
$\left(\frac{3}{2}\pi, \frac{1}{2}\pi\right)$	1	0	1	1	loc. min.
$\left(\frac{3}{2}\pi, \frac{3}{2}\pi\right)$	-1	0	-1	1	loc. max.

$f\left(\frac{1}{2}\pi, \frac{1}{2}\pi\right) = f\left(\frac{3}{2}\pi, \frac{3}{2}\pi\right) = 1; \quad f\left(\frac{1}{2}\pi, \frac{3}{2}\pi\right) = f\left(\frac{3}{2}\pi, \frac{1}{2}\pi\right) = -1$

25. (a) $\nabla f = (2x + ky)\,\mathbf{i} + (2y + kx)\,\mathbf{j}$ and $\nabla f(0,0) = \mathbf{0}$ independent of the value of k.

(b) $f_{xx} = 2, \quad f_{xy} = k, \quad f_{yy} = 2, \quad D = 4 - k^2$. Thus, $D < 0$ for $|k| > 2$ and $(0,0)$ is a saddle point

(c) $D = 4 - k^2 > 0$ for $|k| < 2$. Since $A = f_{xx} = 2 > 0$, $(0,0)$ is a local minimum.

(d) The test is inconclusive when $D = 4 - k^2 = 0$ i.e., for $k = \pm 2$. (If $k = \pm 2$, $f(x,y) = (x \pm y)^2$ and $(0, 0)$ is a minimum.)

27. Let $P(x,y,z)$ be a point in the plane. We want to find the minimum of $f(x,y,z) = \sqrt{x^2 + y^2 + z^2}$. However, it is sufficient to minimize the square of the distance: $F(x,y,z) = x^2 + y^2 + z^2$. It is clear that F has a minimum value, but no maximum value. Since P lies in the plane, $2x - y + 2z = 16$ which implies $y = 2x + 2z - 16 = 2(x + z - 8)$. Thus, we want to find the minimum value of

$$F(x, z) = x^2 + 4(x + z - 8)^2 + z^2$$

Now,

$$\nabla F = [2x + 8(x + z - 8)]\,\mathbf{i} + [8(x + z - 8) + 2z]\,\mathbf{k}$$

The gradient is $\mathbf{0}$ when

$$2x + 8(x + z - 8) = 0 \quad \text{and} \quad 8(x + z - 8) + 2z = 0$$

The only solution to this pair of equations is: $x = z = \dfrac{32}{9}$, from which it follows that $y = -\dfrac{16}{9}$.

The point in the plane that is closest to the origin is $P\left(\frac{32}{9}, -\frac{16}{9}, \frac{32}{9}\right)$.

The distance from the origin to the plane is: $F(P) = \frac{16}{3}$.

Check using (13.6.5): $d(P,0) = \dfrac{|2 \cdot 0 - 0 + 2 \cdot 0 - 16|}{\sqrt{2^2 + (-1)^2 + 2^2}} = \dfrac{16}{3}$.

29. $f(x, y) = (x - 1)^2 + (y - 2)^2 + z^2 = (x - 1)^2 + (y - 2)^2 + x^2 + 2y^2$ $\left[\text{since } z = \sqrt{x^2 + 2y^2}\right]$

$\nabla f = [2(x - 1) + 2x]\,\mathbf{i} + [2(y - 2) + 4y]\,\mathbf{j} = \mathbf{0} \implies x = \dfrac{1}{2}, \; y = \dfrac{2}{3}$.

$f_{xx} = 4 > 0, \quad f_{xy} = 0, \quad f_{yy} = 6, \quad D = 24 > 0$. Thus, f has a local minimum at $(1/2, 2/3)$.

The shortest distance from $(1, 2, 0)$ to the cone is $\sqrt{f\left(\frac{1}{2}, \frac{2}{3}\right)} = \frac{1}{6}\sqrt{114}$

31. (a)

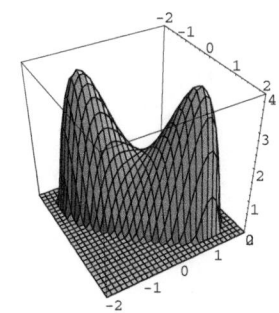

(b)

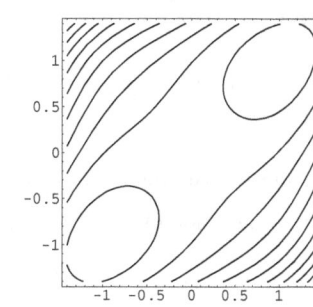

(c) $\nabla f = (4y - 4x^3)\,\mathbf{i} + (4x - 4y^3)\,\mathbf{j} = \mathbf{0}$ at $(0,0)$, $(1,1)$, $(-1,-1)$.

$f_{xx} = -12x^2$, $f_{xy} = 4$, $f_{yy} = -12y^2$, $D = 144x^2y^2 - 16$

point	A	B	C	D	result
$(0,0)$	0	4	0	-16	saddle
$(1,1)$	-12	4	-12	128	loc. max.
$(-1,-1)$	-12	4	-12	128	loc. max.

$f(1,1) = f(-1,-1) = 3$

33. (a)

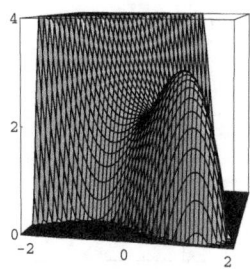

(b)

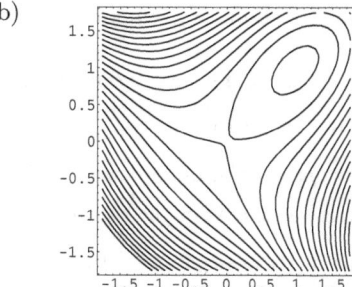

$f(1,1) = 3$ is a local max.; f has a saddle at $(0,0)$.

35. (a)

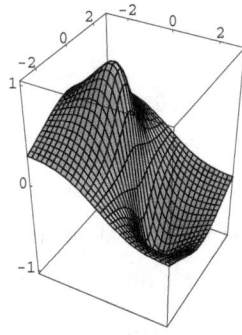

(b)

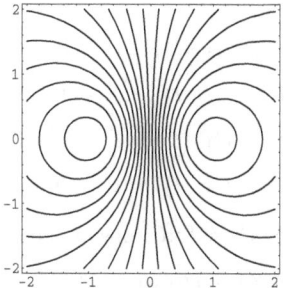

$f(1,0) = -1$ is a local min.; $f(-1,0) = 1$ is a loc. max.

SECTION 16.6

1. $\nabla f = (4x - 4)\,\mathbf{i} + (2y - 2)\,\mathbf{j} = \mathbf{0}$ at $(1,1)$ in D;

    $f(1,1) = -1$

    Next we consider the boundary of D. We
    parametrize each side of the triangle:

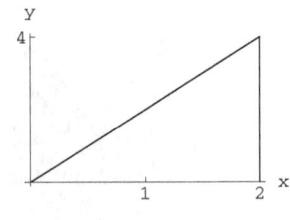

$$C_1 : \ \mathbf{r}_1(t) = t\,\mathbf{i}, \quad t \in [0,2],$$
$$C_2 : \ \mathbf{r}_2(t) = 2\,\mathbf{i} + t\,\mathbf{j}, \quad t \in [0,4],$$
$$C_3 : \ \mathbf{r}_3(t) = t\,\mathbf{i} + 2t\,\mathbf{j}, \quad t \in [0,2],$$

Now,

$$f_1(t) = f(\mathbf{r}_1(t)) = 2(t-1)^2, \quad t \in [0,2]; \quad \text{critical number: } t = 1,$$
$$f_2(t) = f(\mathbf{r}_2(t)) = (t-1)^2 + 1, \quad t \in [0,4]; \quad \text{critical number: } t = 1,$$
$$f_3(t) = f(\mathbf{r}_3(t)) = 6t^2 - 8t + 2, \quad t \in [0,2]; \quad \text{critical number: } t = \tfrac{2}{3}.$$

Evaluating these functions at the endpoints of their domains and at the critical numbers, we find that:

$$f_1(0) = f_3(0) = f(0,0) = 2; \qquad f_1(1) = f(1,0) = 0; \qquad f_1(2) = f_2(0) = f(2,0) = 2;$$
$$f_2(1) = f(2,1) = 1; \qquad f_2(4) = f_3(2) = f(2,4) = 10; \qquad f_3(2/3) = f(2/3, 4/3) = -\tfrac{2}{3}.$$

f takes on its absolute maximum of 10 at $(2,4)$ and its absolute minimum of -1 at $(1,1)$.

3. $\nabla f = (2x + y - 6)\,\mathbf{i} + (x + 2y)\,\mathbf{j} = \mathbf{0}$ at $(4,-2)$ in

    D; $f(4,-2) = -13$

    Next we consider the boundary of D. We
    parametrize each side of the rectangle:

$$C_1 : \ \mathbf{r}_1(t) = -t\,\mathbf{j}, \quad t \in [0,3]$$
$$C_2 : \ \mathbf{r}_2(t) = t\,\mathbf{i} - 3\,\mathbf{j}, \quad t \in [0,5]$$
$$C_3 : \ \mathbf{r}_3(t) = 5\,\mathbf{i} - t\,\mathbf{j}, \quad t \in [0,3]$$
$$C_4 : \ \mathbf{r}_4(t) = t\,\mathbf{i}, \quad t \in [0,5]$$

Now,

$$f_1(t) = f(\mathbf{r}_1(t)) = t^2 - 1, \quad t \in [0,3]; \quad \text{no critical numbers}$$
$$f_2(t) = f(\mathbf{r}_2(t)) = t^2 - 9t + 8, \quad t \in [0,5]; \quad \text{critical number: } t = \tfrac{9}{2}$$
$$f_3(t) = f(\mathbf{r}_3(t)) = t^2 - 5t - 6, \quad t \in [0,3]; \quad \text{critical number: } t = \tfrac{5}{2}$$
$$f_4(t) = f(\mathbf{r}_4(t)) = t^2 - 6t - 1, \quad t \in [0,5]; \quad \text{critical number: } t = 3$$

Evaluating these functions at the endpoints of their domains and at the critical numbers, we find that:

$$f_1(0) = f_4(0) = f(0,0) = -1; \qquad f_1(-3) = f_2(0) = f(0,-3) = 8; \qquad f_2(9/2) = f(9/2, -3) = -\tfrac{49}{4};$$
$$f_2(5) = f_3(3) = f(5,-3) = -12; \qquad f_3(5/2) = f(5,-5/2) = -\tfrac{49}{4}; \qquad f_3(0) = f_4(5) = f(5,0) = -6.$$
$$f_4(3) = f(3,0) = -10$$

f takes on its absolute maximum of 8 at $(0,-3)$ and its absolute minimum of -13 at $(4,-2)$.

5. $\nabla f = (2x + 3y)\,\mathbf{i} + (2y + 3x)\,\mathbf{j} = \mathbf{0}$ at $(0,0)$ in $D;$ $f(0,0) = 2$

Next we consider the boundary of $D.$ We parametrize the circle by:

$$C:\ \mathbf{r}(t) = 2\cos t\,\mathbf{i} + 2\sin t\,\mathbf{j}, \quad t \in [\,0, 2\pi\,]$$

The values of f on the boundary are given by the function

$$F(t) = f(\mathbf{r}(t)) = 6 + 12\sin t\,\cos t, \quad t \in [\,0, 2\pi\,]$$

$$F'(t) = 12\cos^2 t - 12\sin^2 t: \quad F'(t) = 0 \implies \cos t = \pm\sin t \implies t = \tfrac{1}{4}\pi,\ \tfrac{3}{4}\pi,\ \tfrac{5}{4}\pi,\ \tfrac{7}{4}\pi$$

Evaluating F at the endpoints and critical numbers, we have:

$$F(0) = F(2\pi) = f(2,0) = 6; \quad F\left(\tfrac{1}{4}\pi\right) = F\left(\tfrac{5}{4}\pi\right) = f\left(\sqrt{2}, \sqrt{2}\right) = f\left(\left(-\sqrt{2}, -\sqrt{2}\right)\right) = 12;$$

$$F\left(\tfrac{3}{4}\pi\right) = f\left(-\sqrt{2}, \sqrt{2}\right) = F\left(\tfrac{7}{4}\pi\right) = f\left(\sqrt{2}, -\sqrt{2}\right) = 0$$

f takes on its absolute maximum of 12 at $\left(\sqrt{2}, \sqrt{2}\right)$ and at $\left(-\sqrt{2}, -\sqrt{2}\right);$ f takes on its absolute minimum of 0 at $\left(-\sqrt{2}, \sqrt{2}\right)$ and at $\left(\sqrt{2}, -\sqrt{2}\right).$

7. $\nabla f = 2(x - 1)\mathbf{i} + 2(y - 1)\,\mathbf{j} = \mathbf{0}$ only at $(1,1)$ in $D.$ As the sum of two squares, $f(x,y) \geq 0.$ Thus, $f(1,1) = 0$ is a minimum. To examine the behavior of f on the boundary of D, we note that f represents the square of the distance between (x,y) and $(1,1)$. Thus, f is maximal at the point of the boundary furthest from $(1,1)$. This is the point $\left(-\sqrt{2}, -\sqrt{2}\right);$ the maximum value of f is $f\left(-\sqrt{2}, -\sqrt{2}\right) = 6 + 4\sqrt{2}.$

9. $\nabla f = \dfrac{2x^2 - 2y^2 - 2}{(x^2 + y^2 + 1)^2}\,\mathbf{i} + \dfrac{4xy}{(x^2 + y^2 + 1)^2}\,\mathbf{j} = \mathbf{0}$ at $(1,0)$ and $(-1,0)$ in $D;$ $f(1,0) = -1,$ $f(-1,0) = 1.$

Next we consider the boundary of $D.$ We parametrize each side of the square:

$$C_1:\ \mathbf{r}_1(t) = -2\,\mathbf{i} + t\,\mathbf{j}, \quad t \in [-2,2]$$
$$C_2:\ \mathbf{r}_2(t) = t\,\mathbf{i} + 2\,\mathbf{j}, \quad t \in [-2,2]$$
$$C_3:\ \mathbf{r}_3(t) = 2\,\mathbf{i} + t\,\mathbf{j}, \quad t \in [-2,2]$$
$$C_4:\ \mathbf{r}_4(t) = t\,\mathbf{i}, \quad t \in [-2,2]$$

Now,

$$f_1(t) = f(\mathbf{r}_1(t)) = \frac{4}{t^2 + 5}, \quad t \in [-2,2]; \quad \text{critical number: } t = 0$$

$$f_2(t) = f(\mathbf{r}_2(t)) = \frac{-2t}{t^2 + 5}, \quad t \in [-2,2]; \quad \text{no critical numbers}$$

$$f_3(t) = f(\mathbf{r}_3(t)) = \frac{-4}{t^2 + 5}, \quad t \in [-2,2]; \quad \text{critical number: } t = 0$$

$$f_4(t) = f(\mathbf{r}_4(t)) = \frac{-2t}{t^2 + 5}, \quad t \in [-2,2]; \quad \text{no critical numbers}$$

Evaluating these functions at the endpoints of their domains and at the critical numbers, we find that:

$$f_1(-2) = f_4(-2) = f(-2,-2) = \tfrac{4}{9}; \quad f_1(0) = f(-2,0) = \tfrac{4}{5}; \quad f_1(2) = f_2(-2) = f(-2,2) = \tfrac{4}{9};$$
$$f_4(2) = f_3(-2) = f(2,-2) = -\tfrac{4}{9}; \quad f_3(0) = f(2,0) = -\tfrac{4}{5}; \quad f_2(2) = f_3(2) = f(2,2) = -\tfrac{4}{9}.$$

f takes on its absolute maximum of 1 at $(-1,0)$ and its absolute minimum of -1 at $(1,0)$.

11. $\nabla f = (4 - 4x)\cos y\,\mathbf{i} - (4x - 2x^2)\sin y\,\mathbf{j} = \mathbf{0}$ at $(1,0)$ in D: $f(1,0) = 2$

Next we consider the boundary of D. We parametrize each side of the rectangle:

$$C_1 : \mathbf{r}_1(t) = t\,\mathbf{j}, \quad t \in \left[-\tfrac{1}{4}\pi, \tfrac{1}{4}\pi\right]$$
$$C_2 : \mathbf{r}_2(t) = t\,\mathbf{i} - \tfrac{1}{4}\pi\,\mathbf{j}, \quad t \in [0, 2]$$
$$C_3 : \mathbf{r}_3(t) = 2\,\mathbf{i} + t\,\mathbf{j}, \quad t \in \left[-\tfrac{1}{4}\pi, \tfrac{1}{4}\pi\right]$$
$$C_4 : \mathbf{r}_4(t) = t\,\mathbf{i} + \tfrac{1}{4}\pi\,\mathbf{j}, \quad t \in [0, 2]$$

Now,

$$f_1(t) = f(\mathbf{r}_1(t)) = 0;$$
$$f_2(t) = f(\mathbf{r}_2(t)) = \frac{\sqrt{2}}{2}(4t - 2t^2), \quad t \in [0, 2]; \quad \text{critical number: } t = 1;$$
$$f_3(t) = f(\mathbf{r}_3(t)) = 0;$$
$$f_4(t) = f(\mathbf{r}_4(t)) = \frac{\sqrt{2}}{2}(4t - 2t^2), \quad t \in [0, 2]; \quad \text{critical number: } t = 1;$$

f at the vertices of the rectangle has the value 0; $\quad f_2(1) = f_4(1) = f\left(1, -\tfrac{1}{4}\pi\right) = f\left(1, \tfrac{1}{4}\pi\right) = \sqrt{2}.$

f takes on its absolute maximum of 2 at $(1,0)$ and its absolute minimum of 0 along the lines $x = 0$ and $x = 2$.

13. $\nabla f = (3x^2 - 3y)\,\mathbf{i} + (-3x - 3y^2)\,\mathbf{j} = \mathbf{0}$ at $(-1, 1)$ in D;

$f(-1, 1) = 1$

Next we consider the boundary of D. We parametrize each side of the triangle:

$$C_1 : \mathbf{r}_1(t) = -2\,\mathbf{i} + t\,\mathbf{j}, \quad t \in [-2, 2],$$
$$C_2 : \mathbf{r}_2(t) = t\,\mathbf{i} + t\,\mathbf{j}, \quad t \in [-2, 2],$$
$$C_3 : \mathbf{r}_3(t) = t\,\mathbf{i} + 2\,\mathbf{j}, \quad t \in [-2, 2],$$

and evaluate f:

$$f_1(t) = f(\mathbf{r}_1(t)) = -8 + 6t - t^3, \quad t \in [-2, 2]; \quad \text{critical numbers: } t = \pm\sqrt{2},$$
$$f_2(t) = f(\mathbf{r}_2(t)) = -3t^2, \quad t \in [-2, 2]; \quad \text{critical number: } t = 0,$$
$$f_3(t) = f(\mathbf{r}_3(t)) = t^3 - 6t - 8, \quad t \in [-2, 2]; \quad \text{critical numbers: } t = \pm\sqrt{2}.$$

Evaluating these functions at the endpoints of their domains and at the critical numbers, we find that:

$$f_1(-2) = f_2(-2) = f(-2, -2) = -12; \qquad f_1(-\sqrt{2}) = f(-2, -\sqrt{2}) = -8 - 4\sqrt{2} \cong -13.66;$$
$$f_1(\sqrt{2}) = f(-2, \sqrt{2}) = -8 + 4\sqrt{2} \cong -2.34; \qquad f_1(2) = f_3(-2) = f(-2, 2) = -4;$$
$$f_2(0) = f(0, 0) = 0; \qquad f_2(2) = f_3(2) = f(2, 2) = -12;$$
$$f_3(-\sqrt{2}) = f(-\sqrt{2}, 2) = -8 + 4\sqrt{2}; \qquad f_3(\sqrt{2}) = f(\sqrt{2}, 2) = -8 - 4\sqrt{2}$$

f takes on its absolute maximum of 1 at $(-1, 1)$ and its absolute minimum of $-8 - 4\sqrt{2}$ at $(\sqrt{2}, 2)$ and $(-2, -\sqrt{2})$.

15. $\nabla f = \dfrac{4xy}{(x^2 + y^2 + 1)^2}\,\mathbf{i} + \dfrac{2y^2 - 2x^2 - 2}{(x^2 + y^2 + 1)^2}\,\mathbf{j} = \mathbf{0}$ at $(0, 1)$ and $(0, -1)$ in D;

$$f(0, 1) = -1, \quad f(0, -1) = 1$$

Next we consider the boundary of D.   We parametrize the circle by:

$$C : \ \mathbf{r}(t) = 2\cos t\,\mathbf{i} + 2\sin t\,\mathbf{j}, \quad t \in [0, 2\pi]$$

The values of f on the boundary are given by the function

$$F(t) = f(\mathbf{r}(t)) = -\tfrac{4}{5}\sin t, \quad t \in [0, 2\pi]$$

$$F'(t) = -\tfrac{4}{5}\cos t : \quad F'(t) = 0 \implies \cos t = 0 \implies t = \tfrac{1}{2}\pi, \ \tfrac{3}{2}\pi.$$

Evaluating F at the endpoints and critical numbers, we have:

$$F(0) = F(2\pi) = f(2, 0) = 0; \quad F\left(\tfrac{1}{2}\pi\right) = f(0, 2) = -\tfrac{4}{5}; \quad F\left(\tfrac{3}{2}\pi\right) = f(0, -2) = \tfrac{4}{5}$$

f takes on its absolute maximum of 1 at $(0, -1)$ and its absolute minimum of -1 at $(0, 1)$.

17. $\nabla f = 2(x - y)\mathbf{i} - 2(x - y)\,\mathbf{j} = \mathbf{0}$ at each point of the line segment $y = x$ from $(0, 0)$ to $(4, 4)$. Since $f(x, x) = 0$ and $f(x, y) \geq 0$, f takes on its minimum of 0 at each of these points.

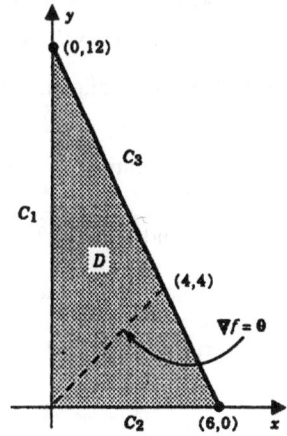

Next we consider the boundary of D. We parametrize each side of the triangle:

$$C_1 : \ \mathbf{r}_1(t) = t\mathbf{j}, \quad t \in [0, 12]$$
$$C_2 : \ \mathbf{r}_2(t) = t\mathbf{i}, \quad t \in [0, 6]$$
$$C_3 : \ \mathbf{r}_3(t) = t\mathbf{i} + (12 - 2t)\,\mathbf{j}, \quad t \in [0, 6]$$

and observe from

$$f(\mathbf{r}_1(t)) = t^2, \quad t \in [0, 12]$$
$$f(\mathbf{r}_2(t)) = t^2, \quad t \in [0, 6]$$
$$f(\mathbf{r}_3(t)) = (3t - 12)^2, \quad t \in [0, 6]$$

that f takes on its maximum of 144 at the point $(0, 12)$.

19. Using the hint, we want to find the maximum value of $f(x, y) = 18xy - x^2 y - xy^2$ in the triangular region. The gradient of f is:

$$\nabla D = \left(18y - 2xy - y^2\right)\mathbf{i} + \left(18x - x^2 - 2xy\right)\mathbf{j}$$

The gradient is $\mathbf{0}$ when

$$18y - 2xy - y^2 = 0 \quad \text{and} \quad 18x - x^2 - 2xy = 0$$

The solution set of this pair of equations is: $(0, 0)$, $(18, 0)$, $(0, 18)$, $(6, 6)$.

It is easy to verify that f is a maximum when $x = y = 6$. The three numbers that satisfy $x + y + z = 18$ and maximize the product xyz are: $x = 6$, $y = 6$, $z = 6$.

21. $f(x, y) = xy(1 - x - y)$, $0 \le x \le 1$, $0 \le y \le 1 - x$.

[dom (f) is the triangle with vertices $(0,0)$, $(1,0)$, $(0,1)$.]

$\nabla f = (y - 2xy - y^2)\mathbf{i} + (x - 2xy - x^2)\mathbf{j} = \mathbf{0} \implies x = y = 0$, $x = 1$, $y = 0$, $x = 0$, $y = 1$, $x = y = \frac{1}{3}$.

(Note that $[0,0]$ is not an interior point of the domain of f.)

$f_{xx} = -2y$, $f_{xy} = 1 - 2x - 2y$, $f_{yy} = -2x$.

At $\left(\frac{1}{3}, \frac{1}{3}\right)$, $D = \frac{1}{3} > 0$ and $A < 0$ so we have a local max; the value is $1/27$.

Since $f(x, y) = 0$ at each point on the boundary of the domain, the local max of $1/27$ is also the absolute max.

23. (a) $\nabla f = \frac{1}{2}x\,\mathbf{i} - \frac{2}{9}y\,\mathbf{j} = \mathbf{0}$ only at $(0,0)$.

(b) The difference

$$f(h, k) - f(0,0) = \frac{1}{4}h^2 - \frac{1}{9}k^2$$

does not keep a constant sign for all small h and k; $(0,0)$ is a saddle point. The function has no local extreme values.

(c) Being the difference of two squares, f can be maximized by maximizing $\frac{1}{4}x^2$ and minimizing $\frac{1}{9}y^2$; $(1,0)$ and $(-1,0)$ give absolute maximum value $\frac{1}{4}$.   Similarly, $(0,1)$ and $(0,-1)$ give absolute minimum value $-\frac{1}{9}$.

25. Let x, y and z be the length, width and height of the box. The surface area is given by

$$S = 2xy + 2xz + 2yz, \quad \text{so} \quad z = \frac{S - 2xy}{2(x + y)}, \quad \text{where S is a constant, and } x, y, z > 0.$$

Now, the volume $V = xyz$ is given by:

$$V(x, y) = xy\left[\frac{S - 2xy}{2(x + y)}\right]$$

and

$$\nabla V = \left\{ y\left[\frac{S - 2xy}{2(x + y)}\right] + xy\,\frac{2(x + y)(-2y) - (S - 2xy)(2)}{4(x + y)^2} \right\} \mathbf{i}$$

$$+ \left\{ x\left[\frac{S - 2xy}{2(x + y)}\right] + xy\,\frac{2(x + y)(-2x) - (S - 2xy)(2)}{4(x + y)^2} \right\} \mathbf{j}$$

Setting $\dfrac{\partial V}{\partial x} = \dfrac{\partial V}{\partial y} = 0$ and simplifying, we get the pair of equations

$$2S - 4x^2 - 8xy = 0$$

$$2S - 4y^2 - 8xy = 0$$

from which it follows that $x = y = \sqrt{S/6}$. From practical considerations, we conclude that V has a maximum value at $(\sqrt{S/6}, \sqrt{S/6})$. Substituting these values into the equation for z, we get $z = \sqrt{S/6}$ and so the box of maximum volume is a cube.

27.
$$f(x,y) = \sum_{i=1}^{3} \left[(x - x_i)^2 + (y - y_i)^2\right]$$

$$\nabla f(x,y) = 2\left[(3x - x_1 - x_2 - x_3)\,\mathbf{i} + (3y - y_1 - y_2 - y_3)\,\mathbf{j}\right]$$

$$\nabla f = \mathbf{0} \quad \text{only at} \quad \left(\frac{x_1 + x_2 + x_3}{3}, \frac{y_1 + y_2 + y_3}{3}\right) = (x_0, y_0).$$

The difference $\quad f(x_0 + h,\, y_0 + k) - f(x_0, y_0)$

$$= \sum_{i=1}^{3} \left[(x_0 + h - x_i)^2 + (y_0 + k - y_i)^2 - (x_0 - x_i)^2 - (y_0 - y_i)^2\right]$$

$$= \sum_{i=1}^{3} \left[2h\,(x_0 - x_i) + h^2 + 2k\,(y_0 - y_i) + k^2\right]$$

$$= 2h\,(3x_0 - x_1 - x_2 - x_3) + 2k\,(3y_0 - y_1 - y_2 - y_3) + 3h^2 + 3k^2$$

$$= 3h^2 + 3k^2$$

is nonnegative for all h and k. Thus, f has its absolute minimum at (x_0, y_0).

29.

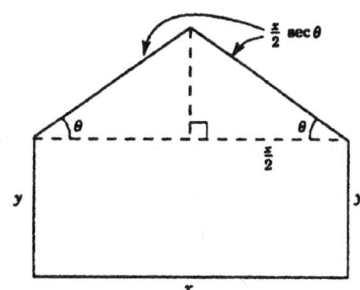

$$A = xy + \frac{1}{2}x\left(\frac{x}{2}\tan\theta\right),$$

$$P = x + 2y + 2\left(\frac{x}{2}\sec\theta\right), \quad y = \tfrac{1}{2}(P - x - x\sec\theta)$$

$$0 < \theta < \frac{1}{2}\pi, \quad 0 < x < \frac{P}{1 + \sec\theta}.$$

$$A(x,\theta) = \tfrac{1}{2}x(P - x - x\sec\theta) + \tfrac{1}{4}x^2\tan\theta,$$

$$\nabla A = \left(\frac{P}{2} - x - x\sec\theta + \frac{x}{2}\tan\theta\right)\mathbf{i} + \left(\frac{x^2}{4}\sec^2\theta - \frac{x^2}{2}\sec\theta\tan\theta\right)\mathbf{j},$$

(Here $\mathbf{j}$ is the unit vector in the direction of increasing θ.)

$$\nabla A = \frac{1}{2}[P + x(\tan\theta - 2\sec\theta - 2)]\,\mathbf{i} + \frac{x^2}{4}\sec\theta\,(\sec\theta - 2\tan\theta)\,\mathbf{j}.$$

From $\dfrac{\partial A}{\partial \theta} = 0$ we get $\theta = \tfrac{1}{6}\pi$ and then from $\dfrac{\partial A}{\partial x} = 0$ we get

$$P + x\left(\tfrac{1}{3}\sqrt{3} - \tfrac{4}{3}\sqrt{3} - 2\right) = 0 \quad \text{so that} \quad x = (2 - \sqrt{3})P.$$

Next,

$$A_{xx} = \tfrac{1}{2}(\tan\theta - 2\sec\theta - 2),$$

$$A_{x\theta} = \frac{x}{2}\sec\theta\,(\sec\theta - 2\tan\theta),$$

$$A_{\theta\theta} = \frac{x^2}{2}\sec\theta\,(\sec\theta\tan\theta - \sec^2\theta - \tan^2\theta).$$

Apply the second-partials test:

$$A = -\tfrac{1}{2}(2 + \sqrt{3}), \quad B = 0, \quad C = -\tfrac{1}{3}P^2\sqrt{3}\,(2 - \sqrt{3})^2, \quad D < 0.$$

Since, $D > 0$ and $A < 0$, the area is a maximum when $\theta = \tfrac{1}{6}\pi$, $x = (2 - \sqrt{3})\,P$ and $y = \tfrac{1}{6}(3 - \sqrt{3})P.$

31. From
$$x = \tfrac{1}{2}y = \tfrac{1}{3}z = t \quad \text{and} \quad x = y - 2 = z = s$$
we take
$$(t, 2t, 3t) \quad \text{and} \quad (s, 2 + s, s)$$
as arbitrary points on the lines. It suffices to minimize the square of the distance between these points:
$$f(t, s) = (t - s)^2 + (2t - 2 - s)^2 + (3t - s)^2$$
$$= 14t^2 - 12ts + 3s^2 - 8t + 4s + 4, \quad t, s \text{ real.}$$

Let $\mathbf{i}$ and $\mathbf{j}$ be the unit vectors in the direction of increasing t and s, respectively.
$$\nabla f = (28t - 12s - 8)\mathbf{i} + (-12t + 6s + 4)\mathbf{j}; \quad \nabla f = \mathbf{0} \implies t = 0, \ s = -2/3.$$
$$f_{tt} = 28, \quad f_{ts} = -12, \quad f_{ss} = 6, \quad D = 6(28) - (-12)^2 = -24 < 0.$$

By the second-partials test, the distance is a minimum when $t = 0$, $s = -2/3$; the nature of the problem tells us the minimum is absolute. The distance is $\sqrt{f(0, -2/3)} = \tfrac{2}{3}\sqrt{6}$.

33. (a) Let x and y be the cross-sectional measurements of the box, and let l be its length. Then
$$V = xyl, \quad \text{where} \quad 2x + 2y + l \le 108, \quad x, y > 0$$
To maximize V we will obviously take $2x + 2y + l = 108$. Therefore, $V(x, y) = xy(108 - 2x - 2y)$ and
$$\nabla V = [y(108 - 2x - 2y) - 2xy]\mathbf{i} + [x(108 - 2x - 2y) - 2xy]\mathbf{j}$$
Setting $\dfrac{\partial V}{\partial x} = \dfrac{\partial V}{\partial y} = 0$, we get the pair of equations
$$\frac{\partial V}{\partial x} = 108y - 4xy - 2y^2 = 0$$
$$\frac{\partial V}{\partial y} = 108x - 4xy - 2x^2 = 0$$
from which it follows that $x = y = 18 \implies l = 36$.
Now, at $(18, 18)$, we have
$$A = V_{xx} = -4y = -72 < 0, \quad B = V_{xy} = 108 - 4x - 4y = -36,$$
$$C = V_{yy} = -4x = -72, \quad \text{and} \quad D = (36)^2 - (72)^2 < 0.$$

Thus, V is a maximum when $x = y = 18$ inches and $l = 36$ inches.

(b) Let r be the radius of the tube and let l be its length. Then
$$V = \pi r^2 l, \quad \text{where} \quad 2\pi r + l \le 108, \quad r > 0$$
To maximize V we take $2\pi r + l = 108$. Then $V(r) = \pi r^2(108 - 2\pi r) = 108\pi r^2 - 2\pi^2 r^3$. Now
$$\frac{dV}{dr} = 216\pi r - 6\pi^2 r^2$$

Setting $\dfrac{dV}{dr} = 0$, we get

$$216\pi\, r - 6\pi^2 r^2 = 0 \quad \Longrightarrow \quad r = \frac{36}{\pi} \quad \Longrightarrow \quad l = 36$$

Now, at $r = 36/\pi$, we have

$$\frac{d^2 V}{dr^2} = 216\pi - 12\pi^2\,\frac{36}{\pi} = -\,216\pi < 0$$

Thus, V is a maximum when $r = 36/\pi$ inches and $l = 36$ inches.

35. Let S denote the cross-sectional area. Then

$$S = \frac{1}{2}\,(12 - 2x + 12 - 2x + 2x\,\cos\theta)\,x\,\sin\theta = 12x\,\sin\theta - 2x^2\,\sin\theta + \frac{1}{2}\,x^2\sin 2\theta,$$

where $\quad 0 < x < 6, \;\; 0 < \theta < \pi/2$

Now, with $\mathbf{j}$ in the direction of increasing θ,

$$\nabla S = (12\,\sin\theta - 4x\,\sin\theta + x\,\sin 2\theta)\,\mathbf{i} + (12x\,\cos\theta - 2x^2\,\cos\theta + x^2\,\cos 2\theta)\,\mathbf{j}$$

Setting $\dfrac{\partial S}{\partial x} = \dfrac{\partial S}{\partial \theta} = 0$, we get the pair of equations

$$12\,\sin\theta - 4x\,\sin\theta + x\,\sin 2\theta = 0$$

$$12x\,\cos\theta - 2x^2\,\cos\theta + x^2\,\cos 2\theta = 0$$

from which it follows that $x = 4, \theta = \pi/3$.

Now, at $(4, \pi/3)$, we have

$$A = S_{xx} = -4\,\sin\theta + \sin 2\theta = -\frac{3}{2}\,\sqrt{3}, \quad B = S_{x\theta} = 12\,\cos\theta - 4x\,\cos\theta + 2x\,\cos 2\theta = -6,$$

$$C = S_{\theta\theta} = -12x\,\sin\theta + 2x^2\,\sin\theta - 2x^2\,\sin 2\theta = -24\sqrt{3} \quad \text{and} \quad D = 108 - 36 > 0.$$

Thus, S is a maximum when $x = 4$ inches and $\theta = \pi/3$.

37. (a) $f(m, b) = [2 - b]^2 + [-5 - (m + b)]^2 + [4 - (2m + b)]^2$.

$\quad f_m = 10m + 6b - 6, \quad f_b = 6m + 6b - 2; \qquad f_m = f_b = 0 \quad \Longrightarrow \quad m = 1, \;\; b = -\frac{2}{3}.$

$\quad f_{mm} = 10, \quad f_{mb} = 6, \quad f_{bb} = 6, \quad D = 24 > 0 \quad \Longrightarrow \quad$ a minimum.

Answer: the line $y = x - \frac{2}{3}$.

(b) $f(\alpha, \beta) = [2 - \beta]^2 + [-5 - (\alpha + \beta)]^2 + [4 - (4\alpha + \beta)]^2$.

$\quad f_\alpha = 34\alpha + 10\beta - 22, \quad f_\beta = 10\alpha + 6\beta - 2; \qquad f_\alpha = f_\beta = 0 \quad \Longrightarrow \quad \left[\begin{array}{l} \alpha = \frac{14}{13} \\[4pt] \beta = -\frac{19}{13} \end{array}\right].$

$\quad f_{\alpha\alpha} = 34, \quad f_{\alpha\beta} = 10, \quad f_{\beta\beta} = 6, \quad D = 104 > 0 \quad \Longrightarrow \quad$ a minimun.

Answer: the parabola $y = \frac{1}{13}\left(14x^2 - 19\right).$

SECTION 16.7

1.
$$f(x,y) = x^2 + y^2, \qquad g(x,y) = xy - 1$$
$$\nabla f = 2x\mathbf{i} + 2y\mathbf{j}, \qquad \nabla g = y\mathbf{i} + x\mathbf{j}.$$
$$\nabla f = \lambda \nabla g \implies 2x = \lambda y \text{ and } 2y = \lambda x.$$

Multiplying the first equation by x and the second equation by y, we get

$$2x^2 = \lambda xy = 2y^2.$$

Thus, $x = \pm y$. From $g(x,y) = 0$ we conclude that $x = y = \pm 1$. The points $(1,1)$ and $(-1,-1)$ clearly give a minimum, since f represents the square of the distance of a point on the hyperbola from the origin. The minimum is 2.

3.
$$f(x,y) = xy, \qquad g(x,y) = b^2x^2 + a^2y^2 - a^2b^2$$
$$\nabla f = y\mathbf{i} + x\mathbf{j}, \qquad \nabla g = 2b^2x\mathbf{i} + 2a^2y\mathbf{j}.$$
$$\nabla f = \lambda \nabla g \implies y = 2\lambda b^2 x \text{ and } x = 2\lambda a^2 y.$$

Multiplying the first equation by a^2y and the second equation by b^2x, we get

$$a^2y^2 = 2\lambda a^2 b^2 xy = b^2 x^2.$$

Thus, $ay = \pm bx$. From $g(x,y) = 0$ we conclude that $x = \pm\frac{1}{2}a\sqrt{2}$ and $y = \pm\frac{1}{2}b\sqrt{2}$.

Since f is continuous and the ellipse is closed and bounded, the minimum exists. It occurs at $\left(\frac{1}{2}a\sqrt{2}, -\frac{1}{2}b\sqrt{2}\right)$ and $\left(-\frac{1}{2}a\sqrt{2}, \frac{1}{2}b\sqrt{2}\right)$;   the minimum is $-\frac{1}{2}ab$.

5. Since f is continuous and the ellipse is closed and bounded, the maximum exists.

$$f(x,y) = xy^2, \qquad g(x,y) = b^2x^2 + a^2y^2 - a^2b^2$$
$$\nabla f = y^2\mathbf{i} + 2xy\mathbf{j}, \qquad \nabla g = 2b^2x\mathbf{i} + 2a^2y\mathbf{j}.$$
$$\nabla f = \lambda \nabla g \implies y^2 = 2\lambda b^2 x \text{ and } 2xy = 2\lambda a^2 y.$$

Multiplying the first equation by a^2y and the second equation by b^2x, we get

$$a^2y^3 = 2\lambda a^2 b^2 xy = 2b^2 x^2 y.$$

We can exclude $y = 0$;   it clearly cannot produce the maximum. Thus,

$$a^2y^2 = 2b^2x^2 \text{ and, from } g(x,y) = 0, \ 3b^2x^2 = a^2b^2.$$

This gives us $x = \pm\frac{1}{3}\sqrt{3}\,a$ and $y = \pm\frac{1}{3}\sqrt{6}\,b$. The maximum occurs at $x = \frac{1}{3}\sqrt{3}\,a$, $y = \pm\frac{1}{3}\sqrt{6}\,b$;   the value there is $\frac{2}{9}\sqrt{3}\,ab^2$.

7. The given curve is closed and bounded. Since $x^2 + y^2$ represents the square of the distance from points on this curve to the origin, the maximum exists.

$$f(x, y) = x^2 + y^2, \qquad g(x, y) = x^4 + 7x^2y^2 + y^4 - 1$$

$$\nabla f = 2x\mathbf{i} + 2y\mathbf{j}, \qquad \nabla g = \left(4x^3 + 14xy^2\right)\mathbf{i} + \left(4y^3 + 14x^2y\right)\mathbf{j}.$$

We use the cross-product equation (16.7.4):

$$2x(4y^3 + 14x^2y) - 2y(4x^3 + 14xy^2) = 0,$$

$$20x^3y - 20xy^3 = 0,$$

$$xy(x^2 - y^2) = 0.$$

Thus, $x = 0$, $y = 0$, or $x = \pm y$. From $g(x, y) = 0$ we conclude that the points to examine are

$$(0, \pm 1), \quad (\pm 1, 0), \quad \left(\pm \tfrac{1}{3}\sqrt{3}, \pm \tfrac{1}{3}\sqrt{3}\right).$$

The value of f at each of the first four points is 1; the value at the last four points is $2/3$. The maximum is 1.

9. The maximum exists since xyz is continuous and the ellipsoid is closed and bounded.

$$f(x, y, z) = xyz, \qquad g(x, y, z) = \frac{x^2}{a^2} + \frac{y^2}{b^2} + \frac{z^2}{c^2} - 1$$

$$\nabla f = yz\mathbf{i} + xz\mathbf{j} + xy\mathbf{k}, \qquad \nabla g = \frac{2x}{a^2}\mathbf{i} + \frac{2y}{b^2}\mathbf{j} + \frac{2z}{c^2}\mathbf{k}.$$

$$\nabla f = \lambda \nabla g \implies yz = \frac{2x}{a^2}\lambda, \quad xz = \frac{2y}{b^2}\lambda, \quad xy = \frac{2z}{c^2}\lambda.$$

We can assume x, y, z are non-zero, for otherwise $f(x, y, z) = 0$, which is clearly not a maximum. Then from the first two equations

$$\frac{yza^2}{x} = 2\lambda = \frac{xzb^2}{y} \quad \text{so that} \quad a^2y^2 = b^2x^2 \quad \text{or} \quad \frac{x^2}{a^2} = \frac{y^2}{b^2}.$$

Similarly from the second and third equations we get

$$b^2z^2 = c^2y^2 \quad \text{or} \quad \frac{y^2}{b^2} = \frac{z^2}{c^2}.$$

From $g(x, y, z) = 0$, we get $\dfrac{3x^2}{a^2} = 1 \implies x \pm \dfrac{a}{\sqrt{3}}$, from which it follows that $y = \pm \dfrac{b}{\sqrt{3}}$, $z = \pm \dfrac{c}{\sqrt{3}}$. The maximum value is $\tfrac{1}{9}\sqrt{3}\,abc$.

11. Since the sphere is closed and bounded and $2x + 3y + 5z$ is continuous, the maximum exists.

$$f(x, y, z) = 2x + 3y + 5z, \qquad g(x, y, z) = x^2 + y^2 + z^2 - 19$$

$$\nabla f = 2\mathbf{i} + 3\mathbf{j} + 5\mathbf{k}, \qquad \nabla g = 2x\mathbf{i} + 2y\mathbf{j} + 2z\mathbf{k}.$$

$$\nabla f = \lambda \nabla g \implies 2 = 2\lambda x, \quad 3 = 2\lambda y, \quad 5 = 2\lambda z.$$

Since $\lambda \neq 0$ here, we solve the equations for x, y and z:

$$x = \frac{1}{\lambda}, \quad y = \frac{3}{2\lambda}, \quad z = \frac{5}{2\lambda},$$

and substitute these results in $g(x, y, z) = 0$ to obtain

$$\frac{1}{\lambda^2} + \frac{9}{4\lambda^2} + \frac{25}{4\lambda^2} - 19 = 0, \quad \frac{38}{4\lambda^2} - 19 = 0, \quad \lambda = \pm\frac{1}{2}\sqrt{2}.$$

The positive value of λ will produce positive values for x, y, z and thus the maximum for f. We get $x = \sqrt{2}, y = \frac{3}{2}\sqrt{2}, z = \frac{5}{2}\sqrt{2}$, and $2x + 3y + 5z = 19\sqrt{2}$.

13.

$$f(x, y, z) = xyz, \qquad g(x, y, z) = \frac{x}{a} + \frac{y}{b} + \frac{z}{c} - 1$$

$$\nabla f = yz\mathbf{i} + xz\mathbf{j} + xy\mathbf{k}, \qquad \nabla g = \frac{1}{a}\mathbf{i} + \frac{1}{b}\mathbf{j} + \frac{1}{c}\mathbf{k}.$$

$$\nabla f = \lambda \nabla g \implies yz = \frac{\lambda}{a}, \quad xz = \frac{\lambda}{b}, \quad xy = \frac{\lambda}{c}.$$

Multiplying these equations by x, y, z respectively, we obtain

$$xyz = \frac{\lambda x}{a}, \quad xyz = \frac{\lambda y}{b}, \quad xyz = \frac{\lambda z}{c}.$$

Adding these equations and using the fact that $g(x, y, z) = 0$, we have

$$3xyz = \lambda\left(\frac{x}{a} + \frac{y}{b} + \frac{z}{c}\right) = \lambda.$$

Since x, y, z are non-zero,

$$yz = \frac{\lambda}{a} = \frac{3xyz}{a}, \quad 1 = \frac{3x}{a}, \quad x = \frac{a}{3}.$$

Similarly, $y = \frac{b}{3}$ and $z = \frac{c}{3}$. The maximum is $\frac{1}{27}abc$.

15. It suffices to minimize the square of the distance from $(0, 1)$ to a point on the parabola. Clearly, the minimum exists.

$$f(x, y) = x^2 + (y - 1)^2, \qquad g(x, y) = x^2 - 4y$$

$$\nabla f = 2x\mathbf{i} + 2(y - 1)\mathbf{j}, \qquad \nabla g = 2x\mathbf{i} - 4\mathbf{j}.$$

We use the cross-product equation (16.7.4):

$$2x(-4) - 2x(2y - 2) = 0, \quad 4x + 4xy = 0, \quad x(y + 1) = 0.$$

Since $y \geq 0$, we have $x = 0$ and thus $y = 0$. The minimum is 1.

17. It suffices to maximize and minimize the square of the distance from $(2, 1, 2)$ to a point on the sphere. Clearly, these extreme values exist.

$$f(x, y, z) = (x - 2)^2 + (y - 1)^2 + (z - 2)^2, \qquad g(x, y, z) = x^2 + y^2 + z^2 - 1$$

$$\nabla f = 2(x - 2)\mathbf{i} + 2(y - 1)\mathbf{j} + 2(z - 2)\mathbf{k}, \qquad \nabla g = 2x\mathbf{i} + 2y\mathbf{j} + 2z\mathbf{k}.$$

$$\nabla f = \lambda \nabla g \implies 2(x - 2) = 2x\lambda, \quad 2(y - 1) = 2y\lambda, \quad 2(z - 2) = 2z\lambda$$

Thus,

$$x = \frac{2}{1-\lambda}, \quad y = \frac{1}{1-\lambda}, \quad z = \frac{2}{1-\lambda}.$$

Using the fact that $x^2 + y^2 + z^2 = 1$, we have

$$\left(\frac{2}{1-\lambda}\right)^2 + \left(\frac{1}{1-\lambda}\right)^2 + \left(\frac{2}{1-\lambda}\right)^2 = 1 \implies \lambda = -2, 4$$

At $\lambda = -2$, $(x, y, z) = (2/3, 1/3, 2/3)$ and $f(2/3, 1/3, 2/3) = 4$

At $\lambda = 4$, $(x, y, z) = (-2/3, -1/3, -2/3)$ and $f(-2/3, -1/3, -2/3) = 16$

Thus, $(2/3, 1/3, 2/3)$ is the closest point and $(-2/3, -1/3, -2/3)$ is the furthest point.

19.
$$f(x, y, z) = 3x - 2y + z, \qquad g(x, y, z) = x^2 + y^2 + z^2 - 14$$

$$\nabla f = 3\mathbf{i} - 2\mathbf{j} + \mathbf{k}, \qquad \nabla g = 2x\,\mathbf{i} + 2y\,\mathbf{j} + 2z\,\mathbf{k}.$$

$$\nabla f = \lambda \nabla g \implies 3 = 2x\lambda, \quad -2 = 2y\lambda, \quad 1 = 2z\lambda.$$

Thus,

$$x = \frac{3}{2\lambda}, \quad y = -\frac{1}{\lambda}, \quad z = \frac{1}{2\lambda}.$$

Using the fact that $x^2 + y^2 + z^2 = 14$, we have

$$\left(\frac{3}{2\lambda}\right)^2 + \left(-\frac{1}{\lambda}\right)^2 + \left(\frac{1}{2\lambda}\right)^2 = 14 \implies \lambda = \pm\frac{1}{2}.$$

At $\lambda = \frac{1}{2}$, $(x, y, z) = (3, -2, 1)$ and $f(3, -2, 1) = 14$

At $\lambda = -\frac{1}{2}$, $(x, y, z) = (-3, 2, -1)$ and $f(-3, 2, -1) = -14$

Thus, the maximum value of f on the sphere is 14.

21. It's easier to work with the square of the distance; the minimum certainly exists.

$$f(x, y, z) = x^2 + y^2 + z^2, \qquad g(x, y, z) = Ax + By + Cz + D$$

$$\nabla f = 2x\mathbf{i} + 2y\mathbf{j} + 2z\mathbf{k}, \qquad \nabla g = A\mathbf{i} + B\mathbf{j} + C\mathbf{k}.$$

$$\nabla f = \lambda \nabla g \implies 2x = A\lambda, \quad 2y = B\lambda, \quad 2z = C\lambda.$$

Substituting these equations in $g(x, y, z) = 0$, we have

$$\frac{1}{2}\lambda\left(A^2 + B^2 + C^2\right) + D = 0, \quad \lambda = \frac{-2D}{A^2 + B^2 + C^2}.$$

Thus, in turn,

$$x = \frac{-DA}{A^2 + B^2 + C^2}, \quad y = \frac{-DB}{A^2 + B^2 + C^2}, \quad z = \frac{-DC}{A^2 + B^2 + C^2}$$

so the minimum value of $\sqrt{x^2 + y^2 + z^2}$ is $|D|\left(A^2 + B^2 + C^2\right)^{-1/2}$.

23.

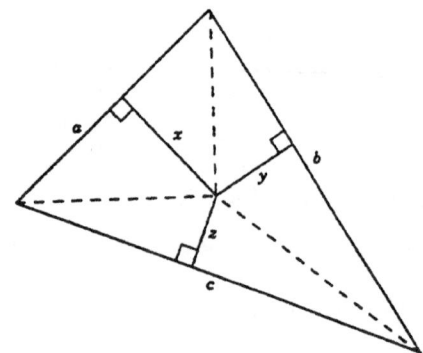

area $A = \frac{1}{2}ax + \frac{1}{2}by + \frac{1}{2}cz$.

The geometry suggests that
$$x^2 + y^2 + z^2$$
has a minimum.

$$f(x,y,z) = x^2 + y^2 + z^2, \qquad g(x,y,z) = ax + by + cz - 2A$$

$$\nabla f = 2x\mathbf{i} + 2y\mathbf{j} + 2z\mathbf{k}, \qquad \nabla g = a\mathbf{i} + b\mathbf{j} + c\mathbf{k}.$$

$$\nabla f = \lambda \nabla g \quad \Longrightarrow \quad 2x = a\lambda, \ \ 2y = b\lambda, \ \ 2z = c\lambda.$$

Solving these equations for x, y, z and substituting the results in $g(x,y,z) = 0$, we have

$$\frac{a^2\lambda}{2} + \frac{b^2\lambda}{2} + \frac{c^2\lambda}{2} - 2A = 0, \quad \lambda = \frac{4A}{a^2 + b^2 + c^2}$$

and thus

$$x = \frac{2aA}{a^2 + b^2 + c^2}, \quad y = \frac{2bA}{a^2 + b^2 + c^2}, \quad z = \frac{2cA}{a^2 + b^2 + c^2}.$$

The minimum is $4A^2(a^2 + b^2 + c^2)^{-1}$.

25. Since the curve is asymptotic to the line $y = x$ as $x \to -\infty$ and as $x \to \infty$, the maximum exists. The distance between the point (x, y) and the line $y - x = 0$ is given by

$$\frac{|y - x|}{\sqrt{1 + 1}} = \frac{1}{2}\sqrt{2}\,|y - x|. \qquad \text{(see Section 1.4)}$$

Since the points on the curve are below the line $y = x$, we can replace $|y - x|$ by $x - y$. To simplify the work we drop the constant factor $\frac{1}{2}\sqrt{2}$.

$$f(x, y) = x - y, \qquad g(x, y) = x^3 - y^3 - 1$$

$$\nabla f = \mathbf{i} - \mathbf{j}, \qquad \nabla g = 3x^2\mathbf{i} - 3y^2\mathbf{j}.$$

We use the cross-product equation (16.7.4):

$$1(-3y^2) - (3x^2)(-1) = 0, \quad 3x^2 - 3y^2 = 0, \quad x = -y \ \ (x \neq y).$$

Now $g(x, y) = 0$ gives us

$$x^3 - (-x)^3 - 1 = 0, \quad 2x^3 = 1, \quad x = 2^{-1/3}.$$

The point is $\left(2^{-1/3}, -2^{-1/3}\right)$.

27. It suffices to show that the square of the area is a maximum when $a = b = c$.

$$f(a, b, c) = s(s - a)(s - b)(s - c), \quad g(a, b, c) = a + b + c - 2s$$

$$\nabla f = -s(s - b)(s - c)\mathbf{i} - s(s - a)(s - c)\mathbf{j} - s(s - a)(s - b)\mathbf{k}, \quad \nabla g = \mathbf{i} + \mathbf{j} + \mathbf{k}.$$

(Here $\mathbf{i}$, $\mathbf{j}$, $\mathbf{k}$ are the unit vectors in the directions of increasing a, b, c.)

$$\nabla f = \lambda \nabla g \implies -s(s - b)(s - c) = -s(s - a)(s - c) = -s(s - a)(s - b) = \lambda.$$

Thus, $s - b = s - a = s - c$ so that $a = b = c$. This gives us the maximum, as no minimum exists. [The area can be made arbitrarily small by taking a close to s.]

29. (a) $f(x, y) = (xy)^{1/2}, \qquad g(x, y) = x + y - k, (x, y \geq 0, \ k$ a nonnegative constant)

$$\nabla f = \frac{y^{1/2}}{2x^{1/2}}\mathbf{i} + \frac{x^{1/2}}{2y^{1/2}}\mathbf{j}, \qquad \nabla g = \mathbf{i} + \mathbf{j}.$$

$$\nabla f = \lambda \nabla g \implies \frac{y^{1/2}}{2x^{1/2}} = \lambda = \frac{x^{1/2}}{2y^{1/2}} \implies x = y = \frac{k}{2}.$$

Thus, the maximum value of f is: $f(k/2, k/2) = \dfrac{k}{2}$.

(b) For all x, y $(x, y \geq 0)$ we have

$$(xy)^{1/2} = f(x, y) \leq f(k/2, k/2) = \frac{k}{2} = \frac{x + y}{2}.$$

31. Simply extend the arguments used in Exercises 29 and 30.

33.
$$S(r, h) = 2\pi r^2 + 2\pi rh, \qquad\qquad g(r, h) = \pi r^2 h - V, \quad (V \text{ constant})$$

$$\nabla S = (4\pi r + 2\pi h)\mathbf{i} + 2\pi r\,\mathbf{j}, \qquad\qquad \nabla g = 2\pi rh\,\mathbf{i} + \pi r^2\,\mathbf{j}.$$

$$\nabla S = \lambda \nabla g \implies 4\pi r + 2\pi h = 2\pi rh\lambda, \quad 2\pi r = \pi r^2 \lambda \implies r = \frac{2}{\lambda}, \quad h = \frac{4}{\lambda}.$$

Now $\pi r^2 h = V, \implies \lambda = \sqrt[3]{\dfrac{16\pi}{V}} \implies r = \sqrt[3]{\dfrac{V}{2\pi}}, \quad h = \sqrt[3]{\dfrac{4V}{\pi}}.$

To minimize the surface area, take $r = \sqrt[3]{\dfrac{V}{2\pi}},$ and $h = \sqrt[3]{\dfrac{4V}{\pi}}.$

35. Same as Exercise 13.

37. Let x, y, z denote the length, width and height of the box. We want to maximize the volume V of the box given that the surface area S is constant. That is:

maximize $V(x, y, z) = xyz$ subject to $S(x, y, z) = 2xy + 2xz + 2yz = S$ constant

Let $g(x, y, z) = 2xy + 2xz + 2yz - S$. Then

$$\nabla V = yz\,\mathbf{i} + xz\,\mathbf{j} + xy\,\mathbf{k}, \qquad \nabla g = (2y + 2z)\mathbf{i} + (2x + 2z)\mathbf{j} + (2x + 2y)\mathbf{k}$$

$\nabla V = \lambda \nabla g$ and the side condition yield the system of equations:

$$yz = \lambda(2y + 2z)$$

$$xz = \lambda(2x + 2z)$$

$$xy = \lambda(2x + 2y)$$

$$xy + 2xz + 2yz = S.$$

Multiply the first equation by x, the second by y and subtract. This gives

$$0 = 2\lambda z(x - y) \implies x = y \quad \text{since} \quad z = 0 \implies V = 0.$$

Multiply the second equation by y, the third by z and subtract. This gives

$$0 = 2\lambda x(y - z) \implies y = z \quad \text{since} \quad x = 0 \implies V = 0.$$

Thus the closed rectangular box of maximum volume is a cube. The cube has side length $x = \sqrt{S/6}$.

39. $S(r, h) = 4\pi r^2 + 2\pi rh, \quad g(r, h) = \dfrac{4}{3}\pi r^3 + \pi r^2 h - 10,000$

$\nabla S = (8\pi r + 2\pi h)\mathbf{i} + 2\pi r\mathbf{j}, \quad \nabla g = (4\pi r^2 + 2\pi rh)\mathbf{i} + \pi r^2\mathbf{j}$

(Here $\mathbf{i}$, $\mathbf{j}$ are the unit vectors in the directions of increasing r and h.)

$\nabla S = \lambda \nabla g \implies 2\pi(4r + h) = 2\pi r\lambda(2r + h), \quad 2\pi r = \lambda\pi r^2 \implies h = 0$

Maximum volume for sphere of radius $r = \sqrt[3]{7500/\pi}$ meters.

41. $f(x, y, z) = 8xyz, \qquad g(x, y, z) = 4x^2 + 9y^2 + 36z^2 - 36.$

$\nabla f(x, y, z) = 8yz\mathbf{i} + 8xz\mathbf{j} + 8xy\mathbf{k}, \qquad \nabla g(x, y, z) = 8x\mathbf{i} + 18y\mathbf{j} + 72z\mathbf{k}.$

$\nabla f = \lambda \nabla g \quad$ gives

$$yz = \lambda x, \quad 4xz = 9\lambda y, \quad xy = 9\lambda z.$$

$$4\frac{xyz}{\lambda} = 4x^2, \quad 4\frac{xyz}{\lambda} = 9y^2, \quad 4\frac{xyz}{\lambda} = 36z^2.$$

Also notice

$$4x^2 + 9y^2 + 36z^2 - 36 = 0$$

We have

$$12\frac{xyz}{\lambda} = 36 \implies x = \sqrt{3}, \quad y = \frac{2}{\sqrt{3}}, \quad z = \frac{1}{\sqrt{3}}.$$

Thus,

$$V = 8xyz = 8 \cdot \sqrt{3} \cdot \frac{2}{\sqrt{3}} \cdot \frac{1}{\sqrt{3}} = \frac{16}{\sqrt{3}}.$$

43. To simplify notation we set $x = Q_1$, $\quad y = Q_2$, $\quad z = Q_3$.

$$f(x,y,z) = 2x + 8y + 24z, \qquad g(x,y,z) = x^2 + 2y^2 + 4z^2 - 4,500,000,000$$

$$\nabla f = 2\mathbf{i} + 8\mathbf{j} + 24\mathbf{k}, \qquad \nabla g = 2x\mathbf{i} + 4y\mathbf{j} + 8z\mathbf{k}.$$

$$\nabla f = \lambda \nabla g \implies 2 = 2\lambda x, \quad 8 = 4\lambda y, \quad 24 = 8\lambda z.$$

Since $\lambda \neq 0$ here, we solve the equations for x, y, z:

$$x = \frac{1}{\lambda}, \quad y = \frac{2}{\lambda}, \quad z = \frac{3}{\lambda},$$

and substitute these results in $g(x,y,z) = 0$ to obtain

$$\frac{1}{\lambda^2} + 2\left(\frac{4}{\lambda^2}\right) + 4\left(\frac{9}{\lambda^2}\right) - 45 \times 10^8 = 0, \quad \frac{45}{\lambda^2} = 45 \times 10^8, \quad \lambda = \pm 10^{-4}.$$

Since x, y, z are non-negative, $\lambda = 10^{-4}$ and

$$x = 10^4 = Q_1, \quad y = 2 \times 10^4 = Q_2, \quad z = 3 \times 10^4 = Q_3.$$

PROJECT 16.7

1. $f(x,y,z) = xy + z^2$, $\quad g(x,y,z) = x^2 + y^2 + z^2 - 4$, $\quad h(x,y,z) = y - x$

$\nabla f = y\mathbf{i} + x\mathbf{j} + 2z\mathbf{k}$, $\quad \nabla g = 2x\mathbf{i} + 2y\mathbf{j} + 2z\mathbf{k}$, $\quad \nabla h = -\mathbf{i} + \mathbf{j}$.

$\nabla f = \lambda \nabla g + \mu \nabla h \implies y = 2\lambda x - \mu, \quad x = 2\lambda y - \mu, \quad 2z = 2\lambda z$

$2z = 2\lambda z \implies \lambda = 0 \quad \text{or} \quad z = 1.$

$\lambda = 0 \implies y = -x$ which contradicts $y = x$.

$z = 1 \implies x^2 + y^2 = 3$, which, with $y = x$ implies $x = \pm\sqrt{3/2}$; $\quad \left(\pm\sqrt{3/2}, \pm\sqrt{3/2}\right)$

Adding the first two equations gives

$$x + y = 2\lambda(x + y) \implies (x+y)[2\lambda - 1] = 0 \implies \lambda = \frac{1}{c} \quad \text{or} \quad x = y = 0.$$

$x = y = 0 \implies z = \pm 2; \quad (0, 0, \pm 2).$

$\lambda = \dfrac{1}{2} \implies z = 0$ and $y = x \implies 2x^4 = 4; \quad x = \pm\sqrt{2}; \quad (\pm\sqrt{2}, \pm\sqrt{2}, 0).$

$f\left(\pm\sqrt{3/2}, \pm\sqrt{3/2}, 1\right) = \dfrac{5}{2}; \quad f(0, 0, \pm 2) = 4; \quad f(\pm\sqrt{2}, \pm\sqrt{2}, 0) = 2.$

The maximum value of f is 4; the minimum value is 2.

3. $f(x,y,z) = x^2 + y^2 + z^2$, $\qquad g(x,y,z) = x + y - z + 1$, $\qquad h(x,y,z) = x^2 + y^2 - z^2$

$\nabla f = 2x\mathbf{i} + 2y\mathbf{j} + 2z\mathbf{k}$, $\qquad \nabla g = \mathbf{i} + \mathbf{j} - \mathbf{k}$, $\qquad \nabla h = 2x\mathbf{i} + 2y\mathbf{j} - 2z\mathbf{k}$.

$$\nabla f = \lambda \nabla g + \mu \nabla h \implies 2x = \lambda + 2x\mu, \quad 2y = \lambda + 2y\mu, \quad 2z = -\lambda - 2z\mu$$

Multiplying the first equation by y, the second equation by x and subtracting, yields

$$\lambda(y - x) = 0.$$

Now $\lambda = 0 \implies \mu = 1 \implies x = y = z = 0$. This is impossible since $x + y - z = -1$.

Therefore, we must have $y = x \implies z = \pm\sqrt{2}\, x$.

Substituting $y = x$, $z = \sqrt{2}\, x$ into the equation $x + y - z + 1 = 0$, we get

$$x = -1 - \frac{\sqrt{2}}{2} \implies y = -1 - \frac{\sqrt{2}}{2},\ z = -1 - \sqrt{2}$$

Substituting $y = x$, $z = -\sqrt{2}\, x$ into the equation $x + y - z + 1 = 0$, we get

$$x = -1 + \frac{\sqrt{2}}{2} \implies y = -1 + \frac{\sqrt{2}}{2},\ z = -1 + \sqrt{2}$$

Since

$$f\left(-1 - \frac{\sqrt{2}}{2}, -1 - \frac{\sqrt{2}}{2}, -1 - \sqrt{2}\right) = 6 + 4\sqrt{2}\ \text{ and}$$

$$f\left(-1 + \frac{\sqrt{2}}{2}, -1 + \frac{\sqrt{2}}{2}, -1 + \sqrt{2}\right) = 6 - 4\sqrt{2},$$

it follows that $\left(-1 + \dfrac{\sqrt{2}}{2}, -1 + \dfrac{\sqrt{2}}{2}, -1 + \sqrt{2}\right)$ is closest to the origin and

$\left(-1 - \dfrac{\sqrt{2}}{2}, -1 - \dfrac{\sqrt{2}}{2}, -1 - \sqrt{2}\right)$ is furthest from the origin.

SECTION 16.8

1. $df = \left(3x^2y - 2xy^2\right)\Delta x + \left(x^3 - 2x^2y\right)\Delta y$

3. $df = (\cos y + y\sin x)\,\Delta x - (x\sin y + \cos x)\,\Delta y$

5. $df = \Delta x - (\tan z)\,\Delta y - \left(y\sec^2 z\right)\Delta z$

7. $df = \dfrac{y(y^2 + z^2 - x^2)}{(x^2 + y^2 + z^2)^2}\,\Delta x + \dfrac{x(x^2 + z^2 - y^2)}{(x^2 + y^2 + z^2)^2}\,\Delta y - \dfrac{2xyz}{(x^2 + y^2 + z^2)^2}\,\Delta z$

9. $df = [\cos(x + y) + \cos(x - y)]\,\Delta x + [\cos(x + y) - \cos(x - y)]\,\Delta y$

11. $df = \left(y^2 z e^{xz} + \ln z\right)\Delta x + 2y e^{xz}\,\Delta y + \left(xy^2 e^{xz} + \dfrac{x}{z}\right)\Delta z$

13.
$$\Delta u = \left[(x + \Delta x)^2 - 3(x + \Delta x)(y + \Delta y) + 2(y + \Delta y)^2\right] - \left(x^2 - 3xy + 2y^2\right)$$
$$= \left[(1.7)^2 - 3(1.7)(-2.8) + 2(-2.8)^2\right] - \left(2^2 - 3(2)(-3) + 2(-3)^2\right)$$
$$= (2.89 + 14.28 + 15.68) - 40 = -7.15$$
$$du = (2x - 3y)\,\Delta x + (-3x + 4y)\,\Delta y$$
$$= (4 + 9)(-0.3) + (-6 - 12)(0.2) = -7.50$$

15.
$$\Delta u = \left[(x + \Delta x)^2(z + \Delta z) - 2(y + \Delta y)(z + \Delta z)^2 + 3(x + \Delta x)(y + \Delta y)(z + \Delta z)\right]$$
$$- \left(x^2 z - 2yz^2 + 3xyz\right)$$
$$= \left[(2.1)^2(2.8) - 2(1.3)(2.8)^2 + 3(2.1)(1.3)(2.8)\right] - \left[(2)^2 3 - 2(1)(3)^2 + 3(2)(1)(3)\right] = 2.896$$
$$du = (2xz + 3yz)\,\Delta x + \left(-2z^2 + 3xz\right)\,\Delta y + \left(x^2 - 4yz + 3xy\right)\,\Delta z$$
$$= [2(2)(3) + 3(1)(3)](0.1) + [-2(3)^2 + 3(2)(3)](0.3) + [2^2 - 4(1)(3) + 3(2)(1)](-0.2) = 2.5$$

17. $f(x, y) = x^{1/2} y^{1/4}$; $x = 121$, $y = 16$, $\Delta x = 4$, $\Delta y = 1$
$$f(x + \Delta x,\, y + \Delta y) \cong f(x, y) + df$$
$$= x^{1/2}\,y^{1/4} + \tfrac{1}{2}x^{-1/2}\,y^{1/4}\,\Delta x + \tfrac{1}{4}x^{1/2}\,y^{-3/4}\,\Delta y$$
$$\sqrt{125}\,\sqrt[4]{17} \cong \sqrt{121}\,\sqrt[4]{16} + \tfrac{1}{2}(121)^{-1/2}\,(16)^{1/4}\,(4) + \tfrac{1}{4}(121)^{1/2}\,(16)^{-3/4}\,(1)$$
$$= 11(2) + \tfrac{1}{2}\left(\tfrac{1}{11}\right)(2)(4) + \tfrac{1}{4}(11)\left(\tfrac{1}{8}\right)$$
$$= 22 + \tfrac{4}{11} + \tfrac{11}{32} = 22\tfrac{249}{352} \cong 22.71$$

19.
$$f(x, y) = \sin x \cos y; \quad x = \pi, \quad y = \frac{\pi}{4}, \quad \Delta x = -\frac{\pi}{7}, \quad \Delta y = -\frac{\pi}{20}$$
$$df = \cos x \cos y\,\Delta x - \sin x \sin y\,\Delta y$$
$$f(x + \Delta x,\, y + \Delta y) \cong f(x, y) + df$$
$$\sin \frac{6}{7}\pi \cos \frac{1}{5}\pi \cong \sin \pi \cos \frac{\pi}{4} + \left(\cos \pi \cos \frac{\pi}{4}\right)\left(-\frac{\pi}{7}\right) - \left(\sin \pi \sin \frac{\pi}{4}\right)\left(-\frac{\pi}{20}\right)$$
$$= 0 + \left(\frac{1}{2}\sqrt{2}\right)\left(\frac{\pi}{7}\right) + 0 = \frac{\pi\sqrt{2}}{14} \cong 0.32$$

21. $f(2.9, 0.01) \cong f(3, 0) + df$,   where df is to be evaluated at $x = 3$, $y = 0$, $\Delta x = -0.1$, $\Delta y = 0.01$.
$$df = \left(2xe^{xy} + x^2 y e^{xy}\right)\,\Delta x + x^3 e^{xy}\,\Delta y = \left[2(3)e^0 + (3)^2(0)e^0\right](-0.1) + 3^3 e^0(0.01) = -0.33$$
Thus, $f(2.9, .01) \cong 3^2 e^0 - 0.33 = 8.67$.

23. $f(2.94, 1.1, 0.92) \cong f(3, 1, 1) + df$,   where df is to be evaluated at $x = 3$, $y = 1$, $z = 1$,
$\Delta x = -0.06$, $\Delta y = 0.1$, $\Delta z = -0.08$
$$df = \tan^{-1} yz\,\Delta x + \frac{xz}{1 + y^2 z^2}\,\Delta y + \frac{xy}{1 + y^2 z^2}\,\Delta z = \frac{\pi}{4}(-0.06) + (1.5)(0.1) + (1.5)(-0.08) \cong -0.0171$$
Thus, $f(2.94, 1.1, 0.92) \cong \tfrac{3}{4}\pi - 0.0171 \cong 2.3391$

25. $df = \dfrac{\partial z}{\partial x}\,\Delta x + \dfrac{\partial z}{\partial y}\,\Delta y = \dfrac{2y}{(x+y)^2}\,\Delta x - \dfrac{2x}{(x+y)^2}\,\Delta y$

With $x = 4, \ y = 2, \ \Delta x = 0.1, \ \Delta y = 0.1,$ we get

$$df = \tfrac{4}{36}(0.1) - \tfrac{8}{36}(0.1) = -\tfrac{1}{90}.$$

The exact change is $\dfrac{4.1 - 2.1}{4.1 + 2.1} - \dfrac{4 - 2}{4 + 2} = \dfrac{2}{6.2} - \dfrac{1}{3} = -\dfrac{1}{93}.$

27. $S = 2\pi r^2 + 2\pi rh; \quad r = 8, \ h = 12, \ \Delta r = -0.3, \ \Delta h = 0.2$

$$dS = \frac{\partial S}{\partial r}\,\Delta r + \frac{\partial S}{\partial h}\,\Delta h = (4\pi r + 2\pi h)\,\Delta r + (2\pi r)\,\Delta h$$

$$= 56\pi(-0.3) + 16\pi(0.2) = -13.6\pi.$$

The area decreases about 13.6π in.^{2}.

29. $S(9.98, 5.88, 4.08) \cong S(10, 6, 4) + dS = 248 + dS,$ where

$dS = (2w + 2h)\,\Delta l + (2l + 2h)\,\Delta w + (2l + 2w)\,\Delta h = 20(-0.02) + 28(-0.12) + 32(0.08) = -1.20$

Thus, $S(9.98, 5.88, 4.08) \cong 248 - 1.20 = 246.80.$

31. (a) $dV = yz\,\Delta x + xz\,\Delta y + xy\,\Delta z = (8)(6)(0.02) + (12)(6)(-0.05) + (12)(8)(0.03) = 0.24$

(b) $\Delta V = (12.02)(7.95)(6.03) - (12)(8)(6) = 0.22077$

33. $T(P) - T(Q) \cong dT = (-2x + 2yz)\,\Delta x + (-2y + 2xz)\,\Delta y + (-2z + 2xy)\,\Delta z$

Letting $x = 1, \ y = 3, \ z = 4, \ \Delta x = 0.15, \ \Delta y = -0.10, \ \Delta z = 0.10,$ we have

$$dT = (22)(0.15) + (2)(-0.10) + (-2)(0.10) = 2.9$$

35. (a) $\pi r^2 h = \pi(r + \Delta r)^2(h + \Delta h) \quad \Longrightarrow \quad \Delta h = \dfrac{r^2 h}{(r + \Delta r)^2} - h = -\dfrac{(2r + \Delta r)h}{(r + \Delta r)^2}\,\Delta r.$

$$df = (2\pi rh)\,\Delta r + \pi r^2\,\Delta h, \qquad df = 0 \quad \Longrightarrow \quad \Delta h = \dfrac{-2h}{r}\,\Delta r.$$

(b) $2\pi r^2 + 2\pi rh = 2\pi(r + \Delta r)^2 + 2\pi(r + \Delta r)(h + \Delta h).$

Solving for $\Delta h,$

$$\Delta h = \dfrac{r^2 + rh - (r + \Delta r)^2}{r + \Delta r} - h = -\dfrac{2r + h + \Delta r}{r + \Delta r}\,\Delta r.$$

$$df = (4\pi r + 2\pi h)\,\Delta r + 2\pi r\,\Delta h, \qquad df = 0 \quad \Longrightarrow \quad \Delta h = -\left(\dfrac{2r + h}{r}\right)\Delta r.$$

37. (a) $A = \dfrac{1}{2}x^2 \sin\theta; \quad \Delta A \cong dA = x\sin\theta\,\Delta x + \dfrac{x^2}{2}\cos\theta\,\Delta\theta$

(b) The area is more sensitive to changes in θ if $x > 2\tan\theta$, otherwise it is more sensitive to changes in x.

39. $s = \dfrac{A}{A - W}$; $A = 9$, $W = 5$, $\Delta A = \pm 0.01$, $\Delta W = \pm 0.02$

$$ds = \frac{\partial s}{\partial A}\,\Delta A + \frac{\partial s}{\partial W}\,\Delta W = \frac{-W}{(A - W)^2}\,\Delta A + \frac{A}{(A - W)^2}\,\Delta W$$

$$= -\frac{5}{16}(\pm 0.01) + \frac{9}{16}(\pm 0.02) \cong \pm 0.014$$

The maximum possible error in the value of s is 0.014 lbs; $2.23 \le s + \Delta s \le 2.27$

SECTION 16.9

1. $\dfrac{\partial f}{\partial x} = xy^2$, $f(x, y) = \frac{1}{2}x^2 y^2 + \phi(y)$, $\dfrac{\partial f}{\partial y} = x^2 y + \phi'(y) = x^2 y$.

Thus, $\phi'(y) = 0$, $\phi(y) = C$,   and $f(x, y) = \frac{1}{2}x^2 y^2 + C$.

3. $\dfrac{\partial f}{\partial x} = y$, $f(x, y) = xy + \phi(y)$, $\dfrac{\partial f}{\partial y} = x + \phi'(y) = x$.

Thus, $\phi'(y) = 0$, $\phi(y) = C$,   and $f(x, y) = xy + C$.

5. No; $\dfrac{\partial}{\partial y}\left(y^3 + x\right) = 3y^2$ whereas $\dfrac{\partial}{\partial x}\left(x^2 + y\right) = 2x$.

7. $\dfrac{\partial f}{\partial x} = \cos x - y \sin x$, $f(x, y) = \sin x + y \cos x + \phi(y)$, $\dfrac{\partial f}{\partial y} = \cos x + \phi'(y) = \cos x$.

Thus, $\phi'(y) = 0$, $\phi(y) = C$,   and $f(x, y) = \sin x + y \cos x + C$.

9. $\dfrac{\partial f}{\partial x} = e^x \cos y^2$, $f(x, y) = e^x \cos y^2 + \phi(y)$, $\dfrac{\partial f}{\partial y} = -2ye^x \sin y^2 + \phi'(y) = -2ye^x \sin y^2$.

Thus, $\phi'(y) = 0$, $\phi(y) = C$,   and $f(x, y) = e^x \cos y^2 + C$.

11. $\dfrac{\partial f}{\partial y} = xe^x - e^{-y}$, $f(x, y) = xye^x + e^{-y} + \phi(x)$, $\dfrac{\partial f}{\partial x} = ye^x + xye^x + \phi'(x) = ye^x(1 + x)$.

Thus, $\phi'(x) = 0$, $\phi(x) = C$,   and $f(x, y) = xye^x + e^{-y} + C$.

13. No; $\dfrac{\partial}{\partial y}\left(xe^{xy} + x^2\right) = x^2 e^{xy}$ whereas $\dfrac{\partial}{\partial x}\left(ye^{xy} - 2y\right) = y^2 e^{xy}$

15. $\dfrac{\partial f}{\partial x} = 1 + y^2 + xy^2$, $f(x, y) = x + xy^2 + \frac{1}{2}x^2 y^2 + \phi(y)$, $\dfrac{\partial f}{\partial y} = 2xy + x^2 y + \phi'(y) = x^2 y + y + 2xy + 1$.

Thus, $\phi'(y) = y + 1$, $\phi(y) = \frac{1}{2}y^2 + y + C$ and $f(x, y) = x + xy^2 + \frac{1}{2}x^2 y^2 + \frac{1}{2}y^2 + y + C$.

17. $\dfrac{\partial f}{\partial x} = \dfrac{x}{\sqrt{x^2 + y^2}}$, $f(x, y) = \sqrt{x^2 + y^2} + \phi(y)$, $\dfrac{\partial f}{\partial y} = \dfrac{y}{\sqrt{x^2 + y^2}} + \phi'(y) = \dfrac{y}{\sqrt{x^2 + y^2}}$.

Thus, $\phi'(y) = 0$, $\phi(y) = C$,   and $f(x, y) = \sqrt{x^2 + y^2} + C$.

19. $\dfrac{\partial f}{\partial x} = x^2 \sin^{-1} y,\quad f(x,y) = \tfrac{1}{3}x^3 \sin^{-1} y + \phi(y),\quad \dfrac{\partial f}{\partial y} = \dfrac{x^3}{3\sqrt{1-y^2}} + \phi'(y) = \dfrac{x^3}{3\sqrt{1-y^2}} - \ln y.$

Thus, $\phi'(y) = -\ln y, \implies \phi(y) = y - y\ln y + C,$ and

$$f(x,y) = \frac{1}{3}\,x^3 \sin^{-1} y + y - y\ln y + C.$$

21. (a) Yes   (b) Yes   (c) No

23. $\dfrac{\partial f}{\partial x} = f(x,y),\quad \dfrac{\partial f/\partial x}{f(x,y)} = 1,\quad \ln f(x,y) = x + \phi(y),\quad \dfrac{\partial f/\partial y}{f(x,y)} = 0 + \phi'(y),\quad \dfrac{\partial f}{\partial y} = f(x,y).$

Thus, $\phi'(y) = 1,\ \phi(y) = y + K,$ and $f(x,y) = e^{x+y+K} = Ce^{x+y}.$

25. (a) $P = 2x,\ Q = z,\ R = y;\qquad \dfrac{\partial P}{\partial y} = 0 = \dfrac{\partial Q}{\partial x},\quad \dfrac{\partial P}{\partial z} = 0 = \dfrac{\partial R}{\partial x},\quad \dfrac{\partial Q}{\partial z} = 1 = \dfrac{\partial R}{\partial y}$

(b), (c), and (d)

$$\frac{\partial f}{\partial x} = 2x,\quad f(x,y,z) = x^2 + g(y,z).$$

$$\frac{\partial f}{\partial y} = 0 + \frac{\partial g}{\partial y}\quad \text{with}\quad \frac{\partial f}{\partial y} = z \implies \frac{\partial g}{\partial y} = z.$$

Then,

$$g(y,z) = yz + h(z) \implies f(x,y,z) = x^2 + yz + h(z),$$

$$\frac{\partial f}{\partial z} = 0 + y + h'(z)\quad \text{and}\quad \frac{\partial f}{\partial z} = y \implies h'(z) = 0.$$

Thus, $h(z) = C$ and $f(x,y,z) = x^2 + yz + C.$

27. The function is a gradient by the test stated before Exercise 25.

Take $P = 2x + y,\quad Q = 2y + x + z,\quad R = y - 2z.$ Then

$$\frac{\partial P}{\partial y} = 1 = \frac{\partial Q}{\partial x},\quad \frac{\partial P}{\partial z} = 0 = \frac{\partial R}{\partial x},\quad \frac{\partial Q}{\partial z} = 1 = \frac{\partial R}{\partial y}.$$

Next, we find f where $\nabla f = P\mathbf{i} + Q\mathbf{j} + R\mathbf{k}.$

$$\frac{\partial f}{\partial x} = 2x + y \implies f(x,y,z) = x^2 + xy + g(y,z).$$

$$\frac{\partial f}{\partial y} = x + \frac{\partial g}{\partial y}\quad \text{with}\quad \frac{\partial f}{\partial y} = 2y + x + z \implies \frac{\partial g}{\partial y} = 2y + z.$$

Then,

$$g(y,z) = y^2 + yz + h(z),$$

$$f(x,y,z) = x^2 + xy + y^2 + yz + h(z).$$

$$\frac{\partial f}{\partial z} = y + h'(z) = y - 2z \implies h'(z) = -2z.$$

Thus, $h(z) = -z^2 + C$ and $f(x,y,z) = x^2 + xy + y^2 + yz - z^2 + C.$

29. The function is a gradient by the test stated before Exercise 25.

Take $P = y^2z^3 + 1$, $Q = 2xyz^3 + y$, $R = 3xy^2z^2 + 1$.   Then

$$\frac{\partial P}{\partial y} = 2yz^3 = \frac{\partial Q}{\partial x}, \quad \frac{\partial P}{\partial z} = 3y^2z^2 = \frac{\partial R}{\partial x}, \quad \frac{\partial Q}{\partial z} = 6xyz^2 = \frac{\partial R}{\partial y}.$$

Next, we find f where $\nabla f = P\mathbf{i} + Q\mathbf{j} + R\mathbf{k}$.

$$\frac{\partial f}{\partial x} = y^2z^3 + 1,$$

$$f(x, y, z) = xy^2z^3 + x + g(y, z).$$

$$\frac{\partial f}{\partial y} = 2xyz^3 + \frac{\partial g}{\partial y} \quad \text{with} \quad \frac{\partial f}{\partial y} = 2xyz^3 + y \quad \Longrightarrow \quad \frac{\partial g}{\partial y} = y.$$

Then,

$$g(y, z) = \tfrac{1}{2}y^2 + h(z),$$

$$f(x, y, z) = xy^2z^3 + x + \tfrac{1}{2}y^2 + h(z).$$

$$\frac{\partial f}{\partial z} = 3xy^2z^2 + h'(z) = 3xy^2z^2 + 1 \quad \Longrightarrow \quad h'(z) = 1.$$

Thus, $h(z) = z + C$ and $f(x, y, z) = xy^2z^3 + x + \tfrac{1}{2}y^2 + z + C.$

31. $\mathbf{F}(\mathbf{r}) = \nabla\left(\dfrac{GmM}{r}\right)$

REVIEW EXERCISES

1. $\nabla f(x, y) = (4x - 4y)\mathbf{i} + (3y^2 - 4x)\mathbf{j}$

3. $\nabla f(x, y) = (ye^{xy}\tan 2x + 2e^{xy}\sec^2 2x)\,\mathbf{i} + xe^{xy}\tan 2x\,\mathbf{j}$

5. $\nabla f(x, y) = 2xe^{-yz}\sec z\,\mathbf{i}, -zx^2e^{-yz}\sec z\mathbf{j} - (x^2ye^{-yz}\sec z - x^2e^{-yz}\sec z\tan z)\mathbf{k}$

7. $\nabla f(x, y) = (2x - 2y)\,\mathbf{i} - 2x\,\mathbf{j}, \quad \nabla f(1, -2) = 6\,\mathbf{i} - 2\,\mathbf{j}; \qquad \mathbf{u_a} = \dfrac{1}{\sqrt{5}}\mathbf{i} + \dfrac{2}{\sqrt{5}}\mathbf{j};$

$f'_{\mathbf{u_a}}(1, -2) = \nabla f(1, -2) \cdot \mathbf{u_a} = \dfrac{2}{\sqrt{5}}.$

9. $\nabla f(x, y, z) = (y^2 + 6xz)\,\mathbf{i} + (2xy + 2z)\,\mathbf{j} + (2y + 3x^2)\,\mathbf{k}, \quad \nabla f(1, -2, 3) = 22\,\mathbf{i} + 2\,\mathbf{j} - \mathbf{k};$

$\mathbf{u_a} = \dfrac{1}{3}\mathbf{i} - \dfrac{2}{3}\mathbf{j} + \dfrac{2}{3}\mathbf{k}; f'_{\mathbf{u_a}}(1, -2, 3) = \nabla f(1, -2, 3) \cdot \mathbf{u_a} = \dfrac{16}{3}.$

11. $\nabla f(x, y) = (6x - 2y^2)\,\mathbf{i} - 4xy\,\mathbf{j}, \quad \nabla f(3, -2) = 10\,\mathbf{i} + 24\,\mathbf{j};$

$\mathbf{a} = (0, 0) - (3, -2) = (-3, 2) = -3\,\mathbf{i} + 2\,\mathbf{j}, \quad \mathbf{u_a} = \dfrac{-3}{\sqrt{13}}\mathbf{i} + \dfrac{2}{\sqrt{13}}\mathbf{j};$

$f'_{\mathbf{u_a}}(3, -2) = \nabla f(3, -2) \cdot \mathbf{u_a} = \dfrac{18}{\sqrt{13}}.$

13. $\nabla f(x,y,z) = \dfrac{1}{\sqrt{x^2+y^2+z^2}}(x\,\mathbf{i}+y\,\mathbf{j}+z\,\mathbf{k})$, $\nabla f(3,-1,4) = \dfrac{1}{\sqrt{26}}(3\,\mathbf{i}-\mathbf{j}+4\,\mathbf{k})$;

$\mathbf{a} = \pm(4\,\mathbf{i}-3\,\mathbf{j}+\mathbf{k})$, $\mathbf{u_a} = \pm\dfrac{1}{\sqrt{26}}(4\,\mathbf{i}-3\,\mathbf{j}+\mathbf{k})$; $f'_{\mathbf{u_a}}(3,-1,4) = \nabla f(3,-1,4) \cdot \mathbf{u_a} = \pm\dfrac{19}{26}$.

15. $\nabla f(x,y,z) = \cos xyz\,(yz\,\mathbf{i}+xz\,\mathbf{j}+xy\,\mathbf{k})$, $\nabla f(\tfrac{1}{2},\tfrac{1}{3},\pi) = \frac{\pi\sqrt{3}}{6}\,\mathbf{i} + \frac{\pi\sqrt{3}}{4}\,\mathbf{j} + \frac{\sqrt{3}}{12}\,\mathbf{k}$;

minimum directional derivative: $f'_{\mathbf{u}} = -\|\nabla f(\tfrac{1}{2},\tfrac{1}{3},\pi)\| = -\frac{\sqrt{39\pi^2+3}}{12}$

17. Let $\mathbf{r}(t) = x(t)\,\mathbf{i} + y(t)\,\mathbf{j}$ be the path of the particle. $\nabla T = -e^{-x}\cos y\,\mathbf{i} - e^{-x}\sin y\,\mathbf{j}$. Then

$$x'(t) = -e^{-x(t)}\cos y(t), \quad y'(t) = -e^{-x(t)}\sin y(t) \quad\Longrightarrow\quad \frac{y'(t)}{x'(t)} = \tan y(t) \quad\Longrightarrow\quad \frac{dy}{dx} = \tan y$$

The solution is $\sin y = Ce^x$. Since $\mathbf{r}(0) = 0$, $C = 0$ and $y = 0$. The particle moves to the right the x-axis.

19. $\nabla f(x,y) = e^x\arctan y\,\mathbf{i} + e^x\dfrac{1}{1+y^2}\,\mathbf{j}$; $\nabla f(0,1) = \dfrac{\pi}{4}\,\mathbf{i} + \dfrac{1}{2}\,\mathbf{j}$.

$\mathbf{u} = \dfrac{\nabla f(0,1)}{\|\nabla f(0,1)\|} = \dfrac{1}{\sqrt{4+\pi^2}}(\pi\,\mathbf{i}+2\,\mathbf{j})$;   rate: $\|\nabla f(0,1)\| = \dfrac{\sqrt{\pi^2+4}}{4}$

21. rate: $\dfrac{df}{dt} = \nabla f \cdot \mathbf{r}' = (4x\,\mathbf{i} - 9y^2\,\mathbf{j}) \cdot \left(\dfrac{1}{2}t^{-1/2}\,\mathbf{i} + 2e^{2t}\,\mathbf{j}\right) = 2 - 18e^{6t}$

23. rate: $\dfrac{df}{dt} = \nabla f \cdot \mathbf{r}' = \left[\left(\dfrac{1}{y}+\dfrac{z}{x^2}\right)\mathbf{i} - \dfrac{x}{y^2}\,\mathbf{j} - \dfrac{1}{x}\,\mathbf{k}\right] \cdot (\cos t\,\mathbf{i} - \sin t\,\mathbf{j} + \sec^2 t\,\mathbf{k}) = \dfrac{1-\sin t}{\cos^2 t}$

25. $\dfrac{du}{dt} = \nabla u \cdot \mathbf{r}' = \left[(3y^2 - 2x)\,\mathbf{i} + 6xy\,\mathbf{j}\right] \cdot \left[(2t+2)\,\mathbf{i} + 3\,\mathbf{j}\right] = 104t^3 + 150t^2 - 8t$

27. area $A = \frac{1}{2}x(t)y(t)\sin\theta(t)$

$\dfrac{dA}{dt} = 0 = \frac{1}{2}y(t)x'(t)\sin\theta(t) + \frac{1}{2}x(t)y'(t)\sin\theta(t) + \frac{1}{2}\theta'(t)x(t)y(t)\cos\theta(t) = 0$

At $x=4$, $y=5$, $\theta = \pi/3$, $\dfrac{dx}{dt} = \dfrac{dy}{dt} = 2$,   we have

$$5\dfrac{d\theta}{dt} + 2\sqrt{3} + \dfrac{5\sqrt{3}}{2} = 0 \quad\Longrightarrow\quad \dfrac{d\theta}{dt} = -\dfrac{9\sqrt{3}}{10}.$$

29. $\dfrac{\partial u}{\partial s} = \dfrac{\partial u}{\partial x} + \dfrac{\partial u}{\partial y}$; $\dfrac{\partial u}{\partial t} = \dfrac{\partial u}{\partial x} - \dfrac{\partial u}{\partial y}$

$\dfrac{\partial u}{\partial s}\dfrac{\partial u}{\partial t} = \left(\dfrac{\partial u}{\partial x} + \dfrac{\partial u}{\partial y}\right)\left(\dfrac{\partial u}{\partial x} - \dfrac{\partial u}{\partial y}\right) = \left(\dfrac{\partial u}{\partial x}\right)^2 - \left(\dfrac{\partial u}{\partial y}\right)^2$

31. $\nabla f(x,y) = (3x^2 - 6xy)\,\mathbf{i} + (-3x^2 + 2y)\,\mathbf{j}$; $\nabla f(1,-1) = \mathbf{N} = 9\,\mathbf{i} - 5\,\mathbf{j}$

normal line: $x = 1 + 9t$, $y = -1 - 5t$;   tangent line: $x = 1 + 5t$, $y = -1 + 9t$

33. Set $f(x, y, z) = x^{1/2} + y^{1/2} - z$

$\nabla f(x, y, z) = \dfrac{1}{2\sqrt{x}}\mathbf{i} + \dfrac{1}{2\sqrt{y}}\mathbf{j} - \mathbf{k}$; $\nabla f(1, 1, 2) = \frac{1}{2}\mathbf{i} + \frac{1}{2}\mathbf{j} - \mathbf{k}$. Take $\mathbf{N} = \mathbf{i} + \mathbf{j} - 2\mathbf{k}$.

tangent plane: $(x - 1) + (y - 1) - 2(z - 2) = 0$; normal line: $x = 1 + t$, $y = 1 + t$, $z = 2 - 2t$

35. Set $f(x, y, z) = z^3 + xyz - 2$.

$\nabla f(x, y, z) = yz\,\mathbf{i} + xz\,\mathbf{j} + (3z^2 + xy)\,\mathbf{k}$; $\nabla f(1, 1, 1) = \mathbf{i} + \mathbf{j} + 4\mathbf{k}$.

tangent plane: $(x - 1) + (y - 1) + 4(z - 1) = 0$; normal line: $x = 1 + t$; $y = 1 + t$; $z = 1 + 4t$

37. The point $(2, 2, 1)$ is on each hyperboloid. Set $f(x, y, z) = x^2 + 2y^2 - 4z^2$, $g(x, y, z) = 4x^2 - y^2 + 2z^2$.

$\nabla f = 2x\,\mathbf{i} + 4y\,\mathbf{j} - 8z\,\mathbf{k}$, $\nabla f(2, 2, 1) = (4, 8, -8)$; $\nabla g = 8x\,\mathbf{i} - 2y\,\mathbf{j} + 4z\,\mathbf{k}$, $\nabla g(2, 2, 1) = (16, -4, 4)$.

Since $\nabla f(2, 2, 1) \cdot \nabla g(2, 2, 1) = 0$, the hyperboloids are mutually perpendicular at $(2, 2, 1)$.

39. $\nabla f(x, y) = (2xy - 2y)\,\mathbf{i} + (x^2 - 2x + 4y - 15)\,\mathbf{j} = \mathbf{0}$ at $(5, 0)$, $(-3, 0)$, $(1, 4)$.

$f_{xx} = 2y$, $f_{xy} = 2x - 2$, $f_{yy} = 4$.

point	A	B	C	D	result
$(5, 0)$	0	8	4	-64	saddle
$(-3, 0)$	0	-8	4	-64	saddle
$(1, 4)$	8	0	4	32	loc. min.

$f(1, 4) = -34$

41. $\nabla f(x, y) = (3x^2 - 18y)\,\mathbf{i} + (3y^2 - 18x)\,\mathbf{j} = \mathbf{0}$ at $(0, 0)$, $(6, 6)$.

$f_{xx} = 6x$, $f_{xy} = -18$, $f_{yy} = 6y$.

point	A	B	C	D	result
$(0, 0)$	0	-18	0	-18^2	saddle
$(6, 6)$	36	-18	36	> 0	loc. min.

$f(6, 6) = -216$

43. $\nabla f(x, y) = (1 - 2xy + y^2)\,\mathbf{i} + (-1 - x^2 + 2xy)\,\mathbf{j} = \mathbf{0}$ at $(1, 1)$, $(-1, -1)$.

$f_{xx} = -2y$, $f_{xy} = -2x + 2y$, $f_{yy} = 2x$.

point	A	B	C	D	result
$(1, 1)$	-2	0	2	-4	saddle
$(-1, -1)$	2	0	-2	-4	saddle

45. $\nabla f = (2x - 2)\,\mathbf{i} + (2y + 2)\,\mathbf{j} = \mathbf{0}$ at $(1, -1)$ in D; $f(1, -1) = 0$

Next we consider the boundary of D. We parametrize the circle by:

$C : \mathbf{r}(t) = 2\cos t\,\mathbf{i} + 2\sin t\,\mathbf{j}$, $t \in [0, 2\pi]$

The values of f on the boundary are given by the function

$$F(t) = f(\mathbf{r}(t)) = 6 - 4\cos t + 4\sin t, \quad t \in [0, 2\pi]$$

$$F'(t) = 4\sin t + 4\cos t: \quad F'(t) = 0 \implies \sin t = -\cos t \implies t = \frac{3}{4}\pi, \frac{7}{4}\pi$$

Evaluating F at the endpoints and critical numbers, we have:

$$F(0) = F(2\pi) = f(2,0) = 2; \quad F\left(\tfrac{3}{4}\pi\right) = f\left(-\sqrt{2}, \sqrt{2}\right) = 6 + 4\sqrt{2};$$

$$F\left(\tfrac{7}{4}\pi\right) = f\left(\sqrt{2}, -\sqrt{2}\right) = 6 - 4\sqrt{2}.$$

f takes on its absolute maximum of $6 + 4\sqrt{2}$ at $(-\sqrt{2}, \sqrt{2})$; f takes on its absolute minimum of 0 at $(1, -1)$.

47. $\nabla f(x, y) = (8x - y)\mathbf{i} + (-x + 2y + 1)\mathbf{j} = \mathbf{0}$ at $(-1/15, -8/15)$ in D; $f(-1/15, -8/15) = -4/15$. On the boundary of D: $x = \cos t$, $y = 2\sin t$. Set

$$F(t) = f(\cos t, 2\sin t) = 4 + 2\sin t - 2\sin t \cos t, \quad 0 \le t \le 2\pi.$$

Then

$$F'(t) = 2\cos t - 4\cos^2 t + 2 = -2(2\cos t + 1)(\cos t - 1); \quad F'(t) = 0 \implies t = \frac{2\pi}{3}, \frac{4\pi}{3}.$$

Evaluating F at the endpoints of the interval and at the critical points, we get

$$F(0) = F(2\pi) = f(1, 0) = 4, \ F(2\pi/3) = f(-1/2, \sqrt{3}) = 4 + \frac{3\sqrt{3}}{2},$$

$$F(4\pi/3) = f(-1/2, -\sqrt{3}) = 4 - \frac{3\sqrt{3}}{2} > -\frac{4}{15}$$

f takes on its absolute maximum of 2 at $(0, 1)$; f takes on its absolute minimum of $-4/15$ at $(-1/15, -8/15)$.

49. Set $f(x, y, z) = D^2 = (x - 1)^2 + (y + 2)^2 + (z - 3)^2$, $g(x, y) = 3x + 2y - z - 5$.

$$\nabla f = 2(x - 1)\mathbf{i} + 2(y + 2)\mathbf{j} + 2(z - 3)\mathbf{k}, \quad \nabla g = 3\mathbf{i} + 2\mathbf{j} - \mathbf{k}.$$

Set $\nabla f = \lambda \nabla g$:

$$2(x - 1) = 3\lambda \implies x = \tfrac{3}{2}\lambda + 1,$$
$$2(y + 2) = 2\lambda \implies y = \lambda - 2,$$
$$2(z - 3) = -\lambda \implies z = -\tfrac{1}{2}\lambda + 3.$$

Substituting these values in $3x + 2y - z = 5$ gives $\lambda = \dfrac{9}{7} \implies x = \dfrac{41}{14}, \ y = -\dfrac{5}{7}, \ z = \dfrac{33}{14}.$

The point on the plane that is closest to $(1, -2, 3)$ is $(41/14, -5/7, 33/14)$. The distance from the point to the plane is $\dfrac{9}{\sqrt{14}}$.

51. Set $f(x, y, z) = x + y - z$, $\quad g(x, y, z) = x^2 + y^2 + 4z^2 - 4$,

$\nabla f = \mathbf{i} + \mathbf{j} - \mathbf{k}$, $\quad \nabla g = 2x\,\mathbf{i} + 2y\,\mathbf{j} + 8z\,\mathbf{k}$.

Set $\nabla f = \lambda \nabla g$:

$$1 = 2\lambda x \implies x = 1/2\lambda, \quad 1 = 2\lambda y \implies y = 1/2\lambda, \quad -1 = 8\lambda z \implies z = -1/8\lambda.$$

Substituting these values in $x^2 + y^2 + 4z^2 = 4$ gives $\lambda = \pm\frac{3}{8} \implies x = 4/3$, $y = 4/3$, $z = -1/3$ or $x = -4/3$, $y = -4/3$, $z = 1/3$. Evaluating f : $f(\frac{4}{3}, \frac{4}{3}, -\frac{1}{3}) = 3$, $f(-\frac{4}{3}, -\frac{4}{3}, \frac{1}{3}) = -3$. The maximum value of f is 3, the minimum value is -3.

53. $df = (9x^2 - 10xy^2 + 2)\,dx + (-10x^2y - 1)\,dy$

55. $df = \dfrac{y^2z + z^2y}{(x + y + z)^2}\,dx + \dfrac{xz^2 + zx^2}{(x + y + z)^2}\,dy + \dfrac{x^2y + y^2x}{(x + y + z)^2}\,dz$

57. Set $f(x, y, z) = e^x \sqrt{y + z^3}$. Then

$$df = e^x \sqrt{y + z^3}\,\Delta x + \frac{e^x}{2}\,\frac{1}{\sqrt{y + z^3}}\,\Delta y + \frac{e^x}{2}\,\frac{3z^2}{\sqrt{y + z^3}}\,\Delta z.$$

With $x = 0$, $y = 15$, $z = 1$, $\Delta x = 0.02$, $\Delta y = 0.2$, $\Delta z = 0.01$, $df = 4\,\Delta x + \frac{1}{8}\,\Delta y + \frac{3}{8}\,\Delta z \cong 0.1088$.

Therefore, $e^{0.02}\sqrt{15.2 + (1.01)^3} \cong e^0\sqrt{15 + 1} + 0.1088 = 4.1088$.

59. $V = \pi r^2 h$; $\quad r = 5\,\text{ft.}$, $\quad h = 22\,\text{ft.}$, $\quad \Delta r = 0.01\,\text{in.} = \frac{1}{1200}\,\text{ft.}$, $\quad \Delta h = 0.01 = \frac{1}{1200}$

$$dV = 2\pi r h\,\Delta r + \pi r^2\,\Delta h$$

Using the values given above,

$$dV = 2\pi(5)(22)\,\frac{1}{1200} + \pi(25)\,\frac{1}{1200} \cong 0.6414\,\text{cu. ft.} \cong 1108.35\,\text{cu. in.}; \quad \frac{1108.35}{231} \cong 4.80.$$

Approximately 4.80 gallons will be needed.

61. $\dfrac{\partial P}{\partial y} = 2x - \sin x = \dfrac{\partial Q}{\partial x}$; the vector function is a gradient.

$\dfrac{\partial f}{\partial x} = 2xy + 3 - y\sin x$, $\quad f(x, y) = x^2y + 3x + y\cos x + \phi(y)$,

$\dfrac{\partial f}{\partial y} = x^2 + \cos x + \phi'(y) = x^2 + 2y + 1 + \cos x$.

Thus, $\quad \phi'(y) = 2y + 1$, $\quad \phi(y) = y^2 + y + C$, $\quad$ and $\quad f(x, y) = x^2y + 3x + y\cos x + y^2 + y + C$.

63. $\dfrac{\partial P}{\partial y} = e^y \sin z = \dfrac{\partial Q}{\partial x}$, $\quad \dfrac{\partial P}{\partial z} = e^y \cos z = \dfrac{\partial R}{\partial x}$, $\quad \dfrac{\partial Q}{\partial z} = xe^y \cos z = \dfrac{\partial R}{\partial y}$;

the vector function is a gradient.

$$f(x,y,z) = \int (e^y \sin z + 2x)\, dx = xe^y \sin z + x^2 + \phi(y,z),$$

$$f_y = xe^y \sin z + \dfrac{\partial \phi}{\partial y} = xe^y \sin z - y^2 \implies \dfrac{\partial \phi}{\partial y} = -y^2 \implies \phi = -\dfrac{1}{3}y^3 + \psi(z),$$

$$f(x,y,z) = xe^y \sin z + x^2 - \dfrac{1}{3}y^3 + \psi(z), \quad f_z = xe^y \cos z + \psi'(z) = xe^y \cos z \implies \psi'(x) = 0 \implies$$
$$\psi(x) = C$$

Therefore $f(x,y,z) = xe^y \sin z + x^2 - \frac{1}{3}y^3 + C$.

CHAPTER 17

SECTION 17.1

1. $\displaystyle\sum_{i=1}^{3}\sum_{j=1}^{3}2^{i-1}3^{j+1} = \left(\sum_{i=1}^{3}2^{i-1}\right)\left(\sum_{j=1}^{3}3^{j+1}\right) = (1+2+4)(9+27+81) = 819$

3. $\displaystyle\sum_{i=1}^{4}\sum_{j=1}^{3}(i^2+3i)(j-2) = \left[\sum_{i=1}^{4}(i^2+3i)\right]\left[\sum_{j=1}^{3}(j-2)\right] = (4+10+18+28)(-1+0+1) = 0$

5. $\displaystyle\sum_{i=1}^{m}\Delta x_i = \Delta x_1 + \Delta x_2 + \cdots + \Delta x_m = (x_1 - x_0) + (x_2 - x_1) + \cdots + (x_m - x_{m-1})$

$$= x_m - x_0 = a_2 - a_1$$

7. $\displaystyle\sum_{i=1}^{m}\sum_{j=1}^{n}\Delta x_i\,\Delta y_j = \left(\sum_{i=1}^{m}\Delta x_i\right)\left(\sum_{j=1}^{n}\Delta y_j\right) = (a_2 - a_1)(b_2 - b_1)$

9. $\displaystyle\sum_{i=1}^{m}(x_i + x_{i-1})\Delta x_i = \sum_{i=1}^{m}(x_i + x_{i-1})(x_i - x_{i-1}) = \sum_{i=1}^{m}(x_i^2 - x_{i-1}^2)$

$$= x_m^2 - x_0^2 = a_2^2 - a_1^2$$

11. $\displaystyle\sum_{i=1}^{m}\sum_{j=1}^{n}(x_i + x_{i-1})\Delta x_i \Delta y_j = \left(\sum_{i=1}^{m}(x_i + x_{i-1})\Delta x_i\right)\left(\sum_{j=1}^{n}\Delta y_j\right)$

$$\text{(Exercise 9)}\underset{\uparrow}{}$$

$$= \left(a_2^2 - a_1^2\right)(b_2 - b_1)$$

13. $\displaystyle\sum_{i=1}^{m}\sum_{j=1}^{n}(2\Delta x_i - 3\Delta y_j) = 2\left(\sum_{i=1}^{m}\Delta x_i\right)\left(\sum_{j=1}^{n}1\right) - 3\left(\sum_{i=1}^{m}1\right)\left(\sum_{j=1}^{n}\Delta y_j\right)$

$$= 2n(a_2 - a_1) - 3m(b_2 - b_1)$$

15. $\displaystyle\sum_{i=1}^{m}\sum_{j=1}^{n}\sum_{k=1}^{q}\Delta x_i\,\Delta y_j\,\Delta z_k = \left(\sum_{i=1}^{m}\Delta x_i\right)\left(\sum_{j=1}^{n}\Delta y_j\right)\left(\sum_{k=1}^{q}\Delta z_k\right)$

$$= (a_2 - a_1)(b_2 - b_1)(c_2 - c_1)$$

17. $\displaystyle\sum_{i=1}^{n}\sum_{j=1}^{n}\sum_{k=1}^{n}\delta_{ijk}a_{ijk} = a_{111} + a_{222} + \cdots + a_{nnn} = \sum_{p=1}^{n}a_{ppp}$

SECTION 17.2

1. $L_f(P) = 2\frac{1}{4}, \quad U_f(P) = 5\frac{3}{4}$

3. (a) $\displaystyle L_f(P) = \sum_{i=1}^{m}\sum_{j=1}^{n}(x_{i-1} + 2y_{j-1})\,\Delta x_i\,\Delta y_j, \quad U_f(P) = \sum_{i=1}^{m}\sum_{j=1}^{n}(x_i + 2y_j)\,\Delta x_i\,\Delta y_j$

(b) $L_f(P) \leq \sum_{i=1}^{m} \sum_{j=1}^{n} \left[\frac{x_{i-1} + x_i}{2} + 2 \left(\frac{y_{j-1} + y_j}{2} \right) \right] \Delta x_i \, \Delta y_j \leq U_f(P).$

The middle expression can be written

$$\sum_{i=1}^{m} \sum_{j=1}^{n} \frac{1}{2} \left(x_i{}^2 - x_{i-1}^2 \right) \Delta y_j + \sum_{i=1}^{m} \sum_{j=1}^{n} \left(y_j{}^2 - y_{j-1}^2 \right) \Delta x_i.$$

The first double sum reduces to

$$\sum_{i=1}^{m} \sum_{j=1}^{n} \frac{1}{2} \left(x_i{}^2 - x_{i-1}^2 \right) \Delta y_j = \frac{1}{2} \left(\sum_{i=1}^{m} \left(x_i{}^2 - x_{i-1}^2 \right) \right) \left(\sum_{j=1}^{n} \Delta y_j \right) = \frac{1}{2} \left(4 - 0 \right) \left(1 - 0 \right) = 2.$$

In like manner the second double sum also reduces to 2. Thus, $I = 4$; the volume of the prism bounded above by the plane $z = x + 2y$ and below by R.

5. $L_f(P) = -7/24, \quad U_f(P) = 7/24$

7. (a) $L_f(P) = \sum_{i=1}^{m} \sum_{j=1}^{n} \left(4 x_{i-1} \, y_{j-1} \right) \Delta x_i \, \Delta y_j, \quad U_f(P) = \sum_{i=1}^{m} \sum_{j=1}^{n} \left(4 x_i \, y_j \right) \Delta x_i \, \Delta y_j$

(b) $L_f(P) \leq \sum_{i=1}^{m} \sum_{j=1}^{n} \left(x_i + x_{i-1} \right) \left(y_j + y_{j-1} \right) \Delta x_1 \, \Delta y_j \leq U_f(P).$

The middle expression can be written

$$\sum_{i=1}^{m} \sum_{j=1}^{n} \left(x_i{}^2 - x_{i-1}^2 \right) \left(y_j{}^2 - y_{j-1}^2 \right) = \left(\sum_{i=1}^{m} x_i{}^2 - x_{i-1}^2 \right) \left(\sum_{j=1}^{n} y_j{}^2 - y_{j-1}^2 \right)$$

$$\text{by } (17.1.5) \underline{\qquad}\uparrow$$

$$= \left(b^2 - 0^2 \right) \left(d^2 - 0^2 \right) = b^2 d^2.$$

It follows that $I = b^2 d^2$.

9. (a) $L_f(P) = \sum_{i=1}^{m} \sum_{j=1}^{n} 3 \left(x_{i-1}^2 - y_j{}^2 \right) \Delta x_i \, \Delta y_j, \quad U_f(P) = \sum_{i=1}^{m} \sum_{j=1}^{n} 3 \left(x_i{}^2 - y_{j-1}^2 \right) \Delta x_i \, \Delta y_j$

(b) $L_f(P) \leq \sum_{i=1}^{m} \sum_{j=1}^{n} \left[\left(x_i{}^2 + x_i x_{i-1} + x_{i-1}^2 \right) - \left(y_j{}^2 + y_j y_{j-1} + y_{j-1}^2 \right) \right] \Delta x_i \, \Delta y_j \leq U_f(P).$

Since in general $\left(A^2 + AB + B^2 \right) \left(A - B \right) = A^3 - B^3,$ the middle expression can be written

$$\sum_{i=1}^{m} \sum_{j=1}^{n} \left(x_i{}^3 - x_{i-1}^3 \right) \Delta y_j - \sum_{i=1}^{m} \sum_{j=1}^{n} \left(y_j{}^3 - y_{j-1}^3 \right) \Delta x_i,$$

which reduces to

$$\left(\sum_{i=1}^{m} x_i{}^3 - x_{i-1}^3 \right) \left(\sum_{j=1}^{n} \Delta y_j \right) - \left(\sum_{i=1}^{m} \Delta x_i \right) \left(\sum_{j=1}^{n} y_j{}^3 - y_{j-1}^3 \right).$$

This can be evaluated as $b^3 d - b d^3 = b d \left(b^2 - d^2 \right).$ It follows that $I = b d \left(b^2 - d^2 \right).$

11. $\displaystyle\iint_\Omega dx dy = \int_a^b \phi(x)\,dx$

13. Suppose $f(x_0, y_0) \neq 0$. Assume $f(x_0, y_0) > 0$. Since f is continuous, there exists a disc Ω_ϵ with radius ϵ centered at (x_0, y_0) such that $f(x, y) > 0$ on Ω_ϵ. Let R be a rectangle contained in Ω_ϵ. Then $\displaystyle\iint_R f(x, y)\,dx dy > 0$, which contradicts the hypothesis.

15. By Exercise 7, Section 17.2, $\displaystyle\iint_R 4xy\,dx dy = 2^2 3^2 = 36$. Thus

$$f_{avg} = \frac{1}{\text{area}\,(R)} \iint_R 4xy\,dx dy = \frac{1}{6}\,(36) = 6$$

17. By Theorem 16.2.10, there exists a point $(x_1, y_1) \in D_r$ such that

$$\iint_{D_r} f(x, y)\,dx dy = f(x_1, y_1) \iint_R dx dy = f(x_1, y_1)\pi r^2 \quad\Longrightarrow\quad f(x_1, y_1) = \frac{1}{\pi r^2} \iint_{D_r} f(x, y)\,dx dy$$

As $r \to 0$, $(x_1, y_1) \to (x_0, y_0)$ and $f(x_1, y_1) \to f(x_0, y_0)$ since f is continuous.

The result follows.

19. $z = \sqrt{4 - x^2 - y^2}$ on $\Omega : x^2 + y^2 \leq 4$, $x \geq 0$, $y \geq 0$; $\displaystyle\iint_\Omega \sqrt{4 - x^2 - y^2}\,dx dy$ is the volume V of one quarter of a hemisphere; $V = \frac{4}{3}\pi$.

21. $z = 6 - 2x - 3y \Rightarrow \dfrac{x}{3} + \dfrac{y}{2} + \dfrac{z}{6} = 1$; the solid is the tetrahedron bounded by the coordinate planes and the plane: $\dfrac{x}{3} + \dfrac{y}{2} + \dfrac{z}{6} = 1$; $V = \frac{1}{6}(3)(2)(6) = 6$

SECTION 17.3

1. $\displaystyle\int_0^1 \int_0^3 x^2\,dy\,dx = \int_0^1 3x^2\,dx = 1$

3. $\displaystyle\int_0^1 \int_0^3 xy^2\,dy\,dx = \int_0^1 x\left[\frac{1}{3}y^3\right]_0^3 dx = \int_0^1 9x\,dx = \frac{9}{2}$

5. $\displaystyle\int_0^1 \int_0^x xy^3\,dy\,dx = \int_0^1 x\left[\frac{1}{4}y^4\right]_0^x dx = \int_0^1 \frac{1}{4}x^5\,dx = \frac{1}{24}$

7. $\displaystyle\int_0^{\pi/2} \int_0^{\pi/2} \sin(x + y)\,dy\,dx = \int_0^{\pi/2} [-\cos(x + y)]_0^{\pi/2}\,dx = \int_0^{\pi/2} \left[\cos x - \cos\left(x + \frac{\pi}{2}\right)\right] dx = 2$

9. $\displaystyle\int_0^{\pi/2} \int_0^{\pi/2} (1 + xy)\,dy\,dx = \int_0^{\pi/2} \left[y + \frac{1}{2}xy^2\right]_0^{\pi/2} dx = \int_0^{\pi/2} \left(\frac{1}{2}\pi + \frac{1}{8}\pi^2 x\right) dx = \frac{1}{4}\pi^2 + \frac{1}{64}\pi^4$

11. $\displaystyle\int_0^1 \int_{y^2}^y \sqrt{xy}\, dx\, dy = \int_0^1 \sqrt{y}\left[\frac{2}{3}x^{3/2}\right]_{y^2}^y dy = \int_0^1 \frac{2}{3}\left(y^2 - y^{7/2}\right) dy = \frac{2}{27}$

13. $\displaystyle\int_{-2}^2 \int_{\frac{1}{2}y^2}^{4-\frac{1}{2}y^2} \left(4 - y^2\right) dx\, dy = \int_{-2}^2 \left(4 - y^2\right)\left[\left(4 - \frac{1}{2}y^2\right) - \left(\frac{1}{2}y^2\right)\right] dy$

$$= 2\int_0^2 \left(16 - 8y^2 + y^4\right) dy = \frac{512}{15}$$

15. 0   by symmetry (integrand odd in y, Ω symmetric about x-axis)

17. $\displaystyle\int_0^2 \int_0^{x/2} e^{x^2} dy\, dx = \int_0^2 \frac{1}{2}x e^{x^2} dx = \left[\frac{1}{4}e^{x^2}\right]_0^2 = \frac{1}{4}\left(e^4 - 1\right)$

19.

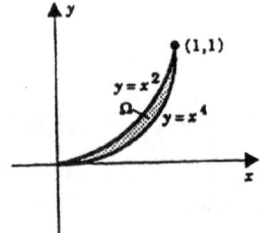

$$\int_0^1 \int_{y^{1/2}}^{y^{1/4}} f(x,y)\, dx\, dy$$

21.

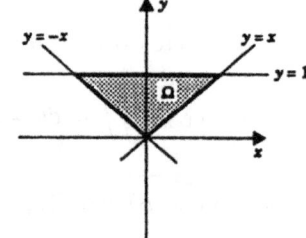

$$\int_{-1}^0 \int_{-x}^1 f(x,y)\, dy\, dx + \int_0^1 \int_x^1 f(x,y)\, dy\, dx$$

23.

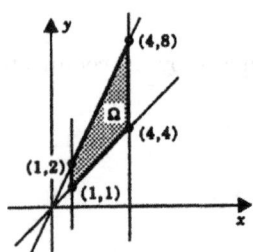

$$\int_1^2 \int_1^y f(x,y)\, dx\, dy + \int_2^4 \int_{y/2}^y f(x,y)\, dx\, dy$$

$$+ \int_4^8 \int_{y/2}^4 f(x,y)\, dx\, dy$$

25. $\displaystyle\int_{-2}^4 \int_{1/4x^2}^{\frac{1}{2}x+2} dy\, dx = \int_{-2}^4 \left[\frac{1}{2}x + 2 - \frac{1}{4}x^2\right] dx = 9$

27. $\displaystyle\int_0^{1/4} \int_{2y^{3/2}}^y dx\, dy = \int_0^{1/4} \left[y - 2y^{3/2}\right] dy = \frac{1}{160}$

29.

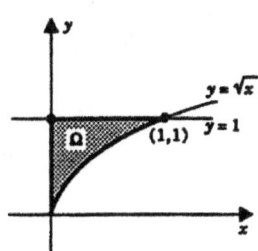

$$\int_0^1 \int_0^{y^2} \sin\left(\frac{y^3+1}{2}\right) dx\, dy = \int_0^1 y^2 \sin\left(\frac{y^3+1}{2}\right) dy$$

$$= \left[-\frac{2}{3}\cos\left(\frac{y^3+1}{2}\right)\right]_0^1$$

$$= \frac{2}{3}\left(\cos\frac{1}{2}-\cos 1\right)$$

31.

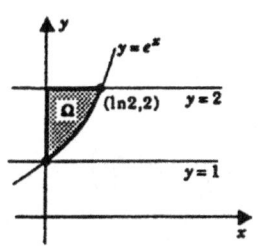

$$\int_0^{\ln 2}\int_{e^x}^2 e^{-x}\, dy\, dx = \int_0^{\ln 2} e^{-x}\left(2-e^x\right) dx$$

$$= \left[-2e^{-x}-x\right]_0^{\ln 2} = 1-\ln 2$$

33. $$\int_1^2\int_{y-1}^{2/y} dx\, dy = \int_1^2\left[\frac{2}{y}-(y-1)\right] dy = \ln 4 - \frac{1}{2}$$

35. $$\int_0^2\int_0^{3-\frac{3}{2}x}\left(4-2x-\frac{4}{3}y\right) dy\, dx = \int_0^3\int_0^{2-\frac{2}{3}y}\left(4-2x-\frac{4}{3}y\right) dx\, dy = 4$$

37. $$\int_0^2\int_0^{1-\frac{1}{2}x} x^3 y\, dy\, dx = \int_0^2\int_0^{2-2y} x^3 y\, dx\, dy = \frac{2}{15}$$

39. $$\int_0^2\int_{-\sqrt{2x-x^2}}^{\sqrt{2x-x^2}}(2x+1)\, dy\, dx = \int_{-1}^1\int_{1-\sqrt{1-y^2}}^{1+\sqrt{1-y^2}}(2x+1)\, dx\, dy$$

$$= \int_{-1}^1\left[x^2+x\right]_{1-\sqrt{1-y^2}}^{1+\sqrt{1-y^2}} dy$$

$$= 6\int_{-1}^1\sqrt{1-y^2}\, dy = 6\left(\frac{\pi}{2}\right) = 3\pi$$

41. $$\int_0^1\int_0^{1-x}(x^2+y^2)\, dy\, dx = \int_0^1\left(2x^2-\frac{4}{3}x^3-x+\frac{1}{3}\right) dx = \frac{1}{6}$$

43. $$\int_0^1\int_{x^2}^x(x^2+3y^2)\, dy\, dx = \int_0^1\left(2x^3-x^4-x^6\right) dx = \frac{11}{70}$$

45. $$\int_0^a\int_0^{\sqrt{a^2-x^2}}\sqrt{a^2-x^2}\, dy\, dx = \int_0^a(a^2-x^2)\, dx = \frac{2}{3}a^3$$

47. $$\int_0^1\int_y^1 e^{y/x}\, dx\, dy = \int_0^1\int_0^x e^{y/x}\, dy\, dx = \int_0^1\left[xe^{y/x}\right]_0^x dx = \int_0^1 x(e-1)\, dx = \frac{1}{2}(e-1)$$

49. $\displaystyle\int_0^1 \int_x^1 x^2 e^{y^4}\, dy\, dx = \int_0^1 \int_0^y x^2 e^{y^4}\, dx\, dy = \int_0^1 \left[\frac{1}{3}x^3 e^{y^4}\right]_0^y dy = \frac{1}{3}\int_0^1 y^3 e^{y^4}\, dy = \frac{1}{12}(e-1)$

51. $\displaystyle f_{avg} = \frac{1}{8}\int_{-1}^1 \int_0^4 x^2 y\, dy\, dx = \frac{1}{8}\int_{-1}^1 8x^2\, dx = \int_{-1}^1 x^2\, dx = \frac{2}{3}$

53. $\displaystyle f_{avg} = \frac{1}{(\ln 2)^2}\int_{\ln 2}^{2\ln 2}\int_{\ln 2}^{2\ln 2}\frac{1}{xy}\, dy\, dx = \frac{1}{(\ln 2)^2}\int_{\ln 2}^{2\ln 2}\frac{1}{x}\ln 2\, dx = 1$

55. $\displaystyle\iint_R f(x)g(y)\, dx\, dy = \int_c^d \int_a^b f(x)g(y)\, dx\, dy = \int_c^d \left(\int_a^b f(x)g(y)\, dx\right) dy$

$$= \int_c^d g(y)\left(\int_a^b f(x)\, dx\right) dy = \left(\int_a^b f(x)\, dx\right)\left(\int_c^d g(y)\, dy\right)$$

57. Note that $\quad \Omega = \{(x,y): 0 \le x \le y,\ 0 \le y \le 1\}.$

Set $\quad \Omega' = \{(x,y): 0 \le y \le x,\ 0 \le x \le 1\}.$

$$\iint_\Omega f(x)f(y)\, dx\, dy = \int_0^1 \int_0^y f(x)f(y)\, dx\, dy$$

$$= \int_0^1 \int_0^x f(y)f(x)\, dy\, dx$$

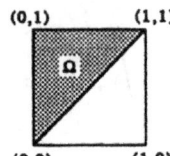 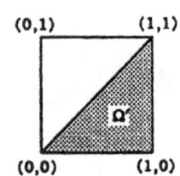

x and y are dummy variables

$$= \int_0^1 \int_0^x f(x)f(y)\, dy\, dx = \iint_{\Omega'} f(x)f(y)\, dx\, dy.$$

Note that Ω and Ω' don't overlap and their union is the unit square

$$R = \{(x,y): 0 \le x \le 1,\ 0 \le y \le 1\}.$$

If $\displaystyle\int_0^1 f(x)\, dx = 0,\quad$ then

$$0 = \left(\int_0^1 f(x)\, dx\right)\left(\int_0^1 f(y)\, dy\right) = \iint_R f(x)f(y)\, dx\, dy$$

by Exercise 55

$$= \iint_\Omega f(x)f(y)\, dx\, dy + \iint_{\Omega'} f(x)f(y)\, dx\, dy$$

$$= 2 \iint_\Omega f(x)f(y)\, dx\, dy$$

and therefore $\displaystyle\iint_\Omega f(x)f(y)\, dx\, dy = 0.$

59. Let M be the maximum value of $|f(x, y)|$ on Ω.

$$\int_{\phi_1(x+h)}^{\phi_2(x+h)} = \int_{\phi_1(x+h)}^{\phi_1(x)} + \int_{\phi_1(x)}^{\phi_2(x)} + \int_{\phi_2(x)}^{\phi_2(x+h)}$$

$$|F(x+h) - F(x)| = \left| \int_{\phi_1(x+h)}^{\phi_2(x+h)} f(x,y)\,dy - \int_{\phi_1(x)}^{\phi_2(x)} f(x,y)\,dy \right|$$

$$= \left| \int_{\phi_1(x+h)}^{\phi_1(x)} f(x,y)\,dy + \int_{\phi_2(x)}^{\phi_2(x+h)} f(x,y)\,dy \right|$$

$$\leq \left| \int_{\phi_1(x+h)}^{\phi_1(x)} f(x,y)\,dy \right| + \left| \int_{\phi_2(x)}^{\phi_2(x+h)} f(x,y)\,dy \right|$$

$$\leq |\phi_1(x) - \phi_1(x+h)|\,M + |\phi_2(x+h) - \phi_2(x)|\,M.$$

The expression on the right tends to 0 as h tends to 0 since ϕ_1 and ϕ_2 are continuous.

61. (a) $\displaystyle\int_1^2 \int_{x^2-2x+2}^{1+\sqrt{x-1}} 1\,dy\,dx = \frac{1}{3}$ (b) $\displaystyle\int_1^2 \int_{y^2-2y+2}^{1+\sqrt{y-1}} 1\,dx\,dy = \frac{1}{3}$

SECTION 17.4

1. $\displaystyle\int_0^{\pi/2} \int_0^{\sin\theta} r\cos\theta\,dr\,d\theta = \int_0^{\pi/2} \frac{1}{2}\sin^2\theta\cos\theta\,d\theta = \left[\frac{1}{6}\sin^3\theta\right]_0^{\pi/2} = \frac{1}{6}$

3. $\displaystyle\int_0^{\pi/2} \int_0^{3\sin\theta} r^2\,dr\,d\theta = \int_0^{\pi/2} 9\sin^3\theta\,d\theta = 9\int_0^{\pi/2}(1-\cos^2\theta)\sin\theta\,d\theta = 9\left[-\cos\theta + \frac{1}{3}\cos^3\theta\right]_0^{\pi/2} = 6$

5. (a) $\Gamma: 0 \leq \theta \leq 2\pi, \quad 0 \leq r \leq 1$

$$\iint_\Gamma (\cos r^2)r\,dr d\theta = \int_0^{2\pi} \int_0^1 (\cos r^2)r\,dr\,d\theta = 2\pi \int_0^1 r\cos r^2\,dr = \pi\sin 1$$

(b) $\Gamma: 0 \leq \theta \leq 2\pi, \quad 1 \leq r \leq 2$

$$\iint_\Gamma (\cos r^2)r\,dr d\theta = \int_0^{2\pi} \int_1^2 (\cos r^2)r\,dr\,d\theta = 2\pi \int_1^2 r\cos r^2\,dr = \pi(\sin 4 - \sin 1)$$

7. (a) $\Gamma: 0 \leq \theta \leq \pi/2, \quad 0 \leq r \leq 1$

$$\iint_\Gamma (r\cos\theta + r\sin\theta)r\,dr d\theta = \int_0^{\pi/2} \int_0^1 r^2(\cos\theta + \sin\theta)\,dr\,d\theta$$

$$= \left(\int_0^{\pi/2} (\cos\theta + \sin\theta)\,d\theta\right)\left(\int_0^1 r^2\,dr\right) = 2\left(\frac{1}{3}\right) = \frac{2}{3}$$

(b) $\Gamma: 0 \leq \theta \leq \pi/2, \quad 1 \leq r \leq 2$

$$\iint_\Gamma (r\cos\theta + r\sin\theta)r\,dr d\theta = \int_0^{\pi/2} \int_1^2 r^2(\cos\theta + \sin\theta)\,dr\,d\theta$$

$$= \left(\int_0^{\pi/2} (\cos\theta + \sin\theta)\,d\theta\right)\left(\int_1^2 r^2\,dr\right) = 2\left(\frac{7}{3}\right) = \frac{14}{3}$$

9. $\displaystyle\int_{-\pi/2}^{\pi/2}\int_0^1 r^2\,dr\,d\theta = \frac{1}{3}\,\pi$

11. $\displaystyle\int_{1/2}^1\int_0^{\sqrt{1-x^2}} dy\,dx = \int_0^{\pi/3}\int_{\frac{1}{2}\sec\theta}^1 r\,dr\,d\theta = \int_0^{\pi/3}\left(\frac{1}{2}-\frac{1}{8}\sec^2\theta\right)d\theta = \frac{1}{6}\pi - \frac{\sqrt{3}}{8}$

13. $\displaystyle\int_0^1\int_0^{\sqrt{1-x^2}}\sin\sqrt{x^2+y^2}\,dy\,dx = \int_0^{\pi/2}\int_0^1\sin(r)\,r\,dr\,d\theta = \int_0^{\pi/2}(\sin 1 - \cos 1)\,d\theta = \frac{\pi}{2}(\sin 1 - \cos 1)$

15. $\displaystyle\int_0^2\int_0^{\sqrt{2x-x^2}} x\,dy\,dx = \int_0^{\pi/2}\int_0^{2\cos\theta} r\cos\theta\,r\,dr\,d\theta = \frac{8}{3}\int_0^{\pi/2}\cos^4\theta\,d\theta = \frac{8}{3}\cdot\frac{3}{4}\cdot\frac{1}{2}\cdot\frac{\pi}{2} = \frac{\pi}{2}$

(See Exercise 62, Section 8.3)

17. $\displaystyle A = \int_0^{\pi/3}\int_0^{3\sin 3\theta} r\,dr\,d\theta = \frac{9}{2}\int_0^{\pi/3}\sin^2 3\theta\,d\theta = \frac{9}{4}\int_0^{\pi/3}(1-6\cos\theta)\,d\theta = \frac{3\pi}{4}$

19. First we find the points of intersection:

$$r = 4\cos\theta = 2 \quad\Longrightarrow\quad \cos\theta = \frac{1}{2}$$
$$\Longrightarrow\quad \theta = \pm\frac{\pi}{3}.$$

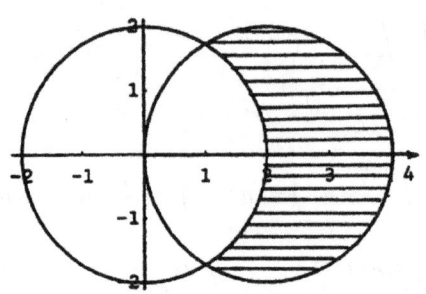

$$A = \int_{-\pi/3}^{\pi/3}\int_2^{4\cos\theta} r\,dr\,d\theta = \int_{-\pi/3}^{\pi/3}(8\cos^2\theta - 2)\,d\theta = \int_{-\pi/3}^{\pi/3}(2+4\cos 2\theta)\,d\theta = \frac{4\pi}{3}+2\sqrt{3}$$

21. $\displaystyle A = 4\int_0^{\pi/4}\int_0^{2\sqrt{\cos 2\theta}} r\,dr\,d\theta = 8\int_0^{\pi/4}\cos 2\theta\,d\theta = 4$

23. $\displaystyle\int_0^{2\pi}\int_0^b (r^2\sin\theta + br)\,dr\,d\theta = \int_0^{2\pi}\left[\frac{1}{3}r^3\sin\theta + \frac{b}{2}r^2\right]_0^b d\theta$

$$= b^3\int_0^{2\pi}\left(\frac{1}{3}\sin\theta + \frac{1}{2}\right)d\theta = b^3\pi$$

25. $\displaystyle 8\int_0^{\pi/2}\int_0^2 \frac{r}{2}\sqrt{12-3r^2}\,dr\,d\theta = 8\int_0^{\pi/2}\left[-\frac{1}{18}(12-3r^2)^{3/2}\right]_0^2 d\theta$

$$= 8\int_0^{\pi/2}\frac{4}{3}\sqrt{3}\,d\theta = \frac{16}{3}\sqrt{3}\,\pi$$

27. $\displaystyle\int_0^{2\pi}\int_0^1 r\sqrt{4-r^2}\,dr\,d\theta = \int_0^{2\pi}\left[-\frac{1}{3}(4-r^2)^{3/2}\right]_0^1 d\theta$

$$= \int_0^{2\pi}\left(\frac{8}{3}-\sqrt{3}\right)d\theta = \frac{2}{3}(8-3\sqrt{3})\pi$$

29. $\int_{-\pi/2}^{\pi/2} \int_0^{2\cos\theta} 2r^2 \cos\theta \, dr \, d\theta = \int_{-\pi/2}^{\pi/2} \left[\frac{2}{3} r^3 \cos\theta\right]_0^{2\cos\theta} d\theta$

$$= \int_{-\pi/2}^{\pi/2} \frac{16}{3} \cos^4\theta \, d\theta = \frac{32}{3} \int_0^{\pi/2} \cos^4\theta \, d\theta = \frac{32}{3}\left(\frac{3}{16}\pi\right) = 2\pi$$

Ex. 46, Sect. 8.3

31. $\frac{b}{a} \int_0^\pi \int_0^{a\sin\theta} r\sqrt{a^2 - r^2} \, dr \, d\theta = \frac{b}{a} \int_0^\pi \left[-\frac{1}{3}\left(a^2 - r^2\right)^{3/2}\right]_0^{a\sin\theta} d\theta$

$$= \frac{1}{3} a^2 b \int_0^\pi \left(1 - \cos^3\theta\right) d\theta = \frac{1}{3}\pi a^2 b$$

33. $A = 2 \int_0^{\pi/4} \int_0^{2\cos 2\theta} r \, dr \, d\theta = 2\int_0^{\pi/4} \frac{1}{2}\left(2\cos 2\theta\right)^2 d\theta = 2\int_0^{\pi/4}\left(1 + \cos 4\theta\right) d\theta = \frac{\pi}{2}$

SECTION 17.5

1. $M = \int_{-1}^1 \int_0^1 x^2 \, dy \, dx = \frac{2}{3}$

$x_M M = \int_{-1}^1 \int_0^1 x^3 \, dy \, dx = 0 \implies x_M = 0$

$y_M M = \int_{-1}^1 \int_0^1 x^2 y \, dy \, dx = \int_{-1}^1 \frac{1}{2} x^2 \, dx = \frac{1}{3} \implies y_M = \frac{1/3}{2/3} = \frac{1}{2}$

3. $M = \int_0^1 \int_{x^2}^1 xy \, dy \, dx = \frac{1}{2}\int_0^1 \left(x - x^5\right) dx = \frac{1}{6}$

$x_M M = \int_0^1 \int_{x^2}^1 x^2 y \, dy \, dx = \frac{1}{2}\int_0^1 \left(x^2 - x^6\right) dx = \frac{2}{21} \implies x_M = \frac{2/21}{1/6} = \frac{4}{7}$

$y_M M = \int_0^1 \int_{x^2}^1 xy^2 \, dy \, dx = \frac{1}{3}\int_0^1 \left(x - x^7\right) dx = \frac{1}{8} \implies y_M = \frac{1/8}{1/6} = \frac{3}{4}$

5. $M = \int_0^8 \int_0^{x^{1/3}} y^2 \, dy \, dx = \frac{1}{3}\int_0^8 x \, dx = \frac{32}{3}$

$x_M M = \int_0^8 \int_0^{x^{1/3}} xy^2 \, dy \, dx = \frac{1}{3}\int_0^8 x^2 \, dx = \frac{512}{9} \implies x_M = \frac{512/9}{32/3} = \frac{16}{3}$

$y_M M = \int_0^8 \int_0^{x^{1/3}} y^3 \, dy \, dx = \frac{1}{4}\int_0^8 x^{4/3} \, dx = \frac{96}{7} \implies y_M = \frac{96/7}{32/3} = \frac{9}{7}$

7. $M = \int_0^1 \int_{2x}^{3x} xy \, dy \, dx = \frac{5}{2}\int_0^1 x^3 \, dx = \frac{5}{8}$

$x_M M = \int_0^1 \int_{2x}^{3x} x^2 y \, dy \, dx = \frac{5}{2}\int_0^1 x^4 \, dx = \frac{1}{2} \implies x_M = \frac{1/2}{5/8} = \frac{4}{5}$

$y_M M = \int_0^1 \int_{2x}^{3x} xy^2 \, dy \, dx = \frac{19}{3}\int_0^1 x^4 \, dx = \frac{19}{15} \implies y_M = \frac{19/15}{5/8} = \frac{152}{75}$

9. $M = \int_0^{2\pi} \int_0^{1+\cos\theta} r^2 \, dr \, d\theta = \frac{1}{3} \int_0^{2\pi} (1 + 3\cos\theta + 3\cos^2\theta + \cos^3\theta) \, d\theta = \frac{5\pi}{3}$

$x_M M = \int_0^{2\pi} \int_0^{1+\cos\theta} r^3 \cos\theta \, dr \, d\theta = \frac{1}{4} \int_0^{2\pi} (1+\cos\theta)^4 \cos\theta \, d\theta$

$$= \frac{1}{4} \int_0^{2\pi} \left[\cos\theta + 4\cos^2\theta + 6\cos^3\theta + 4\cos^4\theta + \cos^5\theta \right] d\theta$$

$$= \frac{7\pi}{4}$$

Therefore, $x_M = \dfrac{7\pi/4}{5\pi/3} = \dfrac{21}{20}$.

$y_M M = \int_0^{2\pi} \int_0^{1+\cos\theta} r^3 \sin\theta \, dr \, d\theta = \frac{1}{4} \int_0^{2\pi} (1+\cos\theta)^4 \sin\theta \, d\theta = \frac{1}{4} \left[\frac{1}{5}(1+\cos\theta)^5 \right]_0^{2\pi} = 0$

Therefore, $y_M = 0$.

11. $\Omega : \ -L/2 \le x \le L/2, \quad -W/2 \le y \le W/2$

$I_x = \iint_\Omega \frac{M}{LW} y^2 \, dxdy = \frac{4M}{LW} \int_0^{W/2} \int_0^{L/2} y^2 \, dx \, dy = \frac{1}{12} MW^2$

$\underset{\text{symmetry}}{\Big\uparrow}$

$I_y = \iint_\Omega \frac{M}{LW} x^2 \, dxdy = \frac{1}{12} ML^2, \quad I_z = \iint_\Omega \frac{M}{LW} (x^2 + y^2) \, dxdy = \frac{1}{12} M (L^2 + W^2)$

$K_x = \sqrt{I_x/M} = \frac{W\sqrt{3}}{6}, \quad K_y = \sqrt{I_y/M} = \frac{L\sqrt{3}}{6}$

$K_z = \sqrt{I_z/M} = \frac{\sqrt{3}\sqrt{L^2+W^2}}{6}$

13. $M = \iint_\Omega k\left(x + \frac{L}{2}\right) dxdy = \iint_\Omega \frac{1}{2} kL \, dxdy = \frac{1}{2} kL (\text{ area of } \Omega) = \frac{1}{2} kL^2 W$

$\underset{\text{symmetry}}{\Big\uparrow}$

$x_M M = \iint_\Omega x \left[k\left(x + \frac{L}{2}\right) \right] dxdy = \iint_\Omega \left(kx^2 + \frac{1}{2} Lx \right) dxdy$

$$= \iint_\Omega kx^2 \, dxdy = 4k \int_0^{W/2} \int_0^{L/2} x^2 \, dx \, dy = \frac{1}{12} kWL^3$$

$\underset{\text{symmetry}}{\Big\uparrow} \qquad\qquad\qquad \underset{\text{symmetry}}{\Big\uparrow}$

$$= \tfrac{1}{6}\left(\tfrac{1}{2} kL^2 W\right) L = \tfrac{1}{6} ML; \quad x_M = \tfrac{1}{6} L$$

$y_M M = \iint_\Omega y \left[k\left(x + \frac{L}{2}\right) \right] dxdy = 0; \quad y_M = 0$

$\underset{\text{by symmetry}}{\Big\uparrow}$

15. $I_x = \displaystyle\iint\limits_\Omega \dfrac{4M}{\pi R^2} y^2\, dxdy = \dfrac{4M}{\pi R^2} \displaystyle\int_0^{\pi/2} \int_0^R r^3 \sin^2\theta\, dr\, d\theta$

$\quad = \dfrac{4M}{\pi R^2} \left(\displaystyle\int_0^{\pi/2} \sin^2\theta\, d\theta \right) \left(\displaystyle\int_0^R r^3\, dr \right) = \dfrac{4M}{\pi R^2} \left(\dfrac{\pi}{4} \right) \left(\dfrac{1}{4} R^4 \right) = \dfrac{1}{4} MR^2$

$I_y = \frac{1}{4} MR^2, \quad I_z = \frac{1}{2} MR^2$

$K_x = K_y = \frac{1}{2} R, \quad K_z = R/\sqrt{2}$

17. I_M, the moment of inertia about the vertical line through the center of mass, is

$$\iint\limits_\Omega \dfrac{M}{\pi R^2} \left(x^2 + y^2 \right) dxdy$$

where Ω is the disc of radius R centered at the origin. Therefore

$$I_M = \dfrac{M}{\pi R^2} \int_0^{2\pi} \int_0^R r^3\, dr\, d\theta = \dfrac{1}{2} MR^2.$$

We need $I_0 = \frac{1}{2} MR^2 + d^2 M$ where d is the distance from the center of the disc to the origin. Solving this equation for d, we have $d = \sqrt{I_0 - \frac{1}{2} MR^2} \,/\, \sqrt{M}$.

19. $\Omega: \ 0 \le x \le a, \quad 0 \le y \le b$

$I_x = \displaystyle\iint\limits_\Omega \dfrac{4M}{\pi ab} y^2\, dxdy = \dfrac{4M}{\pi ab} \displaystyle\int_0^a \int_0^{\frac{b}{a}\sqrt{a^2 - x^2}} y^2\, dy\, dx = \dfrac{1}{4} Mb^2$

$I_y = \displaystyle\iint\limits_\Omega \dfrac{4M}{\pi ab} x^2\, dxdy = \dfrac{4M}{\pi ab} \displaystyle\int_0^a \int_0^{\frac{b}{a}\sqrt{a^2 - x^2}} x^2\, dy\, dx = \dfrac{1}{4} Ma^2$

$I_z = \frac{1}{4} M \left(a^2 + b^2 \right)$

21. $I_x = \displaystyle\int_0^1 \int_{x^2}^1 xy^3\, dy\, dx = \dfrac{1}{4} \int_0^1 (x - x^9)\, dx = \dfrac{1}{10}$

$I_y = \displaystyle\int_0^1 \int_{x^2}^1 x^3 y\, dy\, dx = \dfrac{1}{2} \int_0^1 (x^3 - x^7)\, dx = \dfrac{1}{16}$

$I_z = \displaystyle\int_0^1 \int_{x^2}^1 xy(x^2 + y^2)\, dy\, dx = I_x + I_y = \dfrac{13}{80}$

23. $I_x = \displaystyle\int_0^{2\pi} \int_0^{1+\cos\theta} r^4 \sin^2\theta\, dr\, d\theta = \dfrac{1}{5} \int_0^{2\pi} (1 + \cos\theta)^5 \sin^2\theta\, d\theta = \dfrac{33\pi}{40}$

$I_y = \displaystyle\int_0^{2\pi} \int_0^{1+\cos\theta} r^4 \cos^2\theta\, dr\, d\theta = \dfrac{1}{5} \int_0^{2\pi} (1 + \cos\theta)^5 \cos^2\theta\, d\theta = \dfrac{93\pi}{40}$

$I_z = \displaystyle\int_0^{2\pi} \int_0^{1+\cos\theta} r^4\, dr\, d\theta = I_x + I_y = \dfrac{63\pi}{20}$

25. $\Omega: \quad r_1^2 \leq x^2 + y^2 \leq r_2^2, \quad A = \pi\left(r_2^2 - r_1^2\right)$

   (a) Place the diameter on the x-axis.

$$I_x = \iint\limits_\Omega \frac{M}{A}y^2\,dxdy = \frac{M}{A}\int_0^{2\pi}\int_{r_1}^{r_2}\left(r^2\sin^2\theta\right)r\,dr\,d\theta = \frac{1}{4}M\left(r_2^2 + r_1^2\right)$$

   (b) $\frac{1}{4}M\left(r_2^2 + r_1^2\right) + Mr_1^2 = \frac{1}{4}M\left(r_2^2 + 5r_1^2\right)$ (parallel axis theorem)

   (c) $\frac{1}{4}M\left(r_2^2 + r_1^2\right) + Mr_2^2 = \frac{1}{4}M\left(5r_2^2 + r_1^2\right)$

27. $\Omega: \quad r_1^2 \leq x^2 + y^2 \leq r_2^2, \quad A = \pi\left(r_2^2 - r_1^2\right)$

$$I = \iint\limits_\Omega \frac{M}{A}\left(x^2 + y^2\right)dxdy = \frac{M}{A}\int_0^{2\pi}\int_{r_1}^{r_2}r^3\,dr\,d\theta = \frac{1}{2}M(r_2^2 + r_1^2)$$

29. $M = \iint\limits_\Omega k\left(R - \sqrt{x^2 + y^2}\right)dxdy = k\int_0^\pi\int_0^R\left(Rr - r^2\right)dr\,d\theta = \frac{1}{6}k\pi R^3$

$x_M = 0$ by symmetry

$$y_M M = \iint\limits_\Omega y\left[k\left(R - \sqrt{x^2 + y^2}\right)\right]dxdy = k\int_0^\pi\int_0^R\left(Rr^2 - r^3\right)\sin\theta\,dr\,d\theta = \frac{1}{6}kR^4$$

$y_M = R/\pi$

31. Place P at the origin.

$$M = \iint\limits_\Omega k\sqrt{x^2 + y^2}\,dxdy$$

$$= k\int_0^\pi\int_0^{2R\sin\theta}r^2\,dr\,d\theta = \frac{32}{9}kR^3$$

$x_M = 0$ by symmetry

$$y_M M = \iint\limits_\Omega y\left(k\sqrt{x^2 + y^2}\right)dxdy = k\int_0^\pi\int_0^{2R\sin\theta}r^3\sin\theta\,dr\,d\theta = \frac{64}{15}kR^4$$

$y_M = 6R/5$

Answer: the center of mass lies on the diameter through P at a distance $6R/5$ from P.

33. Suppose Ω, a basic region of area A, is broken up into n basic regions $\Omega_1, \cdots, \Omega_n$ with areas $A_1, \cdots, A_n$. Then

$$\bar{x}A = \iint\limits_\Omega x\,dxdy = \sum_{i=1}^n\left(\iint\limits_{\Omega_i} x\,dxdy\right) = \sum_{i=1}^n \bar{x}_i A_i = \bar{x}_1 A_1 + \cdots + \bar{x}_n A_n.$$

The second formula can be derived in a similar manner.

SECTION 17.6

1. They are equal; they both give the volume of T.

3. $\displaystyle\iiint_\Pi \alpha\, dx\, dy\, dz = \alpha \iiint_\Pi dx\, dy\, dz = \alpha\,(\text{volume of } \Pi) = \alpha(a_2 - a_1)(b_2 - b_1)(c_2 - c_1)$

5. Let $P_1 = \{x_0, \cdots, x_m\}$, $P_2 = \{y_0, \cdots, y_n\}$, $P_3 = \{z_0, \cdots, z_q\}$ be partitions of $[0,a]$, $[0,b]$, $[0,c]$ respectively and let $P = P_1 \times P_2 \times P_3$. Note that

$$x_{i-1}y_{j-1} \le \left(\frac{x_i + x_{i-1}}{2}\right)\left(\frac{y_j + y_{j-1}}{2}\right) \le x_i y_j$$

and therefore

$$x_{i-1}y_{j-1}\,\Delta x_i\,\Delta y_j\,\Delta z_k \le \tfrac{1}{4}\left(x_i{}^2 - x_{i-1}^2\right)\left(y_j{}^2 - y_{j-1}^2\right)\Delta z_k \le x_i y_j\,\Delta x_i\,\Delta y_j \Delta z_k.$$

It follows that

$$L_f(P) \le \frac{1}{4}\sum_{i=1}^{m}\sum_{j=1}^{n}\sum_{k=1}^{q}\left(x_i{}^2 - x_{i-1}^2\right)\left(y_j{}^2 - y_{j-1}^2\right)\Delta z_k \le U_f(P).$$

The middle term can be written

$$\frac{1}{4}\left(\sum_{i=1}^{m} x_i{}^2 - x_{i-1}^2\right)\left(\sum_{j=1}^{n} y_j{}^2 - y_{j-1}^2\right)\left(\sum_{k=1}^{q}\Delta z_k\right) = \frac{1}{4}a^2 b^2 c.$$

7. $\overline{x}_1 = a$, $\overline{y}_1 = b$, $\overline{z}_1 = c$; $\overline{x}_0 = A$, $\overline{y}_0 = B$, $\overline{z}_0 = C$

$$\overline{x}_1 V_1 + \overline{x}\, V = \overline{x}_0 V_0 \quad\Longrightarrow\quad a^2 bc + (ABC - abc)\,\overline{x} = A^2 BC$$

$$\Longrightarrow\quad \overline{x} = \frac{A^2 BC - a^2 bc}{ABC - abc}$$

similarly

$$\overline{y} = \frac{AB^2 C - ab^2 c}{ABC - abc}, \quad \overline{z} = \frac{ABC^2 - abc^2}{ABC - abc}$$

9. $\displaystyle M = \iiint_\Pi Kz\, dx\, dy\, dz$

Let $P_1 = \{x_0, \cdots, x_m\}$, $P_2 = \{y_0, \cdots, y_n\}$, $P_3 = \{z_0, \cdots, z_q\}$ be partitions of $[0,a]$ and let $P = P_1 \times P_2 \times P_3$. Note that

$$z_{k-1} \le \tfrac{1}{2}\left(z_k + z_{k-1}\right) \le z_k$$

and therefore

$$Kz_{k-1}\,\Delta x_i\,\Delta y_j\,\Delta z_k \le \tfrac{1}{2}K\,\Delta x_i\,\Delta y_j\left(z_k{}^2 - z_{k-1}^2\right) \le Kz_k\,\Delta x_i\,\Delta y_j \Delta z_k.$$

It follows that

$$L_f(P) \le \frac{1}{2}\,K\sum_{i=1}^{m}\sum_{j=1}^{n}\sum_{k=1}^{q}\Delta x_i\,\Delta y_j\left(z_k{}^2 - z_{k-1}^2\right) \le U_f(P).$$

The middle term can be written

$$\frac{1}{2}K \left(\sum_{i=1}^{m} \Delta x_i \right) \left(\sum_{j=1}^{n} \Delta y_j \right) \left(\sum_{k=1}^{q} z_k{}^2 - z_{k-1}^2 \right) = \frac{1}{2}K(a)(a)(a^2) = \frac{1}{2}Ka^4.$$

$M = \frac{1}{2}Ka^4$ where K is the constant of proportionality for the density function.

11.
$$I_z = \iiint_{\Pi} Kz\left(x^2 + y^2\right) dx dy dz$$

$$= \underbrace{\iiint_{\Pi} Kx^2 z \, dx dy dz}_{I_1} + \underbrace{\iiint_{\Pi} Ky^2 z \, dx dy dz}_{I_2}.$$

We will calculate I_1 using the partitions we used in doing Exercise 9.   Note that

$$x_{i-1}^2 z_{k-1} \le \left(\frac{x_i{}^2 + x_i x_{i-1} + x_{i-1}^2}{3} \right) \left(\frac{z_k + z_{k-1}}{2} \right) \le x_i{}^2 z_k$$

and therefore

$$Kx_{i-1}^2 z_{k-1} \, \Delta x_i \, \Delta y_j \, \Delta z_k \le \tfrac{1}{6} K\left(x_i{}^3 - x_{i-1}^3\right) \Delta y_j \left(z_k{}^2 - z_{k-1}^2\right) \le Kx_i{}^2 z_k{}^2 \, \Delta x_i \, \Delta y_j \, \Delta z_k.$$

It follows that

$$L_f(P) \le \frac{1}{6}K \sum_{i=1}^{m} \sum_{j=1}^{n} \sum_{k=1}^{q} \left(x_i{}^3 - x_{i-1}^3\right) \Delta y_j \left(z_k{}^2 - z_{k-1}^2\right) \le U_f(P).$$

The middle term can be written

$$\frac{1}{6}K \left(\sum_{i=1}^{m} x_i{}^3 - x_{i-1}^3 \right) \left(\sum_{j=1}^{n} \Delta y_j \right) \left(\sum_{k=1}^{q} z_k{}^2 - z_{k-1}^2 \right) = \frac{1}{6}Ka^3(a)(a^2) = \frac{1}{6}Ka^6.$$

Similarly $I_2 = \frac{1}{6}Ka^6$ and therefore $I_z = \frac{1}{3}Ka^6 = \frac{2}{3}\left(\frac{1}{2}Ka^4\right)a^2 = \frac{2}{3}Ma^2.$

by Exercise 9 $\longrightarrow$ ⌐

SECTION 17.7

1. $\displaystyle\int_0^a \int_0^b \int_0^c dx \, dy \, dz = \int_0^a \int_0^b c \, dy \, dz = \int_0^a bc \, dz = abc$

3. $\displaystyle\int_0^1 \int_1^{2y} \int_0^x (x + 2z) \, dz \, dx \, dy = \int_0^1 \int_1^{2y} \left[xz + z^2 \right]_0^x dx \, dy = \int_0^1 \int_1^{2y} 2x^2 \, dx \, dy$

$$= \int_0^1 \left[\frac{2}{3}x^3 \right]_1^{2y} dy = \int_0^1 \left(\frac{16}{3}y^3 - \frac{2}{3} \right) dy = \frac{2}{3}$$

5. $\displaystyle\int_0^2 \int_{-1}^1 \int_0^3 (z - xy) \, dz \, dy \, dx = \int_0^2 \int_{-1}^1 \left[\frac{1}{2}z^2 - xyz \right]_1^3 dy \, dx$

$$= \int_0^2 \int_{-1}^1 (4 - 2xy) \, dy \, dx = \int_0^2 \left[2y - xy^2 \right]_{-1}^1 dx = \int_0^2 8 \, dy = 16$$

7. $\displaystyle\int_0^{\pi/2}\int_0^1\int_0^{\sqrt{1-x^2}} x\cos z\,dy\,dx\,dz = \int_0^{\pi/2}\int_0^1 [xy\cos z]_0^{\sqrt{1-x^2}}\,dx\,dz$

$\displaystyle = \int_0^{\pi/2}\int_0^1 x\sqrt{1-x^2}\cos z\,dx\,dz = \int_0^{\pi/2}\left[-\frac{1}{3}(1-x^2)^{3/2}\cos z\right]_0^1 dz = \frac{1}{3}\int_0^{\pi/2}\cos z\,dz = \frac{1}{3}$

9. $\displaystyle\int_1^2\int_y^{y^2}\int_0^{\ln x} ye^z\,dz\,dx\,dy = \int_1^2\int_y^{y^2}[ye^z]_0^{\ln x}\,dx\,dy$

$\displaystyle = \int_1^2\int_y^{y^2} y(x-1)\,dx\,dy = \int_1^2\left[\frac{1}{2}x^2y - xy\right]_y^{y^2}dy = \int_1^2\left(\frac{1}{2}y^5 - \frac{3}{2}y^3 + y^2\right)dy = \frac{47}{24}$

11. $\displaystyle\iiint_\Pi f(x)g(y)h(z)\,dxdydz = \int_{c_1}^{c_2}\left[\int_{b_1}^{b_2}\left(\int_{a_1}^{a_2} f(x)g(y)h(z)\,dx\right)dy\right]dz$

$\displaystyle = \int_{c_1}^{c_2}\left[\int_{b_1}^{b_2} g(y)h(z)\left(\int_{a_1}^{a_2} f(x)\,dx\right)dy\right]dz$

$\displaystyle = \int_{c_1}^{c_2}\left[h(z)\left(\int_{a_1}^{a_2} f(x)\,dx\right)\left(\int_{b_1}^{b_2} g(y)\,dy\right)dz\right]$

$\displaystyle = \left(\int_{a_1}^{a_2} f(x)\,dx\right)\left(\int_{b_1}^{b_2} g(y)\,dy\right)\left(\int_{c_1}^{c_2} h(z)\,dz\right)$

13. $\displaystyle\left(\int_0^1 x^2\,dx\right)\left(\int_0^2 y^2\,dy\right)\left(\int_0^3 z^2\,dz\right) = \left(\frac{1}{3}\right)\left(\frac{8}{3}\right)\left(\frac{27}{3}\right) = 8$

15. $\displaystyle x_M M = \iiint_\Pi kx^2yz\,dxdydz = k\left(\int_0^a x^2\,dx\right)\left(\int_0^b y\,dy\right)\left(\int_0^c z\,dz\right)$

$\displaystyle = k\left(\tfrac{1}{3}a^3\right)\left(\tfrac{1}{2}b^2\right)\left(\tfrac{1}{2}c^2\right) = \tfrac{1}{12}ka^3b^2c^2.$

By Exercise 14, $M = \frac{1}{8}ka^2b^2c^2$. Therefore $\overline{x} = \frac{2}{3}a$. Similarly, $\overline{y} = \frac{2}{3}b$ and $\overline{z} = \frac{2}{3}c$.

17.

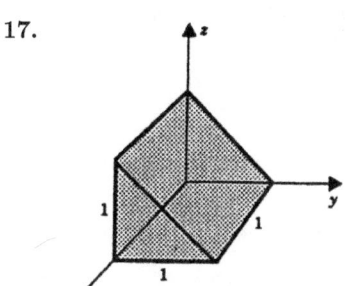

19. center of mass is the centroid

$$\overline{x} = \tfrac{1}{2} \quad \text{by symmetry}$$

$$\overline{y}V = \iiint\limits_{T} y\, dxdydz = \int_0^1 \int_0^1 \int_0^{1-y} y\, dz\, dy\, dx = \int_0^1 \int_0^1 (y - y^2)\, dy\, dx$$

$$= \int_0^1 \left[\frac{1}{2} y^2 - \frac{1}{3} y^3 \right]_0^1 dx = \int_0^1 \frac{1}{6}\, dx = \frac{1}{6}$$

$$\overline{z}V = \iiint\limits_{T} z\, dxdydz = \int_0^1 \int_0^1 \int_0^{1-y} z\, dz\, dy\, dx = \int_0^1 \int_0^1 \frac{1}{2}(1-y)^2\, dy\, dx$$

$$= \frac{1}{2} \int_0^1 \int_0^1 (1 - 2y + y^2)\, dy\, dx = \frac{1}{2} \int_0^1 \left[y - y^2 \frac{1}{3} y^3 \right]_0^1 dx = \frac{1}{2} \int_0^1 \frac{1}{3}\, dx = \frac{1}{6}$$

$$V = \tfrac{1}{2} \text{ (by Exercise 18)}; \quad \overline{y} = \tfrac{1}{3}, \quad \overline{z} = \tfrac{1}{3}$$

21. $$\int_{-r}^{r} \int_{-\phi(x)}^{\phi(x)} \int_{-\psi(x,y)}^{\psi(x,y)} k\left(r - \sqrt{x^2 + y^2 + z^2}\right) dz\, dy\, dx \quad \text{with} \quad \phi(x) = \sqrt{r^2 - x^2},$$

$$\psi(x,y) = \sqrt{r^2 - (x^2 + y^2)}, \quad k \text{ the constant of proportionality}$$

23. $$\int_0^1 \int_{-\sqrt{x-x^2}}^{\sqrt{x-x^2}} \int_{-2x-3y-10}^{1-y^2} dz\, dy\, dx$$

25. $$\int_{-1}^1 \int_{-2\sqrt{2-2x^2}}^{2\sqrt{2-2x^2}} \int_{3x^2+y^2/4}^{4-x^2-y^2/4} k\left(z - 3x^2 - \frac{1}{4} y^2\right) dz\, dy\, dx$$

27. $$\iiint\limits_{T} (x^2 z + y)\, dx\, dy\, dz = \int_0^2 \int_1^3 \int_0^1 (x^2 z + y)\, dx\, dy\, dz = \int_0^2 \int_1^3 \left[\frac{1}{3} x^3 z + xy \right]_0^1 dy\, dz$$

$$= \int_0^2 \int_1^3 \left(\frac{1}{3} z + y \right) dy\, dz = \int_0^2 \left[\frac{1}{3} zy + \frac{1}{2} y^2 \right]_1^3 dz = \int_0^2 \left(\frac{2}{3} z + 4 \right) dz = \frac{28}{3}$$

29. $$\iiint\limits_{T} x^2 y^2 z^2\, dx\, dy\, dz = \int_{-1}^0 \int_0^{y+1} \int_0^1 x^2 y^2 z^2\, dx\, dz\, dy + \int_0^1 \int_0^{1-y} \int_0^1 x^2 y^2 z^2\, dx\, dz\, dy$$

$$= \int_{-1}^0 \int_0^{y+1} \left[\frac{1}{3} x^3 y^2 z^2 \right]_0^1 dz\, dy + \int_0^1 \int_0^{1-y} \left[\frac{1}{3} x^3 y^2 z^2 \right]_0^1 dz\, dy$$

$$= \frac{1}{3} \int_{-1}^0 \int_0^{y+1} y^2 z^2\, dz\, dy + \frac{1}{3} \int_0^1 \int_0^{1-y} [y^2 z^2]_0^1\, dz\, dy$$

$$= \frac{1}{3} \int_{-1}^0 \left[\frac{1}{3} y^2 z^3 \right]_0^{y+1} dy + \frac{1}{3} \int_0^1 \left[\frac{1}{3} y^2 z^3 \right]_0^{1-y} dy$$

$$= \frac{1}{9} \int_{-1}^0 (y^5 + 3y^4 + 3y^3 + y^2)\, dy + \frac{1}{9} \int_0^1 (y^2 - 3y^3 + 3y^4 - y^5)\, dy = \frac{1}{270}$$

31. $\displaystyle \iiint_T y^2\, dx\, dy\, dz = \int_0^3 \int_0^{2-2x/3} \int_0^{6-2x-3y} y^2\, dz\, dy\, dx = \int_0^3 \int_0^{2-2x/3} \left[y^2 z\right]_0^{6-2x-3y} dy\, dx$

$$= \int_0^3 \int_0^{2-2x/3} \left(6y^2 - 2xy^2 - 3y^3\right) dy\, dx$$

$$= \int_0^3 \left[2y^3 - \frac{2}{3}xy^3 - \frac{3}{4}y^4\right]_0^{2-2x/3} dx$$

$$= \frac{1}{4} \int_0^3 \left(2 - \frac{2}{3}x\right) dx = \frac{12}{5}$$

33. $\displaystyle V = \int_0^2 \int_{x^2}^{x+2} \int_0^x dz\, dy\, dx = \int_0^2 \int_{x^2}^{x+2} x\, dy\, dx = \int_0^2 \left(x^2 + 2x - x^3\right) dx = \frac{8}{3}$

$\displaystyle \overline{x}V = \int_0^2 \int_{x^2}^{x+2} \int_0^x x\, dz\, dy\, dx = \int_0^2 \int_{x^2}^{x+2} x^2\, dy\, dx = \int_0^2 \left(x^3 + 2x^2 - x^4\right) dx = \frac{44}{15}$

$\displaystyle \overline{y}V = \int_0^2 \int_{x^2}^{x+2} \int_0^x y\, dz\, dy\, dx = \int_0^2 \int_{x^2}^{x+2} xy\, dy\, dx = \int_0^2 \frac{1}{2}\left(x^3 + 4x^2 + 4x - x^5\right) dx = 6$

$\displaystyle \overline{z}V = \int_0^2 \int_{x^2}^{x+2} \int_0^x z\, dz\, dy\, dx = \int_0^2 \int_{x^2}^{x+2} \frac{1}{2}x^2\, dy\, dx = \int_0^2 \frac{1}{2}\left(x^3 + 2x^2 - x^4\right) dx = \frac{22}{15}$

$\displaystyle \overline{x} = \frac{11}{10}, \quad \overline{y} = \frac{9}{4}, \quad \overline{z} = \frac{11}{20}$

35. $\displaystyle V = \int_{-1}^2 \int_0^3 \int_{2-x}^{4-x^2} dz\, dy\, dx = \frac{27}{2}; \quad (\overline{x}, \overline{y}, \overline{z}) = \left(\frac{1}{2}, \frac{3}{2}, \frac{12}{5}\right)$

37. $\displaystyle V = \int_0^a \int_0^{\phi(x)} \int_0^{\psi(x,y)} dz\, dy\, dx = \frac{1}{6}abc \quad \text{with} \quad \phi(x) = b\left(1 - \frac{x}{a}\right), \quad \psi(x,y) = c\left(1 - \frac{x}{a} - \frac{y}{b}\right)$

$\displaystyle (\overline{x}, \overline{y}, \overline{z}) = \left(\frac{1}{4}a, \frac{1}{4}b, \frac{1}{4}c\right)$

39. $\Pi: 0 \le x \le a, \quad 0 \le y \le b, \quad 0 \le z \le c$

    (a) $\displaystyle I_z = \int_0^a \int_0^b \int_0^c \frac{M}{abc}\left(x^2 + y^2\right) dz\, dy\, dx = \frac{1}{3}M\left(a^2 + b^2\right)$

    (b) $\displaystyle I_M = I_z - d^2 M = \frac{1}{3}M\left(a^2 + b^2\right) - \frac{1}{4}\left(a^2 + b^2\right)M = \frac{1}{12}M\left(a^2 + b^2\right)$

        ⌐ parallel axis theorem (17.5.7)

    (c) $\displaystyle I = I_M + d^2 M = \frac{1}{12}M\left(a^2 + b^2\right) + \frac{1}{4}a^2 M = \frac{1}{3}Ma^2 + \frac{1}{12}Mb^2$

        ⌐ parallel axis theorem (17.5.7)

41. $M = \displaystyle\int_0^1 \int_0^1 \int_0^y k\left(x^2 + y^2 + z^2\right) dz\, dy\, dx = \int_0^1 \int_0^1 k\left(x^2 y + y^3 + \frac{1}{3}\,y^3\right) dy\, dx$

$$= \int_0^1 k\left(\frac{1}{2}\,x^2 + \frac{1}{3}\right) dx = \frac{1}{2}\,k$$

$(x_M, y_M, z_M) = \left(\frac{7}{12}, \frac{34}{45}, \frac{37}{90}\right)$

43. (a) 0 by symmetry

(b) $\displaystyle\iiint_T (a_1 x + a_2 y + a_3 z + a_4)\, dx\, dy\, dz = \iiint_T a_4\, dx\, dy\, dz = a_4 \text{ (volume of ball) } = \frac{4}{3}\,\pi a_4$

by symmetry $\longrightarrow$

45. $V = 8 \displaystyle\int_0^a \int_0^{\sqrt{a^2 - x^2}} \int_0^{\sqrt{a^2 - x^2 - y^2}} dz\, dy\, dx = 8 \int_0^a \int_0^{\sqrt{a^2 - x^2}} \sqrt{a^2 - x^2 - y^2}\, dy\, dx$

polar coordinates $\longrightarrow$

$$= 8 \int_0^{\pi/2} \int_0^a \sqrt{a^2 - r^2}\, r\, dr\, d\theta$$

$$= -4 \int_0^{\pi/2} \left[\frac{2}{3}(a^2 - r^2)^{3/2}\right]_0^a d\theta$$

$$= \frac{8}{3} \int_0^{\pi/2} d\theta = \frac{4}{3}\,\pi\,a^3$$

47. $M = \displaystyle\int_{-2}^2 \int_{-\sqrt{4-x^2}/2}^{\sqrt{4-x^2}/2} \int_{x^2+3y^2}^{4-y^2} k|x|\, dz\, dy\, dx = 4 \int_0^2 \int_0^{\sqrt{4-x^2}/2} \int_{x^2+3y^2}^{4-y^2} kx\, dz\, dy\, dx$

$$= 4k \int_0^2 \int_0^{\sqrt{4-x^2}/2} \left(4x - x^3 - 4xy^2\right) dy\, dx = \frac{4}{3}\,k \int_0^2 x\left(4 - x^2\right)^{3/2} dx = \frac{128}{15}\,k$$

49. $M = \displaystyle\int_{-1}^2 \int_0^3 \int_{2-x}^{4-x^2} k(1+y)\, dz\, dy\, dx = \frac{135}{4}\,k; \quad (x_M, y_M, z_M) = \left(\frac{1}{2}, \frac{9}{5}, \frac{12}{5}\right)$

51. (a) $V = \displaystyle\int_0^6 \int_{z/2}^3 \int_x^{6-x} dy\, dx\, dz$

(b) $V = \displaystyle\int_0^3 \int_0^{2x} \int_x^{6-x} dy\, dz\, dx$

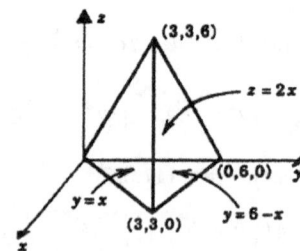

(c) $V = \displaystyle\int_0^6 \int_{z/2}^3 \int_{z/2}^y dx\, dy\, dz + \int_0^6 \int_3^{(12-z)/2} \int_{z/2}^{6-y} dx\, dy\, dz$

53. (a) $V = \iint\limits_{\Omega_{xy}} 2y \, dy dz$

(b) $V = \iint\limits_{\Omega_{xy}} \left(\int_{-y}^{y} dx \right) dy dz$

(c) $V = \int_{0}^{4} \int_{-\sqrt{4-y}}^{\sqrt{4-y}} \int_{-y}^{y} dx \, dz \, dy$

(d) $V = \int_{-2}^{2} \int_{0}^{4-z^2} \int_{-y}^{y} dx \, dy \, dz$

55. (a) $\int_{2}^{4} \int_{3}^{5} \int_{1}^{2} \frac{\ln xy}{z} \, dz \, dy \, dx \cong 6.80703$

(b) $\int_{0}^{4} \int_{1}^{2} \int_{0}^{3} x\sqrt{yz} \, dz \, dy \, dx = \frac{16\sqrt{3}}{3} \left(4\sqrt{2} - 2 \right)$

SECTION 17.8

1. $r^2 + z^2 = 9$ **3.** $z = 2r$ **5.** $4r^2 = z^2$

7. $\int_{0}^{\pi/2} \int_{0}^{2} \int_{0}^{4-r^2} r \, dz \, dr \, d\theta$

$= \int_{0}^{\pi/2} \int_{0}^{2} \left(4r - r^2 \right) dr \, d\theta$

$= \int_{0}^{\pi/2} 4 \, d\theta = 2\pi$

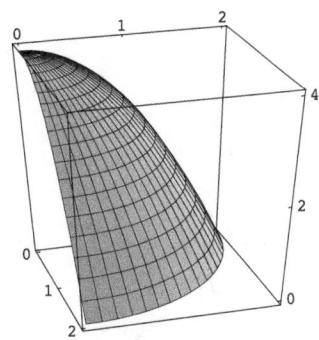

9. $\int_{0}^{2\pi} \int_{0}^{2} \int_{0}^{r^2} r \, dz \, dr \, d\theta$

$= \int_{0}^{2\pi} \int_{0}^{2} r^3 \, dr \, d\theta$

$= \int_{0}^{2\pi} 4 \, d\theta = 8\pi$

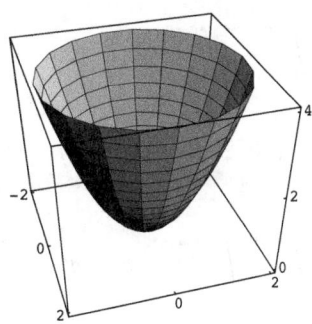

11. $\int_{0}^{1} \int_{0}^{\sqrt{1-x^2}} \int_{0}^{\sqrt{4-(x^2+y^2)}} dz \, dy \, dx = \int_{0}^{\pi/2} \int_{0}^{1} \int_{0}^{\sqrt{4-r^2}} r \, dz \, dr \, d\theta$

$= \int_{0}^{\pi/2} \int_{0}^{1} r\sqrt{4 - r^2} \, dr \, d\theta$

$= \int_{0}^{\pi/2} \left(\frac{8}{3} - \sqrt{3} \right) d\theta = \frac{1}{6} \left(8 - 3\sqrt{3} \right) \pi$

13. $\displaystyle\int_0^3 \int_0^{\sqrt{9-y^2}} \int_0^{\sqrt{9-x^2-y^2}} \frac{1}{\sqrt{x^2+y^2}}\,dz\,dx\,dy = \int_0^{\pi/2} \int_0^3 \int_0^{\sqrt{9-r^2}} \frac{1}{r}\cdot r\,dz\,dr\,d\theta$

$$= \int_0^{\pi/2} \int_0^3 \sqrt{9-r^2}\,dr\,d\theta$$

$$= \int_0^{\pi/2} \left[\frac{r}{2}\sqrt{9-r^2} + \frac{9}{2}\sin^{-1}\frac{r}{3}\right]_0^3 d\theta$$

$$= \frac{9\pi}{4}\int_0^{\pi/2} d\theta = \frac{9}{8}\pi^2$$

15. $\displaystyle\int_0^1 \int_0^{\sqrt{1-x^2}} \int_0^2 \sin(x^2+y^2)\,dz\,dy\,dx = \int_0^{\pi/2} \int_0^1 \int_0^2 \sin(r^2)r\,dz\,dr\,d\theta$

$$= \int_0^{\pi/2} \int_0^1 2r\sin(r^2)\,dr\,d\theta = \tfrac{1}{2}\pi(1-\cos 1) \cong 0.7221$$

17. $(0,1,2) \to (1, \tfrac{1}{2}\pi, 2)$ **19.** $(0,-1,2) \to (1, \tfrac{3}{2}\pi, 2)$

21. $\displaystyle V = \int_{-\pi/2}^{\pi/2} \int_0^{2a\cos\theta} \int_0^r r\,dz\,dr\,d\theta = \int_{-\pi/2}^{\pi/2} \int_0^{2a\cos\theta} r^2\,dr\,d\theta$

$$= \int_{-\pi/2}^{\pi/2} \frac{8}{3}a^3\cos^3\theta\,d\theta = \frac{32}{9}a^3$$

23. $\displaystyle V = \int_{-\pi/2}^{\pi/2} \int_0^{a\cos\theta} \int_0^{a-r} r\,dz\,dr\,d\theta = \int_{-\pi/2}^{\pi/2} \int_0^{a\cos\theta} r(a-r)\,dr\,d\theta$

$$= \int_{-\pi/2}^{\pi/2} a^3\left(\frac{1}{2}\cos^2\theta - \frac{1}{3}\cos^3\theta\right)d\theta = \frac{1}{36}a^3(9\pi - 16)$$

25. $\displaystyle V = \int_{-\pi/2}^{\pi/2} \int_0^{\cos\theta} \int_{r^2}^{r\cos\theta} r\,dz\,dr\,d\theta = \int_{-\pi/2}^{\pi/2} \int_0^{\cos\theta} \left(r^2\cos\theta - r^3\right)dr\,d\theta$

$$= \int_{-\pi/2}^{\pi/2} \frac{1}{12}\cos^4\theta\,d\theta = \frac{1}{32}\pi$$

27. $\displaystyle V = \int_0^{2\pi} \int_0^{1/2} \int_{r\sqrt{3}}^{\sqrt{1-r^2}} r\,dz\,dr\,d\theta = \int_0^{2\pi} \int_0^{1/2} \left(r\sqrt{1-r^2} - r^2\sqrt{3}\right)dr\,d\theta = \frac{1}{3}\pi\left(2 - \sqrt{3}\right)$

29. $\displaystyle V = \int_0^{2\pi} \int_1^3 \int_0^{\sqrt{9-r^2}} r\,dz\,dr\,d\theta = \int_0^{2\pi} \int_1^3 r\sqrt{9-r^2}\,dr\,d\theta = \frac{32}{3}\pi\sqrt{2}$

31. Set the lower base of the cylinder on the xy-plane so that the axis of the cylinder coincides with the z-axis. Assume that the density varies directly as the distance from the lower base.

$$M = \int_0^{2\pi} \int_0^R \int_0^h kzr \, dz \, dr \, d\theta = \frac{1}{2} k\pi R^2 h^2$$

33. $I = I_z = k \int_0^{2\pi} \int_0^R \int_0^h zr^3 \, dr \, d\theta \, dz$

$= \frac{1}{4} k\pi R^4 h^2 = \frac{1}{2} \left(\frac{1}{2} k\pi R^2 h^2 \right) R^2 = \frac{1}{2} MR^2$

$\quad\quad\quad \llcorner$ from Exercise 31

35. Inverting the cone and placing the vertex at the origin, we have

$$V = \int_0^h \int_0^{2\pi} \int_0^{(R/h)z} r \, dr \, d\theta \, dz = \frac{1}{3} \pi R^2 h.$$

37. $I = \dfrac{M}{V} \displaystyle\int_0^h \int_0^{2\pi} \int_0^{(R/h)z} r^3 \, dr \, d\theta \, dz = \dfrac{3}{10} MR^2$

39. $V = \displaystyle\int_0^{2\pi} \int_0^1 \int_0^{1-r^2} r \, dz \, dr \, d\theta = \dfrac{1}{2}\pi$

41. $M = \displaystyle\int_0^{2\pi} \int_0^1 \int_0^{1-r^2} k(r^2 + z^2) r \, dz \, dr \, d\theta = \dfrac{1}{4} k\pi$

SECTION 17.9

1. $\left(\sqrt{3}, \; \frac{1}{4}\pi, \; \cos^{-1}\left[\frac{1}{3}\sqrt{3}\right] \right)$

3. $\left(\frac{3}{4}, \; \frac{3}{4}\sqrt{3}, \; \frac{3}{2}\sqrt{3} \right)$

5. $\rho = \sqrt{2^2 + 2^2 + (2\sqrt{6}/3)^2} = \dfrac{4\sqrt{6}}{3}$

$\phi = \cos^{-1} \left(\dfrac{2\sqrt{6}/3}{4\sqrt{6}/3} \right) = \cos^{-1}(1/2) = \dfrac{\pi}{3}$

$\theta = \tan^{-1}(1) = \dfrac{\pi}{4}$

$(\rho, \theta, \phi) = \left(\dfrac{4\sqrt{6}}{3}, \dfrac{\pi}{4}, \dfrac{\pi}{3} \right)$

7. $x = \rho \sin\phi \cos\theta = 3 \sin 0 \cos(\pi/2) = 0$

$z = \rho \cos\phi = 3 \cos 0 = 3$

$y = \rho \sin\phi \sin\theta = 3 \sin 0 \sin(\pi/2) = 0$

$(x, y, z) = (0, 0, 3)$

9. The circular cylinder $x^2 + y^2 = 1$;   the radius of the cylinder is 1 and the axis is the z-axis.

11. The lower nappe of the circular cone $z^2 = x^2 + y^2$.

13. Horizontal plane one unit above the xy-plane.

15. Sphere of radius 2 centered at the origin:

$$\int_0^{2\pi} \int_0^\pi \int_0^2 \rho^2 \sin\phi \, d\rho \, d\phi \, d\theta = \frac{8}{3} \int_0^{2\pi} \int_0^\pi \sin\phi \, d\phi \, d\theta = \frac{16}{3} \int_0^{2\pi} d\theta = \frac{32\pi}{3}$$

17. The first quadrant portion of the sphere that lies between the x, y-plane and the plane $z = \frac{3}{2}\sqrt{3}$.

$$\int_{\pi/6}^{\pi/2} \int_0^{\pi/2} \int_0^3 \rho^2 \sin\phi \, d\rho \, d\theta \, d\phi = 9 \int_{\pi/6}^{\pi/2} \int_0^{\pi/2} \sin\phi \, d\theta \, d\phi$$

$$= \tfrac{9}{2}\pi \int_{\pi/6}^{\pi/2} \sin\phi \, d\phi$$

$$= \tfrac{9}{2}\pi \left[-\cos\phi \right]_{\pi/6}^{\pi/2} = \tfrac{9}{4}\pi\sqrt{3}$$

19. $\displaystyle \int_0^1 \int_0^{\sqrt{1-x^2}} \int_{\sqrt{x^2+y^2}}^{\sqrt{2-x^2-y^2}} dz \, dy \, dx = \int_0^{\pi/4} \int_0^{\pi/2} \int_0^{\sqrt{2}} \rho^2 \sin\phi \, d\rho \, d\theta \, d\phi$

$$= \tfrac{2}{3}\sqrt{2} \int_0^{\pi/4} \int_0^{\pi/2} \sin\phi \, d\theta \, d\phi$$

$$= \tfrac{\sqrt{2}}{3}\pi \int_0^{\pi/4} \sin\phi \, d\phi = \frac{\sqrt{2}}{6}\pi \left(2 - \sqrt{2}\right)$$

21. $\displaystyle \int_0^3 \int_0^{\sqrt{9-y^2}} \int_0^{\sqrt{9-x^2-y^2}} z\sqrt{x^2 + y^2 + x^2} \, dz \, dx \, dy$

$$= \int_0^{\pi/2} \int_0^{\pi/2} \int_0^3 \rho\cos\phi \cdot \rho \cdot \rho^2 \sin\phi \, d\rho \, d\theta \, d\phi$$

$$= \int_0^{\pi/2} \frac{1}{2}\sin 2\phi \, d\phi \int_0^{\pi/2} d\theta \int_0^3 \rho^4 \, d\rho = \left[-\frac{1}{4}\cos 2\phi\right]_0^{\pi/2} \left(\frac{\pi}{2}\right) \left[\frac{1}{5}\rho^5\right]_0^3$$

$$= \frac{243\pi}{20}$$

23. $\displaystyle V = \int_0^{2\pi} \int_0^\pi \int_0^R \rho^2 \sin\phi \, d\rho \, d\phi \, d\theta = \frac{4}{3}\pi R^3$

25. $\displaystyle V = \int_0^\alpha \int_0^\pi \int_0^R \rho^2 \sin\phi \, d\rho \, d\phi \, d\theta = \frac{2}{3}\alpha R^3$

27. $\displaystyle M = \int_0^{2\pi} \int_0^{\tan^{-1}(r/h)} \int_0^{h\sec\phi} k\rho^3 \sin\phi \, d\rho \, d\phi \, d\theta$

$$= \int_0^{2\pi} \int_0^{\tan^{-1}(r/h)} \frac{kh^4}{4} \tan\phi \sec^3\phi \, d\phi \, d\theta$$

$$= \frac{kh^4}{4} \int_0^{2\pi} \frac{1}{3}\left[\sec^3\phi\right]_0^{\tan^{-1}(r/h)} d\theta = \frac{kh^4}{4} \int_0^{2\pi} \frac{1}{3}\left[\left(\frac{\sqrt{r^2+h^2}}{h}\right)^3 - 1\right] d\theta$$

$$= \frac{1}{6} k\pi h \left(r^2 + h^2\right)^{3/2} - h^3$$

29. center ball at origin; density $= \dfrac{M}{V} = \dfrac{3M}{4\pi R^3}$

(a) $\displaystyle I = \frac{3M}{4\pi R^3} \int_0^{2\pi} \int_0^\pi \int_0^R \rho^4 \sin^3\phi \, d\rho \, d\phi \, d\theta = \frac{2}{5}MR^2$

(b) $I = \frac{2}{5}MR^2 + R^2 M = \frac{7}{5}MR^2$ (parallel axis theorem)

31. center balls at origin; density $= \dfrac{M}{V} = \dfrac{3M}{4\pi \left(R_2{}^3 - R_1{}^3\right)}$

(a) $I = \dfrac{3M}{4\pi \left(R_2{}^3 - R_1{}^3\right)} \displaystyle\int_0^{2\pi} \int_0^\pi \int_{R_1}^{R_2} \rho^4 \sin^3 \phi \, d\rho \, d\phi \, d\theta = \dfrac{2}{5} M \left(\dfrac{R_2{}^5 - R_1{}^5}{R_2{}^3 - R_1{}^3} \right)$

This result can be derived from Exercise 29 without further integration. View the solid as a ball of mass M_2 from which is cut out a core of mass M_1.

$$M_2 = \dfrac{M}{V} V_2 = \dfrac{3M}{4\pi \left(R_2{}^3 - R_1{}^3\right)} \left(\dfrac{4}{3} \pi R_2{}^3 \right) = \dfrac{M R_2{}^3}{R_2{}^3 - R_1{}^3}; \quad \text{similarly} \quad M_1 = \dfrac{M R_1{}^3}{R_2{}^3 - R_1{}^3}.$$

Then

$$I = I_2 - I_1 = \tfrac{2}{5} M_2 R_2{}^2 - \tfrac{2}{5} M_1 R_1{}^2 = \dfrac{2}{5} \left(\dfrac{M R_2{}^3}{R_2{}^3 - R_1{}^3} \right) R_2{}^2 - \dfrac{2}{5} \left(\dfrac{M R_1{}^3}{R_2{}^3 - R_1{}^3} \right) R_1{}^2$$

$$= \dfrac{2}{5} M \left(\dfrac{R_2{}^5 - R_1{}^5}{R_2{}^3 - R_1{}^3} \right).$$

(b) Outer radius R and inner radius R_1 gives

$$\text{moment of inertia} = \dfrac{2}{5} M \left(\dfrac{R^5 - R_1{}^5}{R^3 - R_1{}^3} \right). \qquad [\text{part } (a)]$$

As $R_1 \to R$,

$$\dfrac{R^5 - R_1{}^5}{R^3 - R_1{}^3} = \dfrac{R^4 + R^3 R_1 + R^2 R_1{}^2 + R R_1{}^3 + R_1{}^4}{R^2 + R R_1 + R_1{}^2} \longrightarrow \dfrac{5R^4}{3R^2} = \dfrac{5}{3} R^2.$$

Thus the moment of inertia of spherical shell of radius R is

$$\dfrac{2}{5} M \left(\dfrac{5}{3} R^2 \right) = \dfrac{2}{3} M R^2.$$

(c) $I = \tfrac{2}{3} M R^2 + R^2 M = \tfrac{5}{3} M R^2$ \qquad (parallel axis theorem)

33. $V = \displaystyle\int_0^{2\pi} \int_0^\alpha \int_0^a \rho^2 \sin \phi \, d\rho \, d\phi \, d\theta = \dfrac{2}{3} \pi \left(1 - \cos \alpha\right) a^3$

35. (a) Substituting $\quad x = \rho \sin \phi \cos \theta, \quad y = \rho \sin \phi \sin \theta, \quad z = \rho \cos \phi$
   into $\quad x^2 + y^2 + (z - R)^2 = R^2$
   we have $\quad \rho^2 \sin^2 \phi + (\rho \cos \phi - R)^2 = R^2,$
   which simplifies to $\quad \rho = 2R \cos \phi.$

(b) $0 \le \theta \le 2\pi, \quad 0 \le \phi \le \pi/4, \quad R \sec \phi \le \rho \le 2R \cos \phi$

37.
$$V = \int_0^{2\pi} \int_0^{\pi/4} \int_0^2 \rho^2 \sin \phi \, d\rho \, d\phi \, d\theta + \int_0^{2\pi} \int_{\pi/4}^{\pi/2} \int_0^{2\sqrt{2}\,\cos \phi} \rho^2 \sin \phi \, d\rho \, d\phi \, d\theta$$

$$= \dfrac{1}{3} \left(16 - 6\sqrt{2} \right) \pi$$

39. Encase T in a spherical wedge W. W has spherical coordinates in a box Π that contains S. Define f to be zero outside of T. Then
$$F(\rho, \theta, \phi) = f(\rho \sin \phi \cos \theta, \ \rho \sin \phi \sin \theta, \ \rho \cos \phi)$$
is zero outside of S and
$$\iiint_T f(x,y,z)\,dxdydz = \iiint_W f(x,y,z)\,dxdydz$$
$$= \iiint_\Pi F(\rho,\theta,\phi)\,\rho^2 \sin\phi\,d\rho d\theta d\phi$$
$$= \iiint_S F(\rho,\theta,\phi)\,\rho^2 \sin\phi\,d\rho d\theta d\phi.$$

41. T is the set of all (x,y,z) with spherical coordinates (ρ,θ,ϕ) in the set
$$S: \quad 0 \le \theta \le 2\pi, \quad 0 \le \phi \le \pi/4, \quad R\sec\phi \le \rho \le 2R\cos\phi.$$
T has volume $V = \frac{2}{3}\pi R^3$. By symmetry the $\mathbf{i}, \mathbf{j}$ components of force are zero and
$$\mathbf{F} = \left\{ \frac{3GmM}{2\pi R^3} \iiint_T \frac{z}{(x^2+y^2+z^2)^{3/2}}\,dxdydz \right\} \mathbf{k}$$
$$= \left\{ \frac{3GmM}{2\pi R^3} \iiint_S \left(\frac{\rho\cos\phi}{\rho^3} \right)\rho^2 \sin\phi\,d\rho d\theta d\phi \right\} \mathbf{k}$$
$$= \left\{ \frac{3GmM}{2\pi R^3} \int_0^{2\pi}\int_0^{\pi/4}\int_{R\sec\phi}^{2R\cos\phi} \cos\phi\,\sin\phi\,d\rho\,d\phi\,d\theta \right\} \mathbf{k}$$
$$= \frac{GmM}{R^2}\left(\sqrt{2}-1 \right)\mathbf{k}.$$

SECTION 17.10

1. $ad - bc$ **3.** $2\left(v^2 - u^2\right)$ **5.** $-3u^2v^2$

7. abc **9.** $\rho^2 \sin\phi$

11. $J(\rho,\theta,\phi) = \begin{vmatrix} \sin\phi\cos\theta & \sin\phi\sin\theta & \cos\theta \\ -\rho\sin\phi\sin\theta & \rho\sin\phi\cos\theta & 0 \\ \rho\cos\phi\cos\theta & \rho\cos\phi\sin\theta & -\rho\sin\phi \end{vmatrix} = -\rho^2\sin\phi; \quad |J(\rho,\theta,\phi)| = \rho^2\sin\phi.$

13. Set $u = x+y$, $v = x-y$. Then
$$x = \frac{u+v}{2}, \quad y = \frac{u-v}{2} \quad \text{and} \quad J(u,v) = -\frac{1}{2}.$$
Ω is the set of all (x,y) with uv-coordinates in
$$\Gamma: \quad 0 \le u \le 1, \quad 0 \le v \le 2.$$

Then

$$\iint_\Omega (x^2 - y^2)\, dx\, dy = \iint_\Gamma \frac{1}{2} uv\, du\, dv = \frac{1}{2}\int_0^1\int_0^2 uv\, dv\, du$$

$$= \frac{1}{2}\left(\int_0^1 u\, du\right)\left(\int_0^2 v\, dv\right) = \frac{1}{2}\left(\frac{1}{2}\right)(2) = \frac{1}{2}.$$

15. $\dfrac{1}{2}\displaystyle\int_0^1\int_0^2 u\cos(\pi v)\, dv\, du = \dfrac{1}{2}\left(\int_0^1 u\, du\right)\left(\int_0^2 \cos(\pi v)\, dv\right) = \dfrac{1}{2}\left(\dfrac{1}{2}\right)(0) = 0$

17. Set $u = x - y,\quad v = x + 2y.$ Then

$$x = \frac{2u + v}{3}, \quad y = \frac{v - u}{3}, \quad \text{and} \quad J(u,v) = \frac{1}{3}.$$

Ω is the set of all (x,y) with uv-coordinates in the set

$$\Gamma:\ 0 \le u \le \pi, \quad 0 \le v \le \pi/2.$$

Therefore

$$\iint_\Omega \sin(x - y)\cos(x + 2y)\, dx\, dy = \iint_\Gamma \frac{1}{3}\sin u\cos v\, du\, dv = \frac{1}{3}\int_0^\pi\int_0^{\pi/2}\sin u\cos v\, dv\, du$$

$$= \frac{1}{3}\left(\int_0^\pi \sin u\, du\right)\left(\int_0^{\pi/2}\cos v\, dv\right) = \frac{1}{3}(2)(1) = \frac{2}{3}.$$

19. Set $u = xy,\quad v = y.$ Then

$$x = u/v, \quad y = v \quad \text{and} \quad J(u,v) = 1/v.$$

$$xy = 1, \qquad xy = 4 \quad\Longrightarrow\quad u = 1, \qquad u = 4$$

$$y = x, \qquad y = 4x \quad\Longrightarrow\quad u/v = v, \qquad 4u/v = v \quad\Longrightarrow\quad v^2 = u, \qquad v^2 = 4u$$

Ω is the set of all (x,y) with uv-coordinates in the set

$$\Gamma:\quad 1 \le u \le 4, \quad \sqrt{u} \le v \le 2\sqrt{u}.$$

(a) $\quad A = \displaystyle\iint_\Gamma \frac{1}{v}\, du\, dv = \int_1^4\int_{\sqrt{u}}^{2\sqrt{u}} \frac{1}{v}\, dv\, du = \int_1^4 \ln 2\, du = 3\ln 2$

(b) $\quad \bar{x}A = \displaystyle\int_1^4\int_{\sqrt{u}}^{2\sqrt{u}} \frac{u}{v^2}\, dv\, du = \frac{7}{3};\quad \bar{x} = \frac{7}{9\ln 2}$

$$\bar{y}A = \int_1^4\int_{\sqrt{u}}^{2\sqrt{u}} dv\, du = \frac{14}{3};\quad \bar{y} = \frac{14}{9\ln 2}$$

21. Set $u = x + y,\quad v = 3x - 2y.$ Then

$$x = \frac{2u + v}{5}, \quad y = \frac{3u - v}{5} \quad \text{and} \quad J(u,v) = -\frac{1}{5}.$$

With Γ: $0 \leq u \leq 1$, $0 \leq v \leq 2$

$$M = \int_0^1 \int_0^2 \frac{1}{5} \lambda \, dv \, du = \frac{2}{5} \lambda \quad \text{where} \quad \lambda \text{ is the density.}$$

Then

$$I_x = \int_0^1 \int_0^2 \left(\frac{3u - v}{5} \right)^2 \frac{1}{5} \lambda \, dv \, du = \frac{8\lambda}{375} = \frac{4}{75} \left(\frac{2}{5} \lambda \right) = \frac{4}{75} M,$$

$$I_y = \int_0^1 \int_0^2 \left(\frac{2u + v}{5} \right)^2 \frac{1}{5} \lambda \, dv \, du = \frac{28\lambda}{375} = \frac{14}{75} \left(\frac{2}{5} \lambda \right) = \frac{14}{75} M,$$

$$I_z = I_x + I_y = \frac{18}{75} M.$$

23. Set $u = x - 2y$, $v = 2x + y$.   Then

$$x = \frac{u + 2v}{5}, \quad y = \frac{v - 2u}{5} \quad \text{and} \quad J(u, v) = \frac{1}{5}.$$

Γ is the region between the parabola $v = u^2 - 1$ and the line $v = 2u + 2$.   A sketch of the curves shows that

$$\Gamma : \quad -1 \leq u \leq 3, \quad u^2 - 1 \leq v \leq 2u + 2.$$

Then

$$A = \frac{1}{5} \, (\text{area of } \Gamma) = \frac{1}{5} \int_{-1}^3 \left[(2u + 2) - (u^2 - 1) \right] du = \frac{32}{15}.$$

25. The choice $\theta = \pi/6$ reduces the equation to $13u^2 + 5v^2 = 1$.   This is an ellipse in the uv-plane with area $\pi ab = \pi/\sqrt{65}$.   Since $J(u, v) = 1$, the area of Ω is also $\pi/\sqrt{65}$.

27. $J = abc\rho^2 \sin \phi; \quad V = \int_0^{2\pi} \int_0^\pi \int_0^1 abc\rho^2 \sin \phi \, d\rho \, d\phi \, d\theta = \frac{4}{3} \pi abc$

29.
$$V = \frac{2}{3} \pi abc, \quad \lambda = \frac{M}{V} = \frac{3M}{2\pi abc}$$

$$I_x = \frac{3M}{2\pi abc} \int_0^{2\pi} \int_0^{\pi/2} \int_0^1 \left(b^2 \rho^2 \sin^2 \phi \sin^2 \theta + c^2 \rho^2 \cos^2 \phi \right) abc\rho^2 \sin \phi \, d\rho \, d\phi \, d\theta$$

$$= \tfrac{1}{5} M \left(b^2 + c^2 \right)$$

$$I_y = \tfrac{1}{5} M \left(a^2 + c^2 \right), \quad I_z = \tfrac{1}{5} M \left(a^2 + b^2 \right)$$

PROJECT 17.10

1. (a) $\theta = \tan^{-1} \left[\left(\frac{ay}{bx} \right)^{1/\alpha} \right], \quad r = \left[\left(\frac{x}{a} \right)^{2/\alpha} + \left(\frac{y}{b} \right)^{2/\alpha} \right]^{\alpha/2}$

(b)
$$\left. \begin{array}{c} ar_1 (\cos \theta_1)^\alpha = ar_2 (\cos \theta_2)^\alpha \\ br_1 (\sin \theta_1)^\alpha = br_2 (\sin \theta_2)^\alpha \\ r_1 > 0, \quad 0 < \theta < \tfrac{1}{2}\pi \end{array} \right\} \implies r_1 = r_2, \quad \theta_1 = \theta_2$$

3. (a)

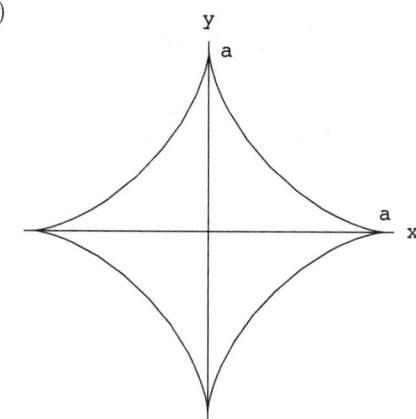

(b) $x = ar\cos^3\theta,\ y = ar\sin^3\theta;\quad x^{\frac{2}{3}} + y^{\frac{2}{3}} = a^{\frac{2}{3}} \Longrightarrow r = 1$ and $x = a\cos^3\theta,\ y = a\sin^3\theta$

$$A = \int_{\frac{\pi}{2}}^{0} y(\theta)x'(\theta)\,d\theta = \int_{\frac{\pi}{2}}^{0} a\sin^3\theta(3a\cos^2\theta[-\sin\theta])\,d\theta$$

$$= 3a^2 \int_{0}^{\frac{\pi}{2}} \sin^4\theta\cos^2\theta\,d\theta = 3a^2 \int_{0}^{\frac{\pi}{2}} (\sin^4\theta - \sin^6\theta)\,d\theta$$

$$= 3a^2 \left[\frac{3 \cdot 1}{4 \cdot 2}\frac{\pi}{2} - \frac{5 \cdot 3 \cdot 1}{6 \cdot 4 \cdot 2}\frac{\pi}{2} \right] \quad \text{(See Exercise 62(b) in 8.3)}$$

$$= \frac{3a^2\pi}{32}$$

(c) Entire area enclosed: $4 \cdot \dfrac{3a^2\pi}{32} = \dfrac{3a^2\pi}{8}$

REVIEW EXERCISES

1. $\displaystyle\int_0^1 \int_y^{\sqrt{y}} xy^2\,dx\,dy = \int_0^1 \left[\frac{1}{2}x^2y^2\right]_y^{\sqrt{y}} dy = \int_0^1 \left(\frac{1}{2}y^3 - \frac{1}{2}y^4\right) dy = \left[\frac{1}{8}y^4 - \frac{1}{10}y^5\right]_0^1 = \frac{1}{40}$

3. $\displaystyle\int_0^1 \int_x^{3x} 2ye^{x^3}\,dy\,dx = \int_0^1 \left[y^2 e^{x^3}\right]_x^{3x} dx = \int_0^1 (9x^2 e^{x^3} - x^2 e^{x^3})\,dx = \left[3e^{x^3} - \frac{1}{3}e^{x^3}\right]_0^1 = \frac{8}{3}e - \frac{8}{3}$

5. $\displaystyle\int_0^{\pi/4} \int_0^{2\sin\theta} r\cos\theta\,dr\,d\theta = \int_0^{\pi/4} \left[\frac{1}{2}r^2\cos\theta\right]_0^{2\sin\theta} d\theta = \int_0^{\pi/4} 2\sin^2\theta\cos\theta\,d\theta = \left[\frac{2}{3}\sin^3\theta\right]_0^{\pi/4} = \frac{\sqrt{2}}{6}$

7. $\displaystyle\int_0^2 \int_0^{2-3x} \int_0^{x+y} x\,dz\,dy\,dx = \int_0^2 \int_0^{2-3x} \left[xz\right]_0^{x+y} dy\,dx = \int_0^2 \int_0^{2-3x} (x^2 + xy)\,dy\,dx$

$$= \int_0^2 \left[x^2 y + \frac{1}{2}xy^2\right]_0^{2-3x} dx = \int_0^2 \left(\frac{3}{2}x^3 - 4x^2 + 2x\right) dx$$

$$= \left[\frac{3}{8}x^4 - \frac{4}{3}x^3 + x^2\right]_0^2 = -\frac{2}{3}$$

9. $\displaystyle\int_{-\frac{\pi}{2}}^{0}\int_{0}^{2\sin\theta}\int_{0}^{r^2} r^2\cos\theta\, dz\, dr\, d\theta = \int_{-\frac{\pi}{2}}^{0}\int_{0}^{2\sin\theta} r^4\cos\theta\, dr\, d\theta = \int_{-\frac{\pi}{2}}^{0}\frac{32}{5}\sin^5\theta\cos\theta\, d\theta = -\frac{16}{15}$

11. $\displaystyle\int_{0}^{1}\int_{y}^{1} e^{x^2}\, dx\, dy = \int_{0}^{1}\int_{0}^{x} e^{x^2}\, dy\, dx = \int_{0}^{1} e^{x^2} y\Big|_{0}^{x}\, dx = \int_{0}^{1} x e^{x^2}\, dx = \frac{1}{2}e^{x^2}\Big|_{0}^{1} = \frac{e-1}{2}$

13. $\displaystyle\int_{0}^{1}\int_{0}^{\sqrt{1-y^2}} \frac{1}{\sqrt{1-y^2}}\, dx\, dy = \int_{0}^{1} dy = 1$

15. $\displaystyle\int_{0}^{1}\int_{0}^{\sqrt{1-x^2}} xy\, dy\, dx = \int_{0}^{1}\left[\frac{1}{2}xy^2\right]_{0}^{\sqrt{1-x^2}} dx = \int_{0}^{1}\left(\frac{1}{2}x - \frac{1}{2}x^3\right) dx = \frac{1}{8}$

17. $\displaystyle\int_{0}^{2}\int_{x}^{3x-x^2} (x^2 - xy)\, dy\, dx = \int_{0}^{2}\left[x^2 y - \frac{1}{2}xy^2\right]_{x}^{3x-x^2} dx = \int_{0}^{2}\left(2x^4 - 2x^3 - \frac{1}{2}x^5\right) dx = -\frac{8}{15}$

19. $\displaystyle\int_{0}^{2}\int_{0}^{y}\int_{0}^{\sqrt{4-y^2}} 2xyz\, dz\, dx\, dy = \int_{0}^{2}\int_{x}^{2} xyz^2\Big|_{0}^{\sqrt{4-y^2}}\, dx\, dy = \int_{0}^{2}\int_{0}^{2} xy(4-y^2)\, dx\, dy$

$$= \int_{0}^{2}\left[\frac{1}{2}y^3(4-y^2)\right] dy = \frac{8}{3}$$

21. $\displaystyle\int_{0}^{2}\int_{0}^{\sqrt{4-x^2}}\int_{0}^{\sqrt{4-x^2-y^2}} xy\, dz\, dy\, dx = \int_{0}^{2}\int_{0}^{\sqrt{4-x^2}}\left[xyz\right]_{0}^{\sqrt{4-x^2-y^2}} dy\, dx$

$$= \int_{0}^{2}\left[-\frac{1}{3}x\sqrt{4-x^2-y^2}\,\right]_{0}^{\sqrt{4-x^2}} dx$$

$$= \int_{0}^{2}\frac{1}{3}x(4-x^2)^{\frac{3}{2}}\, dx = \frac{32}{15}$$

23. $\displaystyle\int_{0}^{2}\int_{0}^{\sqrt{4-y^2}} e^{\sqrt{x^2+y^2}}\, dx\, dy = \int_{0}^{\pi/2}\int_{0}^{2} e^r r\, dr\, d\theta = \frac{\pi}{2}\int_{0}^{2} re^r\, dr = \frac{\pi}{2}\left[re^r - e^r\right]_{0}^{2} = \frac{\pi}{2}(e^2+1)$

25. $\displaystyle V = \int_{0}^{3}\int_{0}^{2\pi} (9-r^2)r\, dr\, d\theta = 2\pi\int_{0}^{3} (9-r^2)r\, dr = \frac{81\pi}{2}$

27. $\displaystyle V = \int_{0}^{1}\int_{0}^{1-x} (x^2+y^2)\, dy\, dx = \int_{0}^{1}\left[x^2 - x^3 + \frac{1}{3}(1-x)^3\right] dx = \left[\frac{1}{3}x^3 - \frac{1}{4}x^4 - \frac{1}{12}(1-x)^4\right]_{0}^{1} = \frac{1}{6}$

29. $\displaystyle M = \int_{-\pi/2}^{\pi/2}\int_{0}^{\cos x} y\, dy\, dx = \int_{-\pi/2}^{\pi/2}\frac{1}{2}\cos^2 x\, dx = \frac{\pi}{4}$

$\displaystyle x_M M = \int_{-\pi/2}^{\pi/2}\int_{0}^{\cos x} xy\, dy\, dx = 0 \text{ by symmetry}$

$\displaystyle y_M M = \int_{-\pi/2}^{\pi/2}\int_{0}^{\cos x} y^2\, dy\, dx = \int_{-\pi/2}^{\pi/2}\frac{1}{3}\cos^3 x\, dx = \frac{4}{9}$

The center of mass is: $\left(0, \frac{16}{9\pi}\right)$

31. $M = \displaystyle\int_0^{\pi/2} \int_r^R u^3 \, du \, d\theta = \frac{\pi}{8}(R^4 - r^4);$ (polar coordinates $[u, \theta]$)

By symmetry, $\bar{x} = \bar{y}$.

$x_M M = \displaystyle\int_0^{\pi/2} \int_r^R u^4 \cos\theta \, du \, d\theta = \frac{1}{5}(R^5 - r^5);$ $x_M = \dfrac{8(R^5 - r^5)}{5\pi(R^4 - r^4)}$

33. Introduce a coordinate system as shown in the figure.

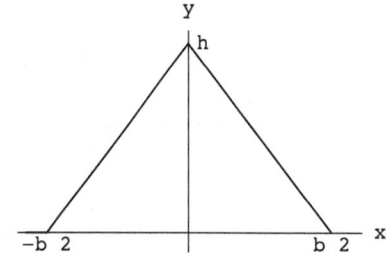

(a) $A = \frac{1}{2}bh;$ by symmetry, $\bar{x} = 0$

$\bar{y}\,A = \displaystyle\int_{-b/2}^0 \int_0^{\frac{2h}{b}(x+\frac{b}{2})} y \, dy \, dx + \int_0^{\frac{b}{2}} \int_0^{-\frac{2h}{b}(x-\frac{b}{2})} y \, dy \, dx$

$= \dfrac{bh^2}{6} \quad \Longrightarrow \quad \bar{y} = \dfrac{h}{3}$

(b) $I = \displaystyle\int_{-b/2}^0 \int_0^{\frac{2h}{b}(x+\frac{b}{2})} \lambda y^2 \, dy \, dx + \int_0^{b/2} \int_0^{-\frac{2h}{b}(x-\frac{b}{2})} \lambda y^2 \, dy \, dx = \dfrac{\lambda b h^3}{12} = \frac{1}{6}Mh^2$

(c) $I = 2\displaystyle\int_0^{b/2} \int_0^{-\frac{2h}{b}(x-\frac{b}{2})} \lambda x^2 \, dx \, dy = \dfrac{1}{48}\lambda h b^3 = \frac{1}{24}Mb^2$

35. $V = \displaystyle\int_0^2 \int_0^x \int_0^{2x+2y+1} dz \, dy \, dx = \int_0^2 \int_0^x (2x + 2y + 1) \, dy \, dx = 10$

37. The curve of intersection of the two surfaces is the circle: $x^2 + y^2 = 4, \; x = 3$

$V = \displaystyle\int_{-2}^2 \int_{-\sqrt{4-x^2}}^{\sqrt{4-x^2}} \int_{2x^2+y^2}^{12-x^2-2y^2} dz \, dy \, dx = \int_{-2}^2 \int_{-\sqrt{4-x^2}}^{\sqrt{4-x^2}} 3\left(4 - x^2 - y^2\right) dy \, dx$

$= 3\displaystyle\int_0^{2\pi} \int_0^2 \left(4 - r^2\right) r \, dr \, d\theta$

$= 3\displaystyle\int_0^{2\pi} \left[2r^2 - \tfrac{1}{4}r^4\right]_0^2 d\theta = 12 \int_0^{2\pi} = 24\pi$

39. $V = \displaystyle\int_0^{2\pi} \int_0^{\pi/3} \int_{\sec\phi}^2 \rho^2 \sin\phi \, d\rho \, d\phi \, d\theta = \int_0^{2\pi} \int_0^{\pi/3} \left[\tfrac{1}{3}\rho^3\right]_{\sec\phi}^2 d\phi \, d\theta$

$= \dfrac{1}{3}\displaystyle\int_0^{2\pi} \int_0^{\pi/3} \left(8 - \sec^3\phi\right) \sin\phi \, d\phi \, d\theta$

$= \dfrac{1}{3}\displaystyle\int_0^{2\pi} \left[-8\cos\phi - \tfrac{1}{2}\sec^2\phi\right]_0^{\pi/3} d\theta$

$= \dfrac{1}{3}\left(\dfrac{5}{2}\right)(2\pi) = \dfrac{5\pi}{3}$

41. $V = \displaystyle\int_0^{2\pi} \int_{\pi/4}^{\pi/2} \int_0^1 \rho^2 \sin\phi \, d\rho \, d\phi \, d\theta = \int_0^{2\pi} \int_{\pi/4}^{\pi/2} \dfrac{1}{3}\sin\phi \, d\phi \, d\theta = \dfrac{\sqrt{2}\pi}{3}$

43. (a) $V = \displaystyle\int_0^1 \int_0^x \int_0^{\sqrt{1-x^2}} dz\, dy\, dx + \int_0^1 \int_0^y \int_0^{\sqrt{1-y^2}} dz\, dx\, dy$

$$= 2 \int_0^1 \int_0^x \sqrt{1-x^2}\, dy\, dx = 2 \int_0^1 x\sqrt{1-x^2}\, dx = \frac{2}{3}$$

By symmetry, $\overline{x} = \overline{y}$.

$$\overline{x}\, V = \int_0^1 \int_0^x \int_0^{\sqrt{1-x^2}} x\, dz\, dy\, dx + \int_0^1 \int_0^y \int_0^{\sqrt{1-y^2}} x\, dz\, dx\, dy$$

For the first integral:

$$\int_0^1 \int_0^x \int_0^{\sqrt{1-x^2}} x\, dz\, dy\, dx = \int_0^1 \int_0^x x\sqrt{1-x^2}\, dy\, dx$$

$$= \int_0^1 x^2\sqrt{1-x^2}\, dx = \int_0^{\pi/2} \sin^2 u\, \cos^2 u\, du = \frac{\pi}{16}$$

$$x = \sin u \underline{\quad\uparrow\quad}$$

For the second integral:

$$\int_0^1 \int_0^y \int_0^{\sqrt{1-y^2}} x\, dz\, dx\, dy = \int_0^1 \int_0^y x\sqrt{1-y^2}\, dx\, dy = \int_0^1 \frac{1}{2} y^2 \sqrt{1-y^2}\, dy = \frac{\pi}{32}$$

Thus, $\overline{x}\, V = \dfrac{3\pi}{32} \implies \overline{x} = \overline{y} = \dfrac{9\pi}{64}$

Now calculate $\overline{z}$:

$$\overline{z}\, V = \int_0^1 \int_0^x \int_0^{\sqrt{1-x^2}} z\, dz\, dy\, dx + \int_0^1 \int_0^y \int_0^{\sqrt{1-y^2}} z\, dz\, dx\, dy;$$

$$\int_0^1 \int_0^x \int_0^{\sqrt{1-x^2}} z\, dz\, dy\, dx = \int_0^1 \int_0^x \frac{1}{2}(1-x^2)\, dy\, dx = \frac{1}{2} \int_0^1 (x - x^3)\, dx = \frac{1}{8}$$

and similarly,

$$\int_0^1 \int_0^y \int_0^{\sqrt{1-y^2}} z\, dz\, dy\, dx = \frac{1}{8}.$$

Therefore, $\overline{z}\, V = \dfrac{1}{4} \implies \overline{z} = \dfrac{3}{8}$

(b) $I_z = \displaystyle\int_0^1 \int_0^x \int_0^{\sqrt{1-x^2}} \lambda \left(\sqrt{x^2+y^2}\right)^2 dz\, dy\, dx + \int_0^1 \int_0^y \int_0^{\sqrt{1-y^2}} \lambda \left(\sqrt{x^2+y^2}\right)^2 dz\, dx\, dy;$

$$\int_0^1 \int_0^x \int_0^{\sqrt{1-x^2}} \lambda \left(\sqrt{x^2+y^2}\right)^2 dz\, dy\, dx = \int_0^{\pi/4} \int_0^{\sec\theta} \int_0^{r\sin\theta} \lambda r^3 dz\, dr\, d\theta = \frac{3}{20}\lambda$$

and $\displaystyle\int_0^1 \int_0^y \int_0^{\sqrt{1-y^2}} \lambda(\sqrt{x^2+y^2})^2 dz\, dx\, dy = \frac{3}{20}\lambda \implies I_z = \frac{3}{10}\lambda$

45. Denote polar coordinates by $[u, \theta]$.

(a) $M = \displaystyle\int_0^{2\pi} \int_0^r \int_0^h u^3 \, dz \, du \, d\theta = 2\pi h \int_0^r u^3 \, du = \dfrac{\pi h r^4}{2}$

(b) By symmetry, $x_M = y_M = 0$

(c) $z_M M = \displaystyle\int_0^{2\pi} \int_0^r \int_0^h u^3 z \, dz \, du \, d\theta = \dfrac{\pi h^2 r^4}{4} \implies z_M = h/2$

47. (a) $M = \displaystyle\int_0^1 \int_0^{2\pi} \int_r^1 r^2 \, dz \, d\theta \, dr = 2\pi \int_0^1 \int_r^1 r^2 \, dz \, dr = 2\pi \int_0^1 r^2(1-r) \, dr = \dfrac{\pi}{6}$

(b) By symmetry, $x_M = y_M = 0$

$z_M M = \displaystyle\int_0^1 \int_0^{2\pi} \int_r^1 r^2 z \, dz \, d\theta \, dr = \pi \int_0^1 r^2(1-r^2) \, dr = \dfrac{2\pi}{15} \implies z_M = \dfrac{4}{5}$

(c) $I_z = \displaystyle\int_0^1 \int_0^{2\pi} \int_r^1 r^4 \, dz \, d\theta \, dr = \dfrac{\pi}{15}$

49. $J(u, v) = \begin{vmatrix} e^u \cos v & e^u \sin v \\ -e^u \sin v & e^u \cos v \end{vmatrix} = e^{2u}$

51. Set $\quad x = \dfrac{v-u}{2}, \quad y = \dfrac{v+u}{2} \implies u = y - x, \quad v = y + x, \quad 1 \leq v \leq 2, \quad J = -\dfrac{1}{2}$

at $x = 0$, $y = u$, $y = v \implies u = v$

at $y = 0$, $-x = u$, $x = v \implies u = -v$

$\displaystyle\iint_\Omega \cos\left(\dfrac{y-x}{y+x}\right) dx \, dy = \int_1^2 \int_{-v}^v \dfrac{1}{2} \cos\left(\dfrac{u}{v}\right) du \, dv = \int_i^2 v \sin 1 \, dv = \dfrac{3}{2} \sin 1$

CHAPTER 18

SECTION 18.1

1. (a) $\mathbf{h}(x,y) = y\,\mathbf{i} + x\,\mathbf{j}; \quad \mathbf{r}(u) = u\,\mathbf{i} + u^2\,\mathbf{j}, \quad u \in [0,1]$

    $x(u) = u, \quad y(u) = u^2; \quad x'(u) = 1, \quad y'(u) = 2u$

    $\mathbf{h}(\mathbf{r}(u)) \cdot \mathbf{r}'(u) = y(u)\,x'(u) + x(u)\,y'(u) = u^2(1) + u(2u) = 3u^2$

    $\displaystyle \int_C \mathbf{h}(\mathbf{r}) \cdot d\mathbf{r} = \int_0^1 3u^2\,du = 1$

    (b) $h(x,y) = y\,\mathbf{i} + x\,\mathbf{j}; \quad \mathbf{r}(u) = u^3\,\mathbf{i} - 2u\,\mathbf{j}, \quad u \in [0,1]$

    $x(u) = u^3, \quad y(u) = -2u; \quad x'(u) = 3u^2, \quad y'(u) = -2$

    $\mathbf{h}(\mathbf{r}(u)) \cdot \mathbf{r}'(u) = y(u)\,x'(u) + x(u)\,y'(u) = (-2u)(3u^2) + u^3(-2) = -8u^3$

    $\displaystyle \int_C \mathbf{h}(\mathbf{r}) \cdot d\mathbf{r} = \int_0^1 -8u^3\,du = -2$

3. $h(x,y) = y\,\mathbf{i} + x\,\mathbf{j}; \quad \mathbf{r}(u) = \cos u\,\mathbf{i} - \sin u\,\mathbf{j}, \quad u \in [0, 2\pi]$

    $x(u) = \cos u, \quad y(u) = -\sin u; \quad x'(u) = -\sin u, \quad y'(u) = -\cos u$

    $\mathbf{h}(\mathbf{r}(u)) \cdot \mathbf{r}'(u) = y(u)\,x'(u) + x(u)\,y'(u) = \sin^2 u - \cos^2 u$

    $\displaystyle \int_C \mathbf{h}(\mathbf{r}) \cdot d\mathbf{r} = \int_0^{2\pi} (\sin^2 u - \cos^2 u)\,du = 0$

5. (a) $\mathbf{r}(u) = (2-u)\,\mathbf{i} + (3-u)\,\mathbf{j}, \quad u \in [0,1]$

    $\displaystyle \int_C \mathbf{h}(\mathbf{r}) \cdot d\mathbf{r} = \int_0^1 (-5 + 5u - u^2)\,du = -\frac{17}{6}$

    (b) $\mathbf{r}(u) = (1+u)\,\mathbf{i} + (2+u)\,\mathbf{j}, \quad u \in [0,1]$

    $\displaystyle \int_C \mathbf{h}(\mathbf{r}) \cdot d\mathbf{r} = \int_0^1 (1 + 3u + u^2)\,du = \frac{17}{6}$

7. $C = C_1 \cup C_2 \cup C_3 \quad$ where,

    $C_1 : \mathbf{r}(u) = (1-u)(-2\,\mathbf{i}) + u(2\,\mathbf{i}) = (4u - 2)\,\mathbf{i}, \quad u \in [0,1]$

    $C_2 : \mathbf{r}(u) = (1-u)(2\,\mathbf{i}) + u(2\,\mathbf{j}) = (2 - 2u)\,\mathbf{i} + 2u\,\mathbf{j}, \quad u \in [0,1]$

    $C_3 : \mathbf{r}(u) = (1-u)(2\,\mathbf{j}) + u(-2\,\mathbf{i}) = -2u\,\mathbf{i} + (2 - 2u)\,\mathbf{j}, \quad u \in [0,1]$

    $\displaystyle \int_C = \int_{C_1} + \int_{C_2} + \int_{C_3} = 0 + (-4) + (-4) = -8$

9.

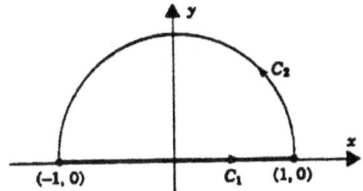

$$C_1 : \mathbf{r}(u) = (-1 + 2u)\,\mathbf{i}, \quad u \in [0, 1]$$

$$C_2 : \mathbf{r}(u) = \cos u\,\mathbf{i} + \sin u\,\mathbf{j}, \quad u \in [0, \pi]$$

$$\int_C = \int_{C_1} + \int_{C_2} = 0 + (-\pi) = -\pi$$

11. (a) $\mathbf{r}(u) = u\,\mathbf{i} + u\,\mathbf{j} + u\,\mathbf{k}, \quad u \in [0, 1]$

$$\int_C \mathbf{h}(\mathbf{r}) \cdot d\mathbf{r} = \int_0^1 3u^2 \, du = 1$$

(b) $\displaystyle \int_C \mathbf{h}(\mathbf{r}) \cdot d\mathbf{r} = \int_0^1 (2u^3 + u^5 + 3u^6)\, du = \frac{23}{21}$

13. (a) $\mathbf{r}(u) = 2u\,\mathbf{i} + 3u\,\mathbf{j} - u\,\mathbf{k}, \quad u \in [0, 1]$

$$\int_C \mathbf{h}(\mathbf{r}) \cdot d\mathbf{r} = \int_0^1 (2\cos 2u + 3\sin 3u + 3u^2)\, du = \left[\sin 2u - \cos 3u + u^3 \right]_0^1 = 2 + \sin 2 - \cos 3$$

(b) $\displaystyle \int_C \mathbf{h}(\mathbf{r}) \cdot d\mathbf{r} = \int_0^1 \left(2u\cos u^2 + 3u^2 \sin u^3 - u^4 \right) du = \left[\sin u^2 - \cos u^3 - \frac{1}{5} u^5 \right]_0^1 = \frac{4}{5} + \sin 1 - \cos 1$

15. $\mathbf{r}(u) = u\,\mathbf{i} + u^2\,\mathbf{j}, \quad u \in [0, 2]$

$$\int_C \mathbf{F}(\mathbf{r}) \cdot d\mathbf{r} = \int_0^2 \left[(u + 2u^2) + (2u + u^2)2u \right] du = \int_0^2 \left(2u^3 + 6u^2 + u \right) du = 26$$

17. $\mathbf{r}(u) = (1 - u)(\mathbf{j} + 4\,\mathbf{k}) + u(\mathbf{i} - 4\,\mathbf{k})$

$$= u\mathbf{i} + (1 - u)\mathbf{j} + (4 - 8u)\mathbf{k}, \quad u \in [0, 1]$$

$$\int_C \mathbf{F}(\mathbf{r}) \cdot d\mathbf{r} = \int_0^1 (-32u + 97u^2 - 64u^3)\, du = \frac{1}{3}$$

19. $\mathbf{r}(u) = \cos u\,\mathbf{i} + \sin u\,\mathbf{j} + u\,\mathbf{k}, \quad u \in [0, 2\pi]$

$$\int_C \mathbf{F}(\mathbf{r}) \cdot d\mathbf{r} = \int_0^{2\pi} \left[-\cos^2 u \sin u + \cos^2 u \sin u + u^2 \right] du = \int_0^{2\pi} u^2 \, du = \frac{8\pi^3}{3}$$

21.
$$\int_C \mathbf{q} \cdot d\mathbf{r} = \int_a^b [\mathbf{q} \cdot \mathbf{r}'(u)]\,du = \int_a^b \frac{d}{du}\,[\mathbf{q} \cdot \mathbf{r}(u)]\,du$$
$$= [\mathbf{q} \cdot \mathbf{r}(b)] - [\mathbf{q} \cdot \mathbf{r}(a)]$$
$$= \mathbf{q} \cdot [\mathbf{r}(b) - \mathbf{r}(a)]$$

$$\int_C \mathbf{r} \cdot d\mathbf{r} = \int_a^b [\mathbf{r}(u) \cdot \mathbf{r}'(u)]\,du$$
$$= \frac{1}{2}\int_a^b \|\mathbf{r}\|\,d\|\mathbf{r}\| \quad \text{(see Exercise 57, Section 14.1)}$$
$$= \frac{1}{2}\left(\|\mathbf{r}(b)\|^2 - \|\mathbf{r}(a)\|^2\right)$$

23. $\displaystyle \int_C \mathbf{f}(\mathbf{r}) \cdot d\mathbf{r} = \int_a^b [\mathbf{f}(\mathbf{r}(u)) \cdot \mathbf{r}'(u)]\,du = \int_a^b [f(u)\,\mathbf{i} \cdot \mathbf{i}]\,du = \int_a^b f(u)\,du$

25. $E : \mathbf{r}(u) = a\cos u\,\mathbf{i} + b\sin u\,\mathbf{j}, \quad u \in [0, 2\pi]$

$$W = \int_0^{2\pi}\left[\left(-\frac{1}{2}b\sin u\right)(-a\sin u) + \left(\frac{1}{2}a\cos u\right)(b\cos u)\right]du = \frac{1}{2}\int_0^{2\pi} ab\,du = \pi ab$$

If the ellipse is traversed in the opposite direction, then $W = -\pi ab$. In both cases $|W| = \pi ab = $ area of the ellipse.

27. $\mathbf{r}(t) = \alpha t\,\mathbf{i} + \beta t^2\,\mathbf{j} + \gamma t^3\,\mathbf{k}$

$\mathbf{r}'(t) = \alpha\,\mathbf{i} + 2\beta t\,\mathbf{j} + 3\gamma t^2\,\mathbf{k}$

force at time $t = m\mathbf{r}''(t) = m(2\beta\mathbf{j} + 6\gamma t\mathbf{k})$

$$W = \int_0^1 [m(2\beta\,\mathbf{j} + 6\gamma t\,\mathbf{k}) \cdot (\alpha\,\mathbf{i} + 2\beta t\mathbf{j} + 3\gamma t^2\,\mathbf{k})]\,dt$$
$$= m\int_0^1 (4\beta^2 t + 18\gamma^2 t^3)\,dt = \left(2\beta^2 + \frac{9}{2}\gamma^2\right)m$$

29. Take $C : \mathbf{r}(t) = r\cos t\,\mathbf{i} + r\sin t\,\mathbf{j}, \quad t \in [0, 2\pi]$

$$\int_C \mathbf{v}(\mathbf{r}) \cdot d\mathbf{r} = \int_0^{2\pi} [\mathbf{v}(\mathbf{r}(t)) \cdot \mathbf{r}'(t)]\,dt$$
$$= \int_0^{2\pi} [f(x(t), y(t))\,\mathbf{r}(t) \cdot \mathbf{r}'(t)]\,dt$$
$$= \int_0^{2\pi} f(x(t), y(t))\,[\mathbf{r}(t) \cdot \mathbf{r}'(t)]\,dt = 0$$

since for the circle $\mathbf{r}(t) \cdot \mathbf{r}'(t) = 0$ identically. The circulation is zero.

31. (a) $\mathbf{r}(u) = (1-u)(\mathbf{i}+2\mathbf{k}) + u(\mathbf{i}+3\mathbf{j}+2\mathbf{k}) = \mathbf{i}+3u\mathbf{j}+2\mathbf{k}, \quad u \in [0,1]$.

$$\int_C \mathbf{F}(\mathbf{r}) \cdot d\mathbf{r} = \int_0^1 \frac{9uk}{(5+9u^2)^{3/2}}\, du = \left[\frac{-k}{\sqrt{5+9u^2}}\right]_0^1 = \frac{k}{\sqrt{5}} - \frac{k}{\sqrt{14}}$$

(b) Let C be an arc on the sphere $\|\mathbf{r}\| = r = 5$.

$$\int_C \mathbf{F}(\mathbf{r}) \cdot d\mathbf{r} = \int_{C_2} \frac{k\mathbf{r}}{\|\mathbf{r}\|^3} \cdot d\mathbf{r}$$

$$= \frac{k}{5^3} \int_{C_2} \mathbf{r} \cdot d\mathbf{r} = \frac{k}{5^3} \int_{C_2} \|\mathbf{r}\|\, d\|\mathbf{r}\| \quad \text{(see Exercise 57, Section 14.1)}$$

$$= \frac{k}{5^3} \left[\frac{1}{2}\|\mathbf{r}\|^2\right]_{(3,4,0)}^{(0,4,3)} = 0$$

33. $\mathbf{r}(u) = u\mathbf{i} + \alpha u(1-u)\mathbf{j}, \quad \mathbf{r}'(u) = \mathbf{i} + \alpha(1-2u)\mathbf{j}, \quad u \in [0,1]$

$$W(\alpha) = \int_C \mathbf{F}(\mathbf{r}) \cdot d\mathbf{r} = \int_0^1 \left[(\alpha^2 u^2(1-u)^2 + 1] + [u + \alpha u(1-u)]\alpha(1-2u)\right] dx$$

$$= \int_0^1 \left[1 + (\alpha+\alpha^2)u - (2\alpha+2\alpha^2)u^2 + \alpha^2 u^4\right] du = 1 - \frac{1}{6}\alpha + \frac{1}{30}\alpha^2$$

$$W'(\alpha) = -\frac{1}{6} + \frac{1}{15}\alpha \quad \Longrightarrow \quad \alpha = \frac{15}{6} = \frac{5}{2}$$

The work done by $\mathbf{F}$ is a minimum when $\alpha = 5/2$.

SECTION 18.2

1. $\mathbf{h}(x,y) = \nabla f(x,y) \quad \text{where} \quad f(x,y) = \frac{1}{2}(x^2+y^2)$

C is closed $\quad \Longrightarrow \quad \displaystyle\int_C \mathbf{h}(\mathbf{r}) \cdot d\mathbf{r} = 0$

3. $\mathbf{h}(x,y) = \nabla f(x,y) \quad \text{where} \quad f(x,y) = x\cos\pi y; \quad \mathbf{r}(0) = \mathbf{0}, \quad \mathbf{r}(1) = \mathbf{i} - \mathbf{j}$

$$\int_C \mathbf{h}(\mathbf{r}) \cdot d\mathbf{r} = \int_C \nabla f(\mathbf{r}) \cdot d\mathbf{r} = f(\mathbf{r}(1)) - f(\mathbf{r}(0)) = f(1,-1) - f(0,0) = -1$$

5. $\mathbf{h}(x,y) = \nabla f(x,y) \quad \text{where} \quad f(x,y) = \frac{1}{2}x^2 y^2; \quad \mathbf{r}(0) = \mathbf{j}, \quad \mathbf{r}(1) = -\mathbf{j}$

$$\int_C \mathbf{h}(\mathbf{r}) \cdot d\mathbf{r} = \int_C \nabla f(\mathbf{r}) \cdot d\mathbf{r} = f(\mathbf{r}(1)) - f(\mathbf{r}(0)) = f(0,-1) - f(0,1) = 0 - 0 = 0$$

7. $\mathbf{h}(x,y) = \nabla f(x,y) \quad \text{where} \quad f(x,y) = x^2 y - xy^2; \quad \mathbf{r}(0) = \mathbf{i}, \, \mathbf{r}(\pi) = -\mathbf{i}$

$$\int_C \mathbf{h}(\mathbf{r}) \cdot d\mathbf{r} = \int_C \nabla f(\mathbf{r}) \cdot d\mathbf{r} = f(\mathbf{r}(\pi)) - f(\mathbf{r}(0)) = f(-1,0) - f(1,0) = 0 - 0 = 0$$

9. $\mathbf{h}(x,y) = \nabla f(x,y)$ where $f(x,y) = (x^2 + y^4)^{3/2}$

$$\int_C \mathbf{h}(\mathbf{r}) \cdot d\mathbf{r} = \int_C \nabla f(\mathbf{r}) \cdot d\mathbf{r} = f(1,0) - f(-1,0) = 1 - 1 = 0$$

11. $\mathbf{h}(x,y)$ is not a gradient, but part of it,

$$2x \cosh y \, \mathbf{i} + (x^2 \sinh y - y)\mathbf{j},$$

is a gradient. Since we are integrating over a closed curve, the contribution of the gradient part is 0. Thus

$$\int_C \mathbf{h}(\mathbf{r}) \cdot d\mathbf{r} = \int_C (-y\mathbf{i}) \cdot d\mathbf{r}.$$

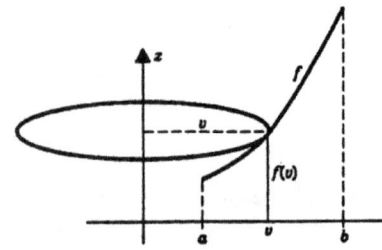

$$C_1 : \mathbf{r}(u) = \mathbf{i} + (-1 + 2u)\mathbf{j}, \quad u \in [0,1]$$
$$C_2 : \mathbf{r}(u) = (1 - 2u)\mathbf{i} + \mathbf{j}, \quad u \in [0,1]$$
$$C_3 : \mathbf{r}(u) = -\mathbf{i} + (1 - 2u)\mathbf{j}, \quad u \in [0,1]$$
$$C_4 : \mathbf{r}(u) = (-1 + 2u)\mathbf{i} - \mathbf{j}, \quad u \in [0,1]$$

$$\int_C \mathbf{h}(\mathbf{r}) \cdot d\mathbf{r} = \int_{C_1} (-y\,\mathbf{i}) \cdot d\mathbf{r} + \int_{C_2} (-y\,\mathbf{i}) \cdot d\mathbf{r} + \int_{C_3} (-y\,\mathbf{i}) \cdot d\mathbf{r} + \int_{C_4} (-y\,\mathbf{i}) \cdot d\mathbf{r}$$

$$= \quad 0 \quad + \int_0^1 -\mathbf{i} \cdot (-2\,\mathbf{i})\,du + \quad 0 \quad + \int_0^1 \mathbf{i} \cdot (2\,\mathbf{i})\,du$$

$$= \quad 0 \quad + \int_0^1 2\,du \quad + \quad 0 \quad + \int_0^1 2\,du$$

$$= \quad 4$$

13. $\mathbf{h}(x,y) = (3x^2 y^3 + 2x)\,\mathbf{i} + (3x^3 y^2 - 4y)\,\mathbf{j}; \quad \dfrac{\partial P}{\partial y} = 9x^2 y^2 = \dfrac{\partial Q}{\partial x}.$ Thus $\mathbf{h}$ is a gradient.

(a) $\mathbf{r}(u) = u\,\mathbf{i} + e^u\,\mathbf{j}, \quad \mathbf{r}'(u) = \mathbf{i} + e^u\,\mathbf{j}, \quad u \in [0,1]$

$$\int_C \mathbf{h}(\mathbf{r}) \cdot d\mathbf{r} = \int_0^1 \left[(3u^2 e^{3u} + 2u) + 3u^3 e^{3u} - 4e^{2u} \right] du = \left[u^3 e^{3u} + u^2 - 2e^{2u} \right]_0^1 = e^3 - 2e^2 + 3$$

(b) $\dfrac{\partial f}{\partial x} = 3x^2 y^3 + 2x \implies f(x,y) = x^3 y^3 + x^2 + g(y);$

$$\dfrac{\partial f}{\partial y} = 3x^3 y^2 + g'(y) = 3x^3 - 4y \implies g'(y) = -4y \implies g(y) = -2y^2$$

Therefore, $f(x,y) = x^3 y^3 + x^2 - 2y^2.$

Now, at $u = 0$, $r(0) = 0\,\mathbf{i} + \mathbf{j} = (0,1)$; at $u = 1$, $r(1) = \mathbf{i} + e\,\mathbf{j} = (1,e)$ and

$$\int_C \mathbf{h}(\mathbf{r}) \cdot d\mathbf{r} = \left[x^3 y^3 + x^2 - 2y^2 \right]_{(0,1)}^{(1,e)} = e^3 - 2e^2 + 3$$

15. $\mathbf{h}(x,y) = (e^{2y} - 2xy)\,\mathbf{i} + (2xe^{2y} - x^2 + 1)\,\mathbf{j}; \quad \dfrac{\partial P}{\partial y} = 2e^{2y} - 2x = \dfrac{\partial Q}{\partial x}.$ Thus $\mathbf{h}$ is a gradient.

(a) $\mathbf{r}(u) = ue^u\,\mathbf{i} + (1+u)\,\mathbf{j}, \quad \mathbf{r}'(u) = (1+u)e^u\,\mathbf{i} + \mathbf{j}, \quad u \in [0,1]$

$$\int_C \mathbf{h}(\mathbf{r}) \cdot d\mathbf{r} = \int_0^1 \left[e^2(3ue^{3u} + e^{3u} - 2u^3e^{2u} - 5u^2e^{2u} - 2ue^{2u} + 1 \right] \, du$$

$$= \left[e^2 u e^{3u} - u^3 e^{2u} - u^2 e^{2u} + u \right]_0^1 = e^5 - 2e^2 + 1$$

(b) $\dfrac{\partial f}{\partial x} = e^{2y} - 2xy \quad \Longrightarrow \quad f(x,y) = xe^{2y} - x^2 y + g(y).$

$$\dfrac{\partial f}{\partial y} = 2xe^{2y} - x^2 + g'(y) = 3x^3 - 4y \quad \Longrightarrow \quad g'(y) = 1 \quad \Longrightarrow \quad g(y) = y$$

Therefore, $f(x,y) = xe^{2y} - x^2 y + y.$

Now, at $u = 0$, $r(0) = 0\,\mathbf{i} + \mathbf{j} = (0,1)$;   at $u = 1$, $r(1) = e\,\mathbf{i} + 2\,\mathbf{j} = (e,2)$ and

$$\int_C \mathbf{h}(\mathbf{r}) \cdot d\mathbf{r} = \left[xe^{2y} - x^2 y + y \right]_{(0,1)}^{(e,2)} = e^5 - 2e^2 + 1$$

17. $\mathbf{h}(x,y,z) = (2xz + \sin y)\,\mathbf{i} + x\cos y\,\mathbf{j} + x^2\,\mathbf{k};$

$$\dfrac{\partial P}{\partial y} = \cos y = \dfrac{\partial Q}{\partial x}, \quad \dfrac{\partial P}{\partial z} = 2x = \dfrac{\partial R}{\partial x}, \quad \dfrac{\partial Q}{\partial z} = 0 = \dfrac{\partial R}{\partial y}. \quad \text{Thus } \mathbf{h} \text{ is a gradient.}$$

$$\dfrac{\partial f}{\partial x} = 2xz + \sin y, \quad \Longrightarrow \quad f(x,y,z) = x^2 z + x\sin y + g(y,z)$$

$$\dfrac{\partial f}{\partial y} = x\cos y + \dfrac{\partial g}{\partial y} = x\cos y, \quad \Longrightarrow \quad g(y,z) = h(z) \quad \Longrightarrow f(x,y,z) = x^2 z + x\sin y + h(z)$$

$$\dfrac{\partial f}{\partial z} = x^2 + h'(z) = x^2 \quad \Longrightarrow \quad h'(z) = 0 \quad \Longrightarrow \quad h(z) = C$$

Therefore, $f(x,y,z) = x^2 z + x\sin y$ (take $C = 0$)

$$\int_C \mathbf{h}(\mathbf{r}) \cdot d\mathbf{r} = \int_C \nabla f \cdot d\mathbf{r} = \left[x^2 z + x\sin y \right]_{\mathbf{r}(0)}^{\mathbf{r}(2\pi)} = \left[x^2 z + x\sin y \right]_{(1,0,0)}^{(1,0,2\pi)} = 2\pi$$

19. $\mathbf{h}(x,y,z) = (2xy + z^2)\,\mathbf{i} + x^2\,\mathbf{j} + 2xz\,\mathbf{k};$

$$\dfrac{\partial P}{\partial y} = 2x = \dfrac{\partial Q}{\partial x}, \quad \dfrac{\partial P}{\partial z} = 2z = \dfrac{\partial R}{\partial x}, \quad \dfrac{\partial Q}{\partial z} = 0 = \dfrac{\partial R}{\partial y}. \quad \text{Thus } \mathbf{h} \text{ is a gradient.}$$

$$\dfrac{\partial f}{\partial x} = 2xy + z^2 \quad \Longrightarrow \quad f(x,y,z) = x^2 y + xz^2 + g(y,z)$$

$$\dfrac{\partial f}{\partial y} = x^2 + \dfrac{\partial g}{\partial y} = x^2 \quad \Longrightarrow \quad g(y,z) = h(z) \quad \Longrightarrow f(x,y,z) = x^2 y + xz^2 + h(z)$$

$$\dfrac{\partial f}{\partial z} = 2xz + h'(z) = 2xz \quad \Longrightarrow \quad h'(z) = 0 \quad \Longrightarrow \quad h(z) = C$$

Therefore, $f(x, y, z) = x^2y + xz^2$ (take $C = 0$)

$$\int_C \mathbf{h}(\mathbf{r}) \cdot d\mathbf{r} = \int_C \nabla f \cdot d\mathbf{r} = \left[x^2y + xz^2 \right]_{\mathbf{r}(0)}^{\mathbf{r}(1)} = \left[x^2y + xz^2 \right]_{(0,2,0)}^{(2,3,-1)} = 14$$

21. $\mathbf{F}(x, y) = (x + e^{2y})\mathbf{i} + (2y + 2xe^{2y})\mathbf{j}$; $\dfrac{\partial P}{\partial y} = 2e^{2y} = \dfrac{\partial Q}{\partial x}$. Thus $\mathbf{F}$ is a gradient.

$$\frac{\partial f}{\partial x} = x + e^{2y} \quad \Longrightarrow \quad f(x, y) = \frac{1}{2}x^2 + xe^{2y} + g(y);$$

$$\frac{\partial f}{\partial y} = 2xe^{2y} + g'(y) = 2y + 2xe^{2y} \quad \Longrightarrow \quad g'(y) = 2y \quad \Longrightarrow g(y) = y^2 \ \text{(take } C = 0)$$

Therefore, $f(x, y) = \dfrac{1}{2}x^2 + xe^{2y} + y^2$.

$$\int_C \mathbf{F}(\mathbf{r}) \cdot d\mathbf{r} = \int_C \nabla f \cdot d\mathbf{r} = \left[\frac{1}{2}x^2 + xe^{2y} + y^2 \right]_{\mathbf{r}(0)}^{\mathbf{r}(2\pi)} = \left[\frac{1}{2}x^2 + xe^{2y} + y^2 \right]_{(3,0)}^{(3,0)} = 0$$

23. Set $f(x, y, z) = g(x)$ and $C : \mathbf{r}(u) = u\,\mathbf{i}, \quad u \in [a, b]$.

In this case

$$\nabla f(\mathbf{r}(u)) = g'(x(u))\mathbf{i} = g'(u)\mathbf{i} \quad \text{and} \quad \mathbf{r}\prime(u) = \mathbf{i},$$

so that

$$\int_C \nabla f(\mathbf{r}) \cdot d\mathbf{r} = \int_a^b \left[\nabla f(\mathbf{r}(u)) \cdot \mathbf{r}\prime(u) \right] du = \int_a^b g'(u)\, du.$$

Since $f(\mathbf{r}(b)) - f(\mathbf{r}(a)) = g(b) - g(a)$,

$$\int_C \nabla f(\mathbf{r}) \cdot d\mathbf{r} = f(\mathbf{r}(b)) - f(\mathbf{r}(a)) \quad \text{gives} \quad \int_a^b g'(u)\, du = g(b) - g(a).$$

25. $\mathbf{F}(\mathbf{r}) = kr\,\mathbf{r} = k\sqrt{x^2 + y^2 + z^2}\,(x\,\mathbf{i} + y\,\mathbf{j} + z\,\mathbf{k}), \quad k > 0$ constant.

$$\frac{\partial P}{\partial y} = \frac{kxy}{\sqrt{x^2 + y^2 + z^2}} = \frac{\partial Q}{\partial x}, \quad \frac{\partial P}{\partial z} = \frac{kxz}{\sqrt{x^2 + y^2 + z^2}} = \frac{\partial R}{\partial x} \quad \frac{\partial Q}{\partial z} = \frac{kyz}{\sqrt{x^2 + y^2 + z^2}} = \frac{\partial R}{\partial y}$$

Therefore, $\mathbf{F}$ is a gradient field.

$$\frac{\partial f}{\partial x} = kx\sqrt{x^2 + y^2 + z^2} \quad \Longrightarrow \quad f(x, y, z) = \frac{k}{3}\left(x^2 + y^2 + z^2\right)^{3/2} + g(y, z).$$

$$\frac{\partial f}{\partial y} = ky\sqrt{x^2 + y^2 + z^2} + \frac{\partial g}{\partial y} = ky\sqrt{x^2 + y^2 + z^2} \quad \Longrightarrow \quad f(x, y, z) = \frac{k}{3}\left(x^2 + y^2 + z^2\right)^{3/2} + h(z)$$

$$\frac{\partial f}{\partial z} = kz\sqrt{x^2 + y^2 + z^2} + h'(z) = kz\sqrt{x^2 + y^2 + z^2} \quad \Longrightarrow \quad h(z) = C, \text{ constant}$$

Therefore, $f(x, y, z) = \frac{k}{3}\left(x^2 + y^2 + z^2\right)^{3/2} + C$.

27. $\mathbf{F}(\mathbf{r}) = \nabla\left(\dfrac{mG}{r}\right);$ $W = \displaystyle\int_C \mathbf{F}(\mathbf{r})\cdot d\mathbf{r} = mG\left(\dfrac{1}{r_2} - \dfrac{1}{r_1}\right)$

29. $\mathbf{F}(x,y,z) = 0\,\mathbf{i} + 0\,\mathbf{j} + \dfrac{-mGr_0^2}{(r_0+z)^2}\,\mathbf{k};$ $\dfrac{\partial P}{\partial y} = 0 = \dfrac{\partial Q}{\partial x},$ $\dfrac{\partial P}{\partial z} = 0 = \dfrac{\partial R}{\partial x},$ $\dfrac{\partial Q}{\partial z} = 0 = \dfrac{\partial R}{\partial y}.$

Therefore, $\mathbf{F}(x,y,z)$ is a gradient.

$\dfrac{\partial f}{\partial x} = 0 \implies f(x,y,z) = g(y,z);$ $\dfrac{\partial f}{\partial y} = \dfrac{\partial g}{\partial y} = 0 \implies g(y,z) = h(z).$

Therefore $f(x,y,z) = h(z).$

Now $\dfrac{\partial f}{\partial z} = h'(z) = \dfrac{-mGr_0^2}{(r_0+z)^2} \implies f(x,y,z) = h(z) = \dfrac{mGr_0^2}{r_0+z}$

SECTION 18.3

1. If f is continuous, then $-f$ is continuous and has antiderivatives u. The scalar fields $U(x,y,z) = u(x)$ are potential functions for $\mathbf{F}$:
$$\nabla U = \frac{\partial U}{\partial x}\,\mathbf{i} + \frac{\partial U}{\partial y}\,\mathbf{j} + \frac{\partial U}{\partial z}\,\mathbf{k} = \frac{du}{dx}\,\mathbf{i} = -f\,\mathbf{i} = -\mathbf{F}.$$

3. The scalar field $U(x,y,z) = \alpha z + d$ is a potential energy function for $\mathbf{F}$. We know that the total mechanical energy remains constant. Thus, for any times t_1 and t_2,
$$\tfrac{1}{2}m[v(t_1)]^2 + U(\mathbf{r}(t_1)) = \tfrac{1}{2}m[v(t_2)]^2 + U(\mathbf{r}(t_2)).$$

This gives

$$\tfrac{1}{2}m[v(t_1)]^2 + \alpha z(t_1) + d = \tfrac{1}{2}m[v(t_2)]^2 + \alpha z(t_2) + d.$$

Solve this equation for $v(t_2)$ and you have the desired formula.

5. (a) We know that $-\nabla U$ points in the direction of maximum decrease of U. Thus $\mathbf{F} = -\nabla U$ attempts to drive objects toward a region where U has lower values.

(b) At a point where u has a minimum, $\nabla U = \mathbf{0}$ and therefore $\mathbf{F} = \mathbf{0}$.

7. (a) By conservation of energy $\tfrac{1}{2}mv^2 + U = E$. Since E is constant and U is constant, v is constant.

(b) ∇U is perpendicular to any surface where U is constant. Obviously so is $\mathbf{F} = -\nabla U$.

9. $f(x,y,z) = -\dfrac{k}{\sqrt{x^2+y^2+z^2}}$ is a potential function for $\mathbf{F}$. The work done by $\mathbf{F}$ moving an object along C is:
$$W = \int_C \mathbf{F}(\mathbf{r})\cdot d\mathbf{r} = \int_a^b \nabla f \cdot d\mathbf{r} = f[\mathbf{r}(b)] - f[\mathbf{r}(a)].$$

Since $\mathbf{r}(a) = (x_0, y_0, z_0)$ and $\mathbf{r}(b) = (x_1, y_1, z_1)$ are points on the unit sphere,

$$f[\mathbf{r}(b)] = f[\mathbf{r}(a)] = -k \quad \text{and so} \quad W = 0$$

SECTION 18.4

1. $\mathbf{r}(u) = u\,\mathbf{i} + 2u\,\mathbf{j}, \quad u \in [0,1]$

$$\int_C (x - 2y)\,dx + 2x\,dy = \int_0^1 \{[x(u) - 2y(u)]x'(u) + 2x(u)\,y'(u)\}\,du = \int_0^1 u\,du = \frac{1}{2}$$

3. $C = C_1 \cup C_2$

$C_1 : \mathbf{r}(u) = u\,\mathbf{i}, \quad u \in [0,1]; \qquad C_2 : \mathbf{r}(u) = \mathbf{i} + 2u\,\mathbf{j}, \quad u \in [0,1]$

$$\int_{C_1} (x - 2y)\,dx + 2x\,dy = \int_{C_1} x\,dx = \int_0^1 x(u)\,x'(u)\,du = \int_0^1 u\,du = \frac{1}{2}$$

$$\int_{C_2} (x - 2y)\,dx + 2x\,dy = \int_{C_2} 2x\,dy = \int_0^1 4\,du = 4$$

$$\int_C = \int_{C_1} + \int_{C_2} = \frac{9}{2}$$

5. $\mathbf{r}(u) = 2u^2\,\mathbf{i} + u\,\mathbf{j}, \quad u \in [0,1]$

$$\int_C y\,dx + xy\,dy = \int_0^1 [y(u)\,x'(u) + x(u)\,y(u)\,y'(u)]\,du = \int_0^1 (4u^2 + 2u^3)\,du = \frac{11}{6}$$

7. $C = C_1 \cup C_2 \qquad C_1 : \mathbf{r}(u) = u\,\mathbf{j}, \quad u \in [0,1]; \qquad C_2 : \mathbf{r}(u) = 2u\,\mathbf{i} + \mathbf{j}, \quad u \in [0,1]$

$$\int_{C_1} y\,dx + xy\,dy = 0$$

$$\int_{C_2} y\,dx + xy\,dy = \int_{C_2} y\,dx = \int_0^1 y(u)\,x'(u)\,du = \int_0^1 2\,du = 2$$

$$\int_C = \int_{C_1} + \int_{C_2} = 2$$

9. $\mathbf{r}(u) = 2u\,\mathbf{i} + 4u\,\mathbf{j}, \quad u \in [0,1]$

$$\int_C y^2\,dx + (xy - x^2)\,dy = \int_0^1 \{y^2(u)x'(u) + [x(u)y(u) - x^2(u)]\,y'(u)\}\,du$$

$$= \int_0^1 [(4u)^2(2) + (8u^2 - 4u^2)(4)]\,du = \int_0^1 48u^2\,du = 16$$

11. $\mathbf{r}(u) = \frac{1}{8}u^2\,\mathbf{i} + u\,\mathbf{j}, \quad u \in [0,4]$

$$\int_C y^2\,dx + (xy - x^2)\,dy = \int_0^4 \{y^2(u)x'(u) + [x(u)y(u) - x^2(u)]\,y'(u)\}\,du$$

$$= \int_0^4 \left[u^2\left(\frac{u}{4}\right) + \left(\frac{u^2}{8}(u) - \left(\frac{u^2}{8}\right)^2 (1)\right) \right]\,du$$

$$= \int_0^4 \left[\frac{3}{8}u^3 - \frac{1}{64}u^4 \right]\,du = \frac{104}{5}$$

13. $\mathbf{r}(u) = u\,\mathbf{i} + u\,\mathbf{j}, \quad u \in [0,1]$

$$\int_C (y^2 + 2x + 1)\,dx + (2xy + 4y - 1)\,dy$$

$$= \int_0^1 \left\{ [y^2(u) + 2x(u) + 1]x'(u) + [2x(u)y(u) + 4y(u) - 1]y'(u) \right\}\,du$$

$$\int_0^1 \left[(u^2 + 2u + 1) + (2u^2 + 4u - 1) \right]\,du = \int_0^1 (3u^2 + 6u)\,du = 4$$

15. $\mathbf{r}(u) = u\,\mathbf{i} + u^3\,\mathbf{j}, \quad u \in [0,1]$

$$\int_C (y^2 + 2x + 1)\,dx + (2xy + 4y - 1)\,dy$$

$$= \int_0^1 \left\{ [y^2(u) + 2x(u) + 1]x'(u) + [2x(u)y(u) + 4y(u) - 1]y'(u) \right\}\,du$$

$$= \int_0^1 \left[(u^6 + 2u + 1) + (2u^4 + 4u^3 - 1)3u^2 \right]\,du = \int_0^1 \left(7u^6 + 12u^5 - 3u^2 + 2u + 1 \right)\,du = 4$$

17. $\mathbf{r}(u) = u\,\mathbf{i} + u\,\mathbf{j} + u\,\mathbf{k}, \quad u \in [0,1]$

$$\int_C y\,dx + 2z\,dy + x\,dz = \int_0^1 \left[y(u)\,x'(u) + 2z(u)\,y'(u) + x(u)\,z'(u) \right]\,du = \int_0^1 4u\,du = 2$$

19. $C = C_1 \cup C_2 \cup C_3$

$C_1 : \mathbf{r}(u) = u\,\mathbf{k}, \quad u \in [0,1]; \quad C_2 : \mathbf{r}(u) = u\,\mathbf{j} + \mathbf{k}, \quad u \in [0,1]; \quad C_3 : \mathbf{r}(u) = u\,\mathbf{i} + \mathbf{j} + \mathbf{k}, \quad u \in [0,1]$

$$\int_{C_1} y\,dx + 2z\,dy + x\,dz = 0$$

$$\int_{C_2} y\,dx + 2z\,dy + x\,dz = \int_{C_2} 2z\,dy = \int_0^1 2z(u)\,y'(u)\,du = \int_0^1 2\,du = 2$$

$$\int_{C_3} y\,dx + 2z\,dy + x\,dz = \int_{C_3} y\,dx = \int_0^1 y(u)\,x'(u)\,du = \int_0^1 du = 1$$

$$\int_C = \int_{C_1} + \int_{C_2} + \int_{C_3} = 3$$

21. $\mathbf{r}(u) = 2u\,\mathbf{i} + 2u\,\mathbf{j} + 8u\,\mathbf{k}, \quad u \in [0,1]$

$$\int_C xy\,dx + 2z\,dy + (y + z)\,dz$$

$$= \int_0^1 \left\{ x(u)y(u)x'(u) + 2z(u)y'(u) + [y(u) + z(u)]z'(u) \right\}\,du$$

$$= \int_0^1 \left[(2u)(2u)(2) + 2(8u)(2) + (2u + 8u)(8) \right]\,du$$

$$= \int_0^1 \left(8u^2 + 112u \right)\,du = \frac{176}{3}$$

23. $\mathbf{r}(u) = u\,\mathbf{i} + u\,\mathbf{j} + 2u^2\,\mathbf{k}, \quad u \in [0,2]$

$$\int_C xy\,dx + 2z\,dy + (y+z)\,dz$$

$$= \int_0^2 \{x(u)y(u)x'(u) + 2z(u)y'(u) + [y(u) + z(u)]z'(u)\}\,du$$

$$= \int_0^2 \left[(u)(u)(1) + 2(2u^2)(1) + (u + 2u^2)(4u) \right]\,du$$

$$= \int_0^2 \left(8u^3 + 9u^2 \right)\,du = 56$$

25. $\mathbf{r}(u) = (u-1)\,\mathbf{i} + (1+2u^2)\,\mathbf{j} + u\,\mathbf{k}, \quad u \in [1,2]$

$$\int_C x^2 y\,dx + y\,dy + xz\,dz$$

$$= \int_1^2 \left[x^2(u)y(u)x'(u) + y(u)y'(u) + x(u)z(u)z'(u) \right]\,du$$

$$= \int_1^2 \left[(u-1)^2(1+2u^2)(1) + (1+2u^2)(4u) + (u-1)u \right]\,du$$

$$= \int_1^2 \left(2u^4 + 4u^3 + 4u^2 + u + 1 \right)\,du = \frac{1177}{30}$$

27. (a) $\dfrac{\partial P}{\partial y} = 6x - 4y = \dfrac{\partial Q}{\partial x}$

$$\frac{\partial f}{\partial x} = x^2 + 6xy - 2y^2 \quad \Longrightarrow \quad f(x,y) = \frac{1}{3}x^3 + 3x^2 y - 2xy^2 + g(y)$$

$$\frac{\partial f}{\partial y} = 3x^2 - 4xy + g'(y) = 3x^2 - 4xy + 2y \quad \Longrightarrow \quad g'(y) = 2y \quad \Longrightarrow \quad g(y) = y^2 + C$$

Therefore, $f(x,y) = \dfrac{1}{3}x^3 + 3x^2 y - 2xy^2 + y^2$ (take $C = 0$)

(b) (i) $\displaystyle\int_C (x^2 + 6xy - 2y^2)\,dx + (3x^2 - 4xy + 2y)\,dy = [f(x,y)]_{(3,0)}^{(0,4)} = 7$

(ii) $\displaystyle\int_C (x^2 + 6xy - 2y^2)\,dx + (3x^2 - 4xy + 2y)\,dy = [f(x,y)]_{(4,0)}^{(0,3)} = -\dfrac{37}{3}$

29. $s'(u) = \sqrt{[x'(u)]^2 + [y'(u)]^2} = a$

(a) $M = \displaystyle\int_C k(x+y)\,ds = k\int_0^{\pi/2} [x(u) + y(u)]\,s'(u)\,du = ka^2 \int_0^{\pi/2} (\cos u + \sin u)\,du = 2ka^2$

$$x_M M = \int_C kx(x+y)\,ds = k\int_0^{\pi/2} x(u)\,[x(u)+y(u)]\,s'(u)\,du$$

$$= ka^3 \int_0^{\pi/2} (\cos^2 u + \cos u \sin u)\,du = \frac{1}{4}ka^3(\pi+2)$$

$$y_M M = \int_C ky(x+y)\,ds = k\int_0^{\pi/2} y(u)\,[x(u)+y(u)]\,s'(u)\,du$$

$$= ka^3 \int_0^{\pi/2} (\sin u \cos u + \sin^2 u)\,du = \frac{1}{4}ka^3(\pi+2)$$

$$x_M = y_M = \tfrac{1}{8}a(\pi+2)$$

(b)

$$I = \int_C k(x+y)y^2\,ds = k\int_0^{\pi/2} \left[x(u)y^2(u) + y^3(u) \right] s'(u)\,du$$

$$= ka^4 \int_0^{\pi/2} \left[\sin^2 u \cos u + \sin^3 u \right] du$$

$$= ka^4 \int_0^{\pi/2} \left[\sin^2 u \cos u + (1 - \cos^2 u)\sin u \right] du$$

$$= ka^4 \left[\tfrac{1}{3}\sin^3 u - \cos u + \tfrac{1}{3}\cos^3 u \right]_0^{\pi/2} = ka^4$$

$I = \tfrac{1}{2}a^2 M.$

31. (a) $I_z = \displaystyle\int_C k(x+y)a^2\,ds = a^2 \int_C k(x+y)\,ds = a^2 M = Ma^2$

(b) The distance from a point (x^*, y^*) to the line $y = x$ is $|y^* - x^*|/\sqrt{2}$.   Therefore

$$I = \int_C k(x+y)\left[\frac{1}{2}(y-x)^2\right] ds = \frac{1}{2}k\int_0^{\pi/2} (a\cos u + a\sin u)(a\sin u - a\cos u)^2 a\,du$$

$$= \frac{1}{2}ka^4 \int_0^{\pi/2} (\sin u - \cos u)^2 \frac{d}{du}(\sin u - \cos u)\,du$$

$$= \frac{1}{2}ka^4 \left[\frac{1}{3}(\sin u - \cos u)^3 \right]_0^{\pi/2} = \frac{1}{3}ka^4.$$

From Exercise 29, $M = 2ka^2$.   Therefore

$$I = \tfrac{1}{6}(2ka^2)a^2 = \tfrac{1}{6}Ma^2.$$

33. (a) $s'(u) = \sqrt{a^2 + b^2}$

$$L = \int_C ds = \int_0^{2\pi} \sqrt{a^2 + b^2}\,du = 2\pi\sqrt{a^2 + b^2}$$

(b) $x_M = 0$, $y_M = 0$ (by symmetry)

$$z_M = \frac{1}{L}\int_C z\,ds = \frac{1}{2\pi\sqrt{a^2+b^2}}\int_0^{2\pi} bu\sqrt{a^2+b^2}\,du = b\pi$$

(c) $I_x = \int_C \frac{M}{L}(y^2+z^2)\,ds = \frac{M}{2\pi}\int_0^{2\pi}(a^2\sin^2 u + b^2u^2)\,du = \frac{1}{6}M(3a^2+8b^2\pi^2)$

$I_y = \frac{1}{6}M(3a^2+8b^2\pi^2)$ similarly

$I_z = Ma^2$ (all the mass is at distance a from the z-axis)

35.
$$M = \int_C k(x^2+y^2+z^2)\,ds$$
$$= k\sqrt{a^2+b^2}\int_0^{2\pi}(a^2+b^2u^2)\,du = \frac{2}{3}\pi k\sqrt{a^2+b^2}\,(3a^2+4\pi^2 b^2)$$

SECTION 18.5

1. (a) $\oint_C xy\,dx + x^2\,dy = \int_{C_1} xy\,dx + x^2\,dy + \int_{C_2} xy\,dx + x^2\,dy + \int_{C_3} xy\,dx + x^2\,dy,$ where

$C_1:\ \mathbf{r}(u) = u\,\mathbf{i} + u\,\mathbf{j},\ \ u\in[0,1];\quad C_2:\ \mathbf{r}(u) = (1-u)\,\mathbf{i}+\mathbf{j},\ \ u\in[0,1]$

$C_3:\ \mathbf{r}(u) = (1-u)\,\mathbf{j},\ \ u\in[0,1].$

$$\int_{C_1} xy\,dx + x^2\,dy = \int_0^1 (u^2+u^2)\,du = \frac{2}{3}$$

$$\int_{C_2} xy\,dx + x^2\,dy = \int_0^1 -(1-u)\,du = -\frac{1}{2}$$

$$\int_{C_3} xy\,dx + x^2\,dy = \int_0^1 0^2(-1)\,du = 0$$

Therefore, $\oint_C xy\,dx + x^2\,dy = \frac{2}{3} - \frac{1}{2} = \frac{1}{6}.$

(b) $\oint_C xy\,dx + x^2\,dy = \iint_\Omega x\,dx\,dy = \int_0^1\int_0^y x\,dx\,dy = \int_0^1\left[\frac{1}{2}x^2\right]_0^y du = \frac{1}{2}\int_0^1 y^2\,dy = \frac{1}{6}$

3. (a) $C:\ \mathbf{r}(u) = 2\cos u\,\mathbf{i} + 3\sin u\,\mathbf{j},\ \ u\in[0,2\pi]$

$$\oint_C (3x^2+y)\,dx + (2x+y^3)\,dy$$

$$= \int_0^{2\pi}\left[(12\cos^2 u + 3\sin u)(-2\sin u) + (4\cos u + 27\sin^3 u)3\cos u\right]\,du$$

$$= \int_0^{2\pi}\left[-24\cos^2 u\,\sin u - 6\sin^2 u + 12\cos^2 u + 81\sin^3 u\,\cos u\right]\,du$$

$$= \left[8\cos^3 u - 3u + \frac{3}{2}\sin 2u + 6u + 3\sin 2u + \frac{81}{4}\sin^4 u\right]_0^{2\pi} = 6\pi$$

(b) $\oint_C (3x^2+y)\,dx + (2x+y^3)\,dy = \iint_\Omega 1\,dx\,dy = $ area of ellipse $\Omega = 6\pi$

5. $\displaystyle\oint_C 3y\,dx + 5x\,dy = \iint_\Omega (5-3)\,dxdy = 2A = 2\pi$

7. $\displaystyle\oint_C x^2\,dy = \iint_\Omega 2x\,dxdy = 2\bar{x}A = 2\left(\frac{a}{2}\right)(ab) = a^2 b$

9.
$$\oint_C (3xy + y^2)\,dx + (2xy + 5x^2)\,dy = \iint_\Omega [(2y + 10x) - (3x + 2y)]\,dxdy$$

$$= \iint_\Omega 7x\,dxdy = 7\bar{x}A = 7(1)(\pi) = 7\pi$$

11. $\displaystyle\oint_C (2x^2 + xy - y^2)\,dx + (3x^2 - xy + 2y^2)\,dy = \iint_\Omega [(6x - y) - (x - 2y)]\,dxdy$

$$= \iint_\Omega (5x + y)\,dxdy = (5\bar{x} + \bar{y})A = (5a + 0)(\pi r^2) = 5a\pi r^2$$

13. $\displaystyle\oint_C e^x \sin y\,dx + e^x \cos y\,dy = \iint_\Omega [e^x \cos y - e^x \cos y]\,dxdy = 0$

15. $\displaystyle\oint_C 2xy\,dx + x^2\,dy = \iint_\Omega [2x - 2x]\,dxdy = 0$

17. $C: \ \mathbf{r}(u) = a\cos u\,\mathbf{i} + a\sin u\,\mathbf{j}; \quad u \in [0, 2\pi]$

$$A = \oint_C -y\,dx = \int_0^{2\pi} (-a\sin u)(-a\sin u)\,du = a^2 \int_0^{2\pi} \sin^2 u\,du = a^2 \left[\frac{1}{2}u - \frac{1}{4}\sin 2u\right]_0^{2\pi} = \pi a^2$$

19. $A = \displaystyle\oint_C x\,dy,$ where $C = C_1 \cup C_2;$

$$C_1: \mathbf{r}(u) = u\,\mathbf{i} + \frac{4}{u}\,\mathbf{j}, \ 1 \le u \le 4; \quad C_2: \mathbf{r}(u) = (4 - 3u)\,\mathbf{i} + (1 + 3u)\,\mathbf{j}, \ 0 \le u \le 1.$$

$$\oint_{C_1} x\,dy = \int_1^4 u\left(\frac{-4}{u^2}\right)du = -4\int_1^4 \frac{1}{u}\,du = -4\ln 4;$$

$$\oint_{C_2} x\,dy = \int_0^1 (4 - 3u)3\,du = \int_0^1 (12 - 9u)\,du = \frac{15}{2}.$$

Therefore, $\quad A = \frac{15}{2} - 4\ln 4.$

21. $\displaystyle\oint_C (ay + b)\,dx + (cx + d)\,dy = \iint_\Omega (c - a)\,dxdy = (c - a)A$

23. We take the arch from $x = 0$ to $x = 2\pi R$. (Figure 9.11.1) Let C_1 be the line segment from $(0,0)$ to $(2\pi R, 0)$ and let C_2 be the cycloidal arch from $(2\pi R, 0)$ back to $(0,0)$. Letting $C = C_1 \cup C_2$, we have

$$A = \oint_C x\,dy = \int_{C_1} x\,dy + \int_{C_2} x\,dy = 0 + \int_{C_2} x\,dy$$

$$= \int_{2\pi}^{0} R(\theta - \sin\theta)(R\sin\theta)\,d\theta$$

$$= R^2 \int_0^{2\pi} (\sin^2\theta - \theta\sin\theta)\,d\theta$$

$$= R^2 \left[\frac{\theta}{2} - \frac{\sin 2\theta}{4} + \theta\cos\theta - \sin\theta \right]_0^{2\pi} = 3\pi R^2.$$

25. Taking Ω to be of type II (see Figure 18.5.2), we have

$$\iint_\Omega \frac{\partial Q}{\partial x}(x,y)\,dxdy = \int_c^d \int_{\psi_1(y)}^{\psi_2(y)} \frac{\partial Q}{\partial x}(x,y)\,dx\,dy$$

$$= \int_c^d \{Q[\psi_2(y),y] - Q[\psi_1(y),y]\}\,dy$$

$$(*) = \int_c^d Q[\psi_2(y),y]\,dy - \int_c^d Q[\psi_1(y),y]\,dy.$$

The graph of $x = \psi_2(y)$ from $x = c$ to $x = d$ is the curve

$$C_4 : \mathbf{r}_4(u) = \psi_2(u)\,\mathbf{i} + u\,\mathbf{j}, \qquad u \in [c,d].$$

The graph of $x = \psi_1(y)$ from $x = c$ to $x = d$ is the curve

$$C_3 : \mathbf{r}_3(u) = \psi_1(u)\,\mathbf{i} + u\,\mathbf{j}, \qquad u \in [c,d].$$

Then

$$\oint_C Q(x,y)\,dy = \int_{C_4} Q(x,y)\,dy - \int_{C_3} Q(x,y)\,dy$$

$$= \int_c^d Q[\psi_2(u),u]\,du - \int_c^d Q[\psi_1(u),u]\,du.$$

Since u is a dummy variable, it can be replaced by y. Comparison with $(*)$ gives the result.

27. Suppose that f is harmonic. By Green's theorem,

$$\int_C \frac{\partial f}{\partial y}\,dx - \frac{\partial f}{\partial x}\,dy = \iint_\Omega \left(-\frac{\partial^2 f}{\partial^2 x} - \frac{\partial^2 f}{\partial^2 y} \right) dxdy = \iint_\Omega 0\,dxdy = 0.$$

29. $\oint_{C_1} = \oint_{C_2} + \oint_{C_3}$

31. $\dfrac{\partial P}{\partial y} = \dfrac{-2xy}{(x^2 + y^2)^2} = \dfrac{\partial Q}{\partial x}$ except at $(0,0)$

(a) If C does not enclose the origin, and Ω is the region enclosed by C, then

$$\oint_C \frac{x}{x^2+y^2}\,dx + \frac{y}{x^2+y^2}\,dy = \iint_\Omega 0\,dx\,dy = 0.$$

(b) If C does enclose the origin, then

$$\oint_C = \oint_{C_a}$$

where $C_a : \mathbf{r}(u) = a\cos u\,\mathbf{i} + a\sin u\,\mathbf{j}, \quad u \in [0, 2\pi]$ is a small circle in the inner region of C.
In this case

$$\oint_C = \int_0^{2\pi}\left[\frac{a\cos u}{a^2}(-a\sin u) + \frac{a\sin u}{a^2}(a\cos u)\right]du = \int_0^{2\pi} 0\,du = 0.$$

The integral is still 0.

33. If Ω is the region enclosed by C, then

$$\oint_C \mathbf{v}\cdot d\mathbf{r} = \oint_C \frac{\partial\phi}{\partial x}\,dx + \frac{\partial\phi}{\partial y}\,dy = \iint_\Omega \left\{\frac{\partial}{\partial x}\left(\frac{\partial\phi}{\partial y}\right) - \frac{\partial}{\partial y}\left(\frac{\partial\phi}{\partial x}\right)\right\}dx\,dy$$

$$= \iint_\Omega 0\,dx\,dy = 0.$$

equality of mixed partials

35. $A = \dfrac{1}{2}\oint_C (-y\,dx + x\,dy)$

$$= \left[\int_{C_1} + \int_{C_2} + \cdots \int_{C_n}\right]$$

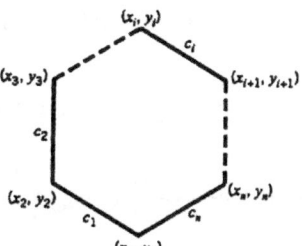

Now

$$\int_{C_i}(-y\,dx + x\,dy) = \int_0^1 \left\{[y_i + u(y_{i+1} - y_i)](x_{i+1} - x_i) + [x_i + u(x_{i+1} - x_i)](y_{i+1} - y_i)\right\}du$$

$$= x_i y_{i+1} - x_{i+1}y_i, \quad i = 1, 2, \ldots, n;\ x_{n+1} = x_1,\ y_{n+1} = y_1$$

Thus, $A = \dfrac{1}{2}\left[(x_1 y_2 - x_2 y_1) + (x_2 y_3 - x_3 y_2) + \cdots + (x_n y_1 - x_1 y_n)\right]$

SECTION 18.6

1. $4[(u^2 - v^2)\mathbf{i} - (u^2 + v^2)\mathbf{j} + 2uv\,\mathbf{k}]$ **3.** $2(\mathbf{j} - \mathbf{i})$

5. $\mathbf{r}(u, v) = 3\cos u\,\cos v\,\mathbf{i} + 2\sin u\,\cos v\,\mathbf{j} + 6\sin v\,\mathbf{k}, \quad u \in [0, 2\pi],\ v \in [0, \pi/2]$

7. $\mathbf{r}(u, v) = 2\cos u\,\cos v\,\mathbf{i} + 2\sin u\,\cos v\,\mathbf{j} + 2\sin v\,\mathbf{k}, \quad u \in [0, 2\pi],\ v \in (\pi/4, \pi/2]$

9. The surface consists of all points of the form $(x, g(x, z), z)$ with $(x, z) \in \Omega$. This set of points is given by

$$\mathbf{r}(u, v) = u\,\mathbf{i} + g(u, v)\,\mathbf{j} + v\,\mathbf{k}, \quad (u, v) \in \Omega.$$

11. $x^2/a^2 + y^2/b^2 + z^2/c^2 = 1$;   ellipsoid

13. $x^2/a^2 - y^2/b^2 = z$;   hyperbolic paraboloid

15. For each $v \in [a, b]$, the points on the surface at level $z = f(v)$ form a circle of radius v.

That circle can be parametrized:
$$\mathbf{R}(u) = v \cos u\,\mathbf{i} + v \sin u\,\mathbf{j} + f(v)\mathbf{k}, \quad u \in [0, 2\pi].$$

Letting v range over $[a, b]$, we obtain the entire surface:
$$\mathbf{r}(u, v) = v \cos u\,\mathbf{i} + v \sin u\,\mathbf{j} + f(v)\mathbf{k}; \quad 0 \leq u \leq 2\pi, \quad a \leq v \leq b.$$

17. Since γ is the angle between p and the xy-plane, γ is the angle between the upper normal to p and $\mathbf{k}$. (Draw a figure.) Therefore, by 18.6.5,

$$\text{area of } \Gamma = \iint_\Omega \sec \gamma \, dx dy = (\sec \gamma) A_\Omega = A_\Omega \quad \sec \gamma.$$

$$\gamma \text{ is constant}$$

19. The surface is the graph of the function
$$f(x, y) = c\left(1 - \frac{x}{a} - \frac{y}{b}\right) = \frac{c}{ab}(ab - bx - ay)$$
defined over the triangle $\Omega : 0 \leq x \leq a, \quad 0 \leq y \leq b(1 - x/a)$. Note that Ω has area $\frac{1}{2}ab$.

$$A = \iint_\Omega \sqrt{[f_x'(x, y)]^2 + [f_y'(x, y)]^2 + 1} \, dx dy$$

$$= \iint_\Omega \sqrt{c^2/a^2 + c^2/b^2 + 1} \, dx dy$$

$$= \frac{1}{ab}\sqrt{a^2 b^2 + a^2 c^2 + b^2 c^2} \iint_\Omega dx\,dy = \frac{1}{2}\sqrt{a^2 b^2 + a^2 c^2 + b^2 c^2}.$$

21. $f(x, y) = x^2 + y^2, \quad \Omega : 0 \leq x^2 + y^2 \leq 4$

$$A = \iint_\Omega \sqrt{4x^2 + 4y^2 + 1} \, dx\,dy \qquad [\text{change to polar coordinates}]$$

$$= \int_0^{2\pi} \int_0^2 \sqrt{4r^2 + 1}\, r\, dr\, d\theta$$

$$= 2\pi \left[\tfrac{1}{12}(4r^2 + 1)^{3/2}\right]_0^2 = \tfrac{1}{6}\pi(17\sqrt{17} - 1)$$

23. $f(x,y) = a^2 - (x^2 + y^2), \quad \Omega : \frac{1}{4}a^2 \le x^2 + y^2 \le a^2$

$$A = \iint_{\Omega} \sqrt{4x^2 + 4y^2 + 1} \; dxdy \qquad [\,\text{change to polar coordinates}\,]$$

$$= \int_0^{2\pi} \int_{a/2}^{a} r\sqrt{4r^2 + 1} \; dr \, d\theta = 2\pi \left[\frac{1}{12}(4r^2 + 1)^{3/2} \right]_{a/2}^{a}$$

$$= \frac{\pi}{6} \left[(4a^2 + 1)^{3/2} - (a^2 + 1)^{3/2} \right]$$

25. $f(x,y) = \frac{1}{3}(x^{3/2} + y^{3/2}), \quad \Omega : 0 \le x \le 1, \quad 0 \le y \le x$

$$A = \iint_{\Omega} \frac{1}{2}\sqrt{x + y + 4} \; dxdy$$

$$= \int_0^1 \int_0^x \frac{1}{2}\sqrt{x + y + 4} \; dy \, dx = \int_0^1 \left[\frac{1}{3}(x + y + 4)^{3/2} \right]_0^x dx$$

$$= \int_0^1 \frac{1}{3} \left[(2x + 4)^{3/2} - (x + 4)^{3/2} \right] dx = \frac{1}{3} \left[\frac{1}{5}(2x + 4)^{5/2} - \frac{2}{5}(x + 4)^{5/2} \right]_0^1$$

$$= \frac{1}{15}(36\sqrt{6} - 50\sqrt{5} + 32)$$

27.   The surface $x^2 + y^2 + z^2 - 4z = 0$ is a sphere of radius 2 centered at $(0, 0, 2)$:

$$x^2 + y^2 + z^2 - 4z = 0 \quad \Longleftrightarrow \quad x^2 + y^2 + (z - 2)^2 = 4.$$

The quadric cone $z^2 = 3(x^2 + y^2)$ intersects the sphere at height $z = 3$:

$$\left. \begin{array}{r} x^2 + y^2 + z^2 - 4z = 0 \\ z^2 = 3(x^2 + y^2) \end{array} \right\} \quad \Longrightarrow \quad \begin{array}{c} 3(x^2 + y^2) + 3z^2 - 12z = 0 \\ 4z^2 - 12z = 0 \\ z = 3. \quad (\text{since } z \ge 2) \end{array}$$

The surface of which we are asked to find the area is a spherical segment of width 1 (from $z = 3$ to $z = 4$) in a sphere of radius 2. The area of the segment is 4π. (Exercise 27, Section 9.9.)

A more conventional solution. The spherical segment is the graph of the function

$$f(x,y) = 2 + \sqrt{4 - (x^2 + y^2)}, \quad \Omega : 0 \le x^2 + y^2 \le 3.$$

Therefore

$$A = \iint_{\Omega} \sqrt{\left(\frac{-x}{\sqrt{4 - x^2 - y^2}} \right)^2 + \left(\frac{-y}{\sqrt{4 - x^2 - y^2}} \right)^2 + 1} \; dxdy$$

$$= \iint_{\Omega} \frac{2}{\sqrt{4 - (x^2 + y^2)}} \; dxdy$$

$$= \int_0^{2\pi} \int_0^{\sqrt{3}} \frac{2r}{\sqrt{4 - r^2}} \; dr \, d\theta \qquad [\,\text{changed to polar coordinates}\,]$$

$$= 2\pi \left[-2\sqrt{4 - r^2} \right]_0^{\sqrt{3}} = 4\pi$$

29. (a)
$$\iint_\Omega \sqrt{\left[\frac{\partial g}{\partial y}(y, z)\right]^2 + \left[\frac{\partial g}{\partial z}(y, z)\right]^2 + 1}\, dydz = \iint_\Omega \sec\left[\alpha(y, z)\right] dydz$$

where α is the angle between the unit normal with positive $\mathbf{i}$ component and the positive x-axis

(b)
$$\iint_\Omega \sqrt{\left[\frac{\partial h}{\partial x}(x, z)\right]^2 + \left[\frac{\partial h}{\partial z}(x, z)\right]^2 + 1}\, dxdz = \iint_\Omega \sec\left[\beta(x, z)\right] dxdz$$

where β is the angle between the unit normal with positive $\mathbf{j}$ component and the positive y-axis

31. (a) $\mathbf{N}(u, v) = v\cos u \sin\alpha\cos\alpha\,\mathbf{i} + v\sin u\sin\alpha\cos\alpha\,\mathbf{j} - v\sin^2\alpha\,\mathbf{k}$

(b)
$$A = \iint_\Omega \|\mathbf{N}(u, v)\|\, dudv = \iint_\Omega v\sin\alpha\, dudv$$

$$= \int_0^{2\pi}\int_0^s v\sin\alpha\, dv\, du = \pi s^2\sin\alpha$$

33. (a) Set $x = a\cos u\cosh v,\quad y = b\sin u\cosh v,\quad z = c\sinh v$. Then,
$$\frac{x^2}{a^2} + \frac{y^2}{b^2} - \frac{z^2}{c^2} = 1.$$

(b)

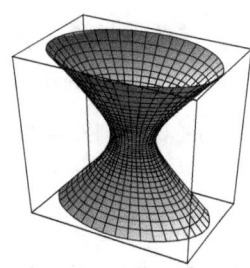

(c) $A = \iint_\Omega \|\mathbf{N}(u, v)\|\, dv\, du$

$$= \int_0^{2\pi}\int_{-\ln 2}^{\ln 2} \sqrt{64\cos^2 u\,\cosh^2 v + 144\sin^2 u\,\cosh^2 v + 36\cosh^2 v\,\sinh^2 v}\, dv\, du$$

35. $A = \sqrt{A_1{}^2 + A_2{}^2 + A_3{}^2}$;   the unit normal to the plane of Ω is a vector of the form
$$\cos\gamma_1\,\mathbf{i} + \cos\gamma_2\,\mathbf{j} + \cos\gamma_3\,\mathbf{k}.$$

Note that
$$A_1 = A\cos\gamma_1,\quad A_2 = A\cos\gamma_2,\quad A_3 = A\cos\gamma_3.$$

Therefore
$$A_1{}^2 + A_2{}^2 + A_3{}^2 = A^2[\cos^2\gamma_1 + \cos^2\gamma_2 + \cos^2\gamma_3] = A^2.$$

37.   (a) (We use Exercise 36.) $f(r, \theta) = r + \theta$; $\Omega : 0 \leq r \leq 1,$ $0 \leq \theta \pi$

$$A = \iint_{\Omega} \sqrt{r^2 \, [f_r'(r, \theta)]^2 + [f_\theta'(r, \theta)]^2 + r^2} \, dr d\theta = \iint_{\Omega} \sqrt{2r^2 + 1} \, dr d\theta$$

$$= \int_0^{\pi} \int_0^1 \sqrt{2r^2 + 1} \, dr \, d\theta = \frac{1}{4} \sqrt{2} \pi \left[\sqrt{6} + \ln \left(\sqrt{2} + \sqrt{3} \right) \right]$$

(b) $f(r, \theta) = re^{\theta}$; $\Omega : 0 \leq r \leq a,$ $0 \leq \theta \leq 2\pi$

$$A = \iint_{\Omega} r \sqrt{2e^{2\theta} + 1} \, dr d\theta = \left(\int_0^{2\pi} \sqrt{2e^{2\theta} + 1} \, d\theta \right) \left(\int_0^a r \, dr \right)$$

$$= \frac{1}{2} a^2 [\sqrt{2e^{4\pi} + 1} - \sqrt{3} + \ln \left(1 + \sqrt{3} \right) - \ln \left(1 + \sqrt{2e^{4\pi} + 1} \right)]$$

SECTION 18.7

For Exercises 1–6 we have $\sec \left[\gamma(x, y) \right] = \sqrt{y^2 + 1}$. $\mathbf{N}(x, y) = -y\mathbf{j} + \mathbf{k}$,   so $\| N(x, y) \| = \sqrt{y^2 + 1}$.

1. $\displaystyle\iint_S d\sigma = \int_0^1 \int_0^1 \sqrt{y^2 + 1} \, dx \, dy = \int_0^1 \sqrt{y^2 + 1} \, dy = \frac{1}{2} [\sqrt{2} + \ln \left(1 + \sqrt{2} \right)]$

3. $\displaystyle\iint_S 3y \, d\sigma = \int_0^1 \int_0^1 3y \sqrt{y^2 + 1} \, dy \, dx = \int_0^1 3y \sqrt{y^2 + 1} \, dy = \left[(y^2 + 1)^{3/2} \right]_0^1 = 2\sqrt{2} - 1$

5. $\displaystyle\iint_S \sqrt{2} z \, d\sigma = \iint_S y \, d\sigma = \frac{1}{3} (2\sqrt{2} - 1)$ (Exercise 3)

7. $\displaystyle\iint_S xy \, d\sigma$; $S : \mathbf{r}(u, v) = (6 - 2u - 3v)\mathbf{i} + u\mathbf{j} + v\mathbf{k},$ $0 \leq u \leq 3 - \frac{3}{2} v,$ $0 \leq v \leq 2$

$$\| \mathbf{N}(u, v) \| = \| (-2\mathbf{i} + \mathbf{j}) \times (-3\mathbf{i} + \mathbf{k}) \| = \sqrt{14}$$

$$\iint_S xy \, d\sigma = \sqrt{14} \iint_{\Omega} x(u, v) y(u, v) \, du \, dv$$

$$= \sqrt{14} \iint_{\Omega} (6 - 2u - 3v) u \, du \, dv$$

$$= \sqrt{14} \int_0^2 \int_0^{3-3v/2} (6u - 2u^2 - 3uv) \, du \, dv$$

$$= \sqrt{14} \left[3 \left(3 - \tfrac{3}{2} v \right)^2 - \tfrac{2}{3} \left(3 - \tfrac{3}{2} v \right)^3 - \tfrac{3}{2} v \left(3 - \tfrac{3}{2} v \right)^2 \right] dv = \frac{9}{2} \sqrt{14}$$

9. $\displaystyle\iint_S x^2 z \, d\sigma;$ $S : \mathbf{r}(u,v) = (\cos u\,\mathbf{i} + v\,\mathbf{j} + \sin u\,\mathbf{k}, \quad 0 \le u \le \pi, \quad 0 \le v \le 2.$

$$\mathbf{N}(u,v) = \begin{vmatrix} \mathbf{i} & \mathbf{j} & \mathbf{k} \\ -\sin u & 0 & \cos u \\ 0 & 1 & 0 \end{vmatrix} = -\cos u\,\mathbf{i} - \sin u\,\mathbf{k} \quad \text{and} \quad \|\mathbf{N}(u,v)\| = 1.$$

$$\iint_S x^2 z \, d\sigma = \iint_\Omega \cos^2 u \, \sin u \, du \, dv = \int_0^2 \int_0^\pi \cos^2 u \, \sin u \, du \, dv = \frac{4}{3}$$

11. $\displaystyle\iint_S (x^2 + y^2) \, d\sigma;$ $S : \mathbf{r}(u,v) = \cos u \cos v\,\mathbf{i} + \cos u \sin v\,\mathbf{j} + \sin u\,\mathbf{k}, \quad 0 \le u \le \pi/2, \quad 0 \le v \le 2\pi.$

$$\mathbf{N}(u,v) = \begin{vmatrix} \mathbf{i} & \mathbf{j} & \mathbf{k} \\ -\sin u \cos v & -\sin u \sin v & \cos u \\ -\cos u \sin v & \cos u \cos v & 0 \end{vmatrix} = -\cos^2 u \cos v\,\mathbf{i} + \cos^2 u \sin v\,\mathbf{j} - \sin u \cos u\,\mathbf{k};$$

$$\|\mathbf{N}(u,v)\| = \cos u.$$

$$\iint_S (x^2 + y^2) \, d\sigma = \iint_\Omega \cos^2 u \, \cos u \, du \, dv = \int_0^{2\pi} \int_0^{\pi/2} \cos^3 u \, du \, dv = \frac{4}{3}\pi$$

13. $\displaystyle M = \iint_S \lambda(x,y,x) \, d\sigma = \int_0^a \int_0^{a-x} k\sqrt{3} \, dy \, dx = \int_0^a k\sqrt{3}\,(a-x) \, dx = \frac{1}{2}a^2 k\sqrt{3}$

15. $\displaystyle M = \iint_S \lambda(x,y,z) \, d\sigma = \int_0^a \int_0^{a-x} kx^2\sqrt{3} \, dy \, dx = \int_0^a k\sqrt{3}x^2(a-x) \, dx = \frac{1}{12}a^4 k\sqrt{3}$

17. $S : \mathbf{r}(u,v) = a\cos u \cos v\,\mathbf{i} + a\sin u \cos v\,\mathbf{j} + a\sin v\,\mathbf{k}$ with $0 \le u \le 2\pi, \quad 0 \le v \le \frac{1}{2}\pi.$ By a previous calculation $\|\mathbf{N}(u,v)\| = a^2 \cos v.$

$\bar{x} = 0, \quad \bar{y} = 0 \qquad$ (by symmetry)

$$\bar{z}A = \iint_S z \, d\sigma = \iint_\Omega z(u,v)\,\|\mathbf{N}(u,v)\| \, du \, dv = \int_0^{2\pi} \int_0^{\pi/2} a^3 \sin v \cos v \, dv \, du = \pi a^3$$

$\bar{z} = \frac{1}{2}a \quad \text{since} \quad A = 2\pi a^2$

19. $\mathbf{N}(u,v) = (\mathbf{i}+\mathbf{j}+2\,\mathbf{k})\cdot(\mathbf{i}-\mathbf{j}) = 2\,\mathbf{i}+2\,\mathbf{j}-2\,\mathbf{k}$

$$\text{flux in the direction of } \mathbf{N} = \iint_S \left(\mathbf{v}\cdot\frac{\mathbf{N}}{\|\mathbf{N}\|}\right) d\sigma = \iint_\Omega [\mathbf{v}(x(u),\,y(u),\,z(u))\cdot\mathbf{N}(u,v)]\,dudv$$

$$= \iint_\Omega [(u+v)\mathbf{i}-(u-v)\mathbf{j}]\cdot[2\,\mathbf{i}+2\,\mathbf{j}-2\,\mathbf{k}]\,dudv.$$

$$= \iint_\Omega 4v\,dudv = 4\int_0^1\int_0^1 v\,dv\,du = 2$$

21. With $\quad \mathbf{v} = z\,\mathbf{k}$

$$\text{flux} = \iint_S (\mathbf{v}\cdot\mathbf{n})\,d\sigma = \frac{1}{a}\iint_S z^2\,d\sigma = \frac{1}{a}\iint_\Omega (a^2\sin^2 v)(a^2\cos v)\,dudv$$

$$= a^3\int_0^{2\pi}\int_{-\pi/2}^{\pi/2}(\sin^2 v\cos v)\,du\,dv = \frac{4}{3}\pi a^3$$

23. With $\quad \mathbf{v} = y\,\mathbf{i}-x\,\mathbf{j}$

$$\text{flux} = \iint_S (\mathbf{v}\cdot\mathbf{n})\,d\sigma = \frac{1}{a}\iint_S \underbrace{(yx-xy)}_{0}\,d\sigma = 0$$

25. With $\quad \mathbf{v} = x\,\mathbf{i}+y\,\mathbf{j}+z\,\mathbf{k}$

$$\text{flux} = \iint_S (\mathbf{v}\cdot\mathbf{n})\,d\sigma = \iint_\Omega (-v_1 f_x' - v_2 f_y' + v_3)\,dxdy$$

$$= \iint_\Omega [-x(-1)-y(-1)+(a-x-y)]\,dxdy = a\iint_\Omega dxdy = aA = \frac{1}{2}\sqrt{3}a^3$$

27. With $\quad \mathbf{v} = x^2\,\mathbf{i}-y^2\,\mathbf{j}$

$$\text{flux} = \iint_S (\mathbf{v}\cdot\mathbf{n})\,d\sigma = \iint_\Omega (-v_1 f_x' - v_2 f_y' + v_3)\,dxdy$$

$$= \iint_\Omega [-x^2(-1)-(-y^2)(-1)+0]\,dxdy = \int_0^a\int_0^{a-x}(x^2-y^2)\,dy\,dx$$

$$= \int_0^a \left[ax^2 - x^3 - \frac{1}{3}(a-x)^3\right]dx = \left[\frac{1}{3}ax^3 - \frac{1}{4}x^4 + \frac{1}{12}(a-x)^4\right]_0^a = 0$$

29. With $\quad \mathbf{v} = xz\,\mathbf{j}-xy\,\mathbf{k}$

$$\text{flux} = \iint_S (\mathbf{v}\cdot\mathbf{n})\,d\sigma = \iint_\Omega (-v_1 f_x' - v_2 f_y' + v_3)\,dxdy$$

$$= \iint_\Omega (-x^3 y - xy)\,dxdy = \int_0^1\int_0^2(-x^3 y - xy)\,dy\,dx$$

$$= \int_0^1 -2(x^3 + x)\,dx = -\frac{3}{2}$$

31. $\mathbf{n} = \dfrac{1}{a}(x\mathbf{i} + y\mathbf{j})$

$$\text{flux} = \iint_S (\mathbf{v}\cdot\mathbf{n})\,d\sigma = \frac{1}{a}\iint_S [(x\mathbf{i} + y\mathbf{j} + z\mathbf{k})\cdot(x\mathbf{i} + y\mathbf{j})]\,d\sigma$$

$$= \frac{1}{a}\iint_S (x^2 + y^2)\,d\sigma = a\iint_S d\sigma = a\,(\text{area of } S) = a\,(2\pi a l) = 2\pi a^2 l$$

33. With $\quad \mathbf{v} = x\mathbf{i} - y\mathbf{j} + \frac{3}{2}z\mathbf{k}$

$$\text{flux} = \iint_S (\mathbf{v}\cdot\mathbf{n})\,d\sigma = \iint_\Omega (-v_1 f_x' - v_2 f_y' + v_3)\,dxdy = \iint_\Omega 2y^{3/2}\,dxdy$$

$$= \int_0^1 \int_0^{1-x} 2y^{3/2}\,dy\,dx = \int_0^1 \frac{4}{5}(1-x)^{5/2}\,dx = \frac{8}{35}$$

35. With $\quad \mathbf{v} = y^2\mathbf{j}$

$$\text{flux} = \iint_S (\mathbf{v}\cdot\mathbf{n})\,d\sigma = \iint_\Omega (-v_1 f_x' - v_2 f_y' + v_3)\,dxdy = \iint_\Omega -y^{5/2}\,d\sigma$$

$$= \int_0^1 \int_0^{1-x} -y^{5/2}\,dy\,dx = \int_0^1 -\frac{2}{7}(1-x)^{7/2}\,dx = -\frac{4}{63}$$

37. $\bar{x} = 0, \quad \bar{y} = 0 \qquad$ by symmetry. You can verify that $\quad \|\mathbf{N}(u,v)\| = v\sin\alpha.$

$$\bar{z}A = \iint_S z\,d\sigma = \iint_\Omega (s\cos\alpha)(v\sin\alpha)\,dudv = \sin\alpha\cos\alpha \int_0^{2\pi}\int_0^s v^2\,dv\,du = \frac{2}{3}\pi\sin\alpha\cos\alpha\,s^3$$

$$\bar{z} = \frac{2}{3}s\cos\alpha \quad \text{since} \quad A = \pi s^2\sin\alpha$$

39. $f(x,y) = \sqrt{x^2 + y^2} \quad \text{on} \quad \Omega: 0 \le x^2 + y^2 \le 1; \quad \lambda(x,y,z) = k\sqrt{x^2 + y^2}$

$$x_M = 0, \quad y_M = 0 \qquad \text{(by symmetry)}$$

$$z_M M = \iint_S z\lambda(x,y,z)\,d\sigma = \iint_\Omega k(x^2 + y^2)\sec[\gamma(x,y)]\,dxdy$$

$$= k\sqrt{2}\iint_\Omega (x^2 + y^2)\,dxdy$$

$$= k\sqrt{2}\int_0^{2\pi}\int_0^1 r^3\,dr\,d\theta = \frac{1}{2}\sqrt{2}\pi k$$

$$z_M = \frac{3}{4} \quad \text{since} \quad M = \frac{2}{3}\sqrt{2}\pi k \qquad \text{(Exercise 38)}$$

41. no answer required

43.
$$x_M M = \iint\limits_{S} x\lambda\,(x,y,z)\,d\sigma = \iint\limits_{S} kx(y^2 + z^2)\,d\sigma$$

$$= 2\sqrt{3}\,k \iint\limits_{\Omega} (u+v)\left[(u-v)^2 + 4u^2\right]\,du\,dv$$

$$= 2\sqrt{3}\,k \int_0^1 \int_0^1 (5u^3 - 2u^2 v + uv^2 + 5u^2 v - 2uv^2 + v^3)\,dv\,du$$

$$= 2\sqrt{3}\,k \int_0^1 \left(5u^3 - u^2 + \frac{1}{3}u + \frac{5}{2}u^2 - \frac{2}{3}u + \frac{1}{4}\right)\,du = \frac{11}{3}\sqrt{3}k$$

$$x_M = \frac{11}{9} \quad \text{since} \quad M = 3\sqrt{3}k \qquad \text{(Exercise 42)}$$

45. Total flux out of the solid is 0. It is clear from a diagram that the outer unit normal to the cylindrical side of the solid is given by $\mathbf{n} = x\,\mathbf{i} + y\,\mathbf{j}$ in which case $\mathbf{v}\cdot\mathbf{n} = 0$. The outer unit normals to the top and bottom of the solid are $\mathbf{k}$ and $-\mathbf{k}$ respectively. So, here as well, $\mathbf{v}\cdot\mathbf{n} = 0$ and the total flux is 0.

47. The surface $z = \sqrt{2 - (x^2 + y^2)}$ is the upper half of the sphere $x^2 + y^2 + z^2 = 2$. The surface intersects the surface $z = x^2 + y^2$ in a circle of radius 1 at height $z = 1$. Thus the upper boundary of the solid, call it S_1, is a segment of width $\sqrt{2} - 1$ on a sphere of radius $\sqrt{2}$. The area of S_1 is therefore $2\pi\sqrt{2}(\sqrt{2} - 1)$. (Exercise 27, Section 9.9). The upper unit normal to S_1 is the vector

$$\mathbf{n} = \frac{1}{\sqrt{2}}(x\,\mathbf{i} + y\,\mathbf{j} + z\,\mathbf{k}).$$

Therefore

$$\text{flux through } S_1 = \iint\limits_{S_1} (\mathbf{v}\cdot\mathbf{n})\,d\sigma = \frac{1}{\sqrt{2}} \iint\limits_{S_1} \overbrace{(x^2 + y^2 + z^2)}^{2}\,d\sigma$$

$$= \sqrt{2} \iint\limits_{S_1} d\sigma = \sqrt{2}\,(\text{area of } S_1) = 4\pi(\sqrt{2} - 1).$$

The lower boundary of the solid, call it S_2, is the graph of the function

$$f(x,y) = x^2 + y^2 \quad \text{on} \quad \Omega : 0 \le x^2 + y^2 \le 1.$$

Taking $\mathbf{n}$ as the lower unit normal, we have

$$\text{flux through } S_2 = \iint\limits_{S_2} (\mathbf{v}\cdot\mathbf{n})\,d\sigma = \iint\limits_{\Omega} (v_1 f_x' + v_2 f_y' - v_3)\,dx\,dy$$

$$= \iint\limits_{\Omega} (x^2 + y^2)\,dx\,dy = \int_0^{2\pi} \int_0^1 r^3\,dr\,d\theta = \frac{1}{2}\pi.$$

The total flux out of the solid is $4\pi(\sqrt{2} - 1) + \frac{1}{2}\pi = (4\sqrt{2} - \frac{7}{2})\pi.$

SECTION 18.8

1. $\nabla \cdot \mathbf{v} = 2, \quad \nabla \times \mathbf{v} = 0$ **3.** $\nabla \cdot \mathbf{v} = 0, \quad \nabla \times \mathbf{v} = 0$ **5.** $\nabla \cdot \mathbf{v} = 6, \quad \nabla \times \mathbf{v} = 0$

7. $\nabla \cdot \mathbf{v} = yz + 1, \quad \nabla \times \mathbf{v} = -x\mathbf{i} + xy\mathbf{j} + (1-x)z\mathbf{k}$

9. $\nabla \cdot \mathbf{v} = 1/r^2, \quad \nabla \times \mathbf{v} = 0$

11. $\nabla \cdot \mathbf{v} = 2(x + y + z)e^{r^2}, \quad \nabla \times \mathbf{v} = 2e^{r^2}\left[(y-z)\mathbf{i} - (x-z)\mathbf{j} + (x-y)\mathbf{k}\right]$

13. $\nabla \cdot \mathbf{v} = f'(x), \quad \nabla \times \mathbf{v} = 0$ **15.** use components.

17. $\nabla \cdot \mathbf{v} = \dfrac{\partial P}{\partial x} + \dfrac{\partial Q}{\partial y} + \dfrac{\partial R}{\partial z} = 2 + 4 - 6 = 0$

19. $\nabla \times \mathbf{F} = \begin{vmatrix} \mathbf{i} & \mathbf{j} & \mathbf{k} \\ \dfrac{\partial}{\partial x} & \dfrac{\partial}{\partial y} & \dfrac{\partial}{\partial z} \\ x & y & -2z \end{vmatrix} = 0$

21. $\nabla^2 f = 12(x^2 + y^2 + z^2)$ **23.** $\nabla^2 f = 2y^3 z^4 + 6x^2 yz^4 + 12x^2 y^3 z^2$

25. $\nabla^2 f = e^r(1 + 2r^{-1})$ **27.** (a) $2r^2$ (b) $-1/r$

29. $\nabla^2 f = \nabla^2 g(r) = \nabla \cdot (\nabla g(r)) = \nabla \cdot \left(g'(r)r^{-1}\mathbf{r}\right)$

$$= \left[(\nabla g'(r)) \cdot r^{-1}\mathbf{r}\right] + g'(r)\left(\nabla \cdot r^{-1}\mathbf{r}\right)$$

$$= \left\{\left[g''(r)r^{-1}\mathbf{r}\right] \cdot r^{-1}\mathbf{r}\right\} + g'(r)(2r^{-1})$$

$$= g''(r) + 2r^{-1}g'(r)$$

31. $\dfrac{\partial f}{\partial x} = 2x + y + 2z, \quad \dfrac{\partial^2 f}{\partial x^2} = 2; \quad \dfrac{\partial f}{\partial y} = 4y + x - 3z, \quad \dfrac{\partial^2 f}{\partial y^2} = 4;$

$\dfrac{\partial f}{\partial z} = -6z + 2x - 3y, \quad \dfrac{\partial^2 f}{\partial z^2} = -6;$

$$\dfrac{\partial^2 f}{\partial x^2} + \dfrac{\partial^2 f}{\partial y^2} + \dfrac{\partial^2 f}{\partial z^2} = 2 + 4 - 6 = 0$$

33. $n = -1$

SECTION 18.9

1. $\displaystyle\iint_S (\mathbf{v} \cdot \mathbf{n})\, d\sigma = \iiint_T (\boldsymbol{\nabla} \cdot \mathbf{v})\, dxdydz = \iiint_T 3\, dxdydz = 3V = 4\pi$

3. $\displaystyle\iint_S (\mathbf{v} \cdot \mathbf{n})\, d\sigma = \iiint_T (\boldsymbol{\nabla} \cdot \mathbf{v})\, dxdydz = \iiint_T 2(x + y + z)\, dxdydz.$

The flux is zero since the function $f(x, y, z) = 2(x + y + z)$ satisfies the relation $f(-x, -y, -z) = -f(x, y, z)$ and T is symmetric about the origin.

5.

face	n	v·n	flux	
$x = 0$	$-\mathbf{i}$	0	0	
$x = 1$	$\mathbf{i}$	1	1	
$y = 0$	$-\mathbf{j}$	0	0	total flux $= 3$
$y = 1$	$\mathbf{j}$	1	1	
$z = 0$	$-\mathbf{k}$	0	0	
$z = 1$	$\mathbf{k}$	1	1	

$$\iiint_T (\boldsymbol{\nabla} \cdot \mathbf{v})\, dxdydz = \iiint_T 3\, dxdydz = 3V = 3$$

7.

face	n	v·n		flux	
$x = 0$	$-\mathbf{i}$	0		0	
$x = 1$	$\mathbf{i}$	1		1	
$y = 0$	$-\mathbf{j}$	xz			
			fluxes add up to 0		total flux $= 2$
$y = 1$	$\mathbf{j}$	$-xz$			
$z = 0$	$-\mathbf{k}$	0		0	
$z = 1$	$\mathbf{k}$	1		1	

$$\iiint_T (\boldsymbol{\nabla} \cdot \mathbf{v})\, dxdydz = \iiint_T 2\,(x + z)\, dxdydz = 2\,(\bar{x} + \bar{z})V = 2\,(\tfrac{1}{2} + \tfrac{1}{2})1 = 2$$

9. flux $= \displaystyle\iiint_T (1 + 4y + 6z)\, dxdydz = (1 + 4\bar{y} + 6\bar{z})V = (1 + 0 + 3)\, 9\pi = 36\pi$

11.

$$\text{flux} = \iiint_T (2x + x - 2x)\, dxdydz \iiint_T x\, dxdydz$$

$$= \int_0^1 \int_0^{1-x} \int_0^{1-x-y} x\, dz\, dy\, dx$$

$$= \int_0^1 \int_0^{1-x} (x - x^2 - xy)\, dy\, dx$$

$$= \int_0^1 \left[xy - x^2 y - \frac{1}{2} xy^2 \right]_0^{1-x} dx$$

$$= \int_0^1 \left(\frac{1}{2} x - x^2 + \frac{1}{2} x^3 \right) dx = \frac{1}{24}$$

13. flux $= \displaystyle\iiint_T 2(x + y + z)\, dxdydz = \int_0^4 \int_0^2 \int_0^{2\pi} 2(r\cos\theta + r\sin\theta + z)r\, d\theta\, dr\, dz$

$$= \int_0^4 \int_0^2 4\pi\, rz\, dr\, dz$$

$$= \int_0^4 8\pi\, z\, dz = 64\pi$$

15. flux $= \displaystyle\iiint_T (2y + 2y + 3y)\, dxdydz = 7\bar{y}V = 0$

17. flux $= \displaystyle\iiint_T (A + B + C)\, dxdydz = (A + B + C)V$

19. Let T be the solid enclosed by S and set $\mathbf{n} = n_1\mathbf{i} + n_2\mathbf{j} + n_3\mathbf{k}$.

$$\iint_S n_1\, d\sigma = \iint_S (\mathbf{i} \cdot \mathbf{n})\, d\sigma = \iiint_T (\boldsymbol{\nabla} \cdot \mathbf{i})\, dxdydz = \iiint_T 0\, dxdydz = 0.$$

Similarly

$$\iint_S n_2\, d\sigma = 0 \quad \text{and} \quad \iint_S n_3\, d\sigma = 0.$$

21. A routine computation shows that $\boldsymbol{\nabla} \cdot (\boldsymbol{\nabla} f \times \boldsymbol{\nabla} g) = 0.$ Therefore

$$\iint_S [(\boldsymbol{\nabla} f \times \boldsymbol{\nabla} g) \cdot \mathbf{n}]\, d\sigma = \iiint_T [\boldsymbol{\nabla} \cdot (\boldsymbol{\nabla} f \times \boldsymbol{\nabla} g)]\, dxdydz = 0.$$

23. Set $\mathbf{F} = F_1\mathbf{i} + F_2\mathbf{j} + F_3\mathbf{k}$.

$$F_1 = \iint_S [\rho(z-c)\mathbf{i}\cdot\mathbf{n}]\,d\sigma = \iiint_T [\boldsymbol{\nabla}\cdot\rho(z-c)\mathbf{i}]\,dxdydz$$

$$= \iiint_T \underbrace{\frac{\partial}{\partial x}[\rho(z-c)]}\,dxdydz = 0.$$

Similarly $F_2 = 0$.

$$F_3 = \iint_S [\rho(z-c)\mathbf{k}\cdot\mathbf{n}]\,d\sigma = \iiint_T [\boldsymbol{\nabla}\cdot\rho(z-c)\mathbf{k}]\,dxdydz$$

$$= \iiint_T \frac{\partial}{\partial z}[\rho(z-c)]\,dxdydz$$

$$= \iiint_T \rho\,dxdydz = W.$$

PROJECT 18.9

1. For $\mathbf{r} \neq \mathbf{0}$, $\quad \boldsymbol{\nabla}\cdot\mathbf{E} = \boldsymbol{\nabla}\cdot qr^{-3}\mathbf{r} = q(-3+3)r^{-3} = 0$ by (17.8.8)

3. On $S_a, \mathbf{n} = \dfrac{\mathbf{r}}{r}$, and thus $\mathbf{E}\cdot\mathbf{n} = q\dfrac{\mathbf{r}}{r^3}\cdot\dfrac{\mathbf{r}}{r} = \dfrac{q}{r^2} = \dfrac{q}{a^2}$

Thus flux of $\mathbf{E}$ out of $S_a = \displaystyle\iint_{S_a}(\mathbf{E}\cdot\mathbf{n})\,d\sigma = \iint_{S_a}\dfrac{q}{a^2}\,d\sigma = \dfrac{q}{a^2}(\text{area of }S_a) = \dfrac{q}{a^2}(4\pi a^2) = 4\pi q$.

SECTION 18.10

For Exercises 1–4: $\quad \mathbf{n} = x\mathbf{i} + y\mathbf{j} + z\mathbf{k} \quad$ and $\quad C : \mathbf{r}(u) = \cos u\,\mathbf{i} + \sin u\,\mathbf{j}, \quad u \in [0, 2\pi]$.

1. (a) $\displaystyle\iint_S [(\boldsymbol{\nabla}\times\mathbf{v})\cdot\mathbf{n}]\,d\sigma = \iint_S (\mathbf{0}\cdot\mathbf{n})\,d\sigma = 0$

(b) S is bounded by the unit circle $\quad C : \mathbf{r}(u) = \cos u\,\mathbf{i} + \sin u\,\mathbf{j}, \quad u \in [0, 2\pi]$.

$\displaystyle\oint_C \mathbf{v}(\mathbf{r})\cdot d\mathbf{r} = 0 \quad$ since $\mathbf{v}$ is a gradient.

3. (a) $\displaystyle\iint_S [(\nabla \times \mathbf{v}) \cdot \mathbf{n}]\, d\sigma = \iint_S [(-3y^2\mathbf{i} + 2z\mathbf{j} + 2\mathbf{k}) \cdot \mathbf{n}]\, d\sigma$

$$= \iint_S (-3xy^2 + 2yz + 2z)\, d\sigma$$

$$= \underbrace{\iint_S (-3xy^2)\, d\sigma}_{0} + \underbrace{\iint_S 2yz\, d\sigma}_{0} + 2 \iint_S z\, d\sigma = 2\bar{z}V = 2(\tfrac{1}{2})2\pi = 2\pi$$

Exercise 17, Section 17.7

(b) $\displaystyle\oint_C \mathbf{v}(\mathbf{r}) \cdot d\mathbf{r} = \oint_C z^2\, dx + 2x\, dy = \oint_C 2x\, dy = \int_0^{2\pi} 2\cos^2 u\, du = 2\pi$

5. (a) $\displaystyle\iint_S [(\nabla \times \mathbf{v}) \cdot \mathbf{n}]\, d\sigma = \iint_S \tfrac{1}{3}\sqrt{3}\, d\sigma = \tfrac{1}{3}\sqrt{3}A = 2$

(b) $\displaystyle\oint_C \mathbf{v}(\mathbf{r}) \cdot d\mathbf{r} = \left(\int_{C_1} + \int_{C_2} + \int_{C_3}\right)\mathbf{v}(\mathbf{r}) \cdot d\mathbf{r} = -2 + 2 + 2 = 2$

7. (a) $\displaystyle\iint_S [(\nabla \times \mathbf{v}) \cdot \mathbf{n}]\, d\sigma = \iint_S (y\mathbf{k} \cdot \mathbf{n})\, d\sigma = \tfrac{1}{3}\sqrt{3} \iint_S y\, d\sigma = \tfrac{1}{3}\sqrt{3}\,\bar{y}A = \tfrac{4}{3}$

(b) $\displaystyle\oint_C \mathbf{v}(\mathbf{r}) \cdot d\mathbf{r} = \left(\int_{C_1} + \int_{C_2} + \int_{C_3}\right)\mathbf{v}(\mathbf{r}) \cdot d\mathbf{r} = \left(\tfrac{4}{3} - \tfrac{32}{5}\right) + \tfrac{32}{5} + 0 = \tfrac{4}{3}$

9. The bounding curve is the set of all (x, y, z) with
$$x^2 + y^2 = 4 \quad \text{and} \quad z = 4.$$
Traversed in the positive sense with respect to $\mathbf{n}$, it is the curve $-C$ where
$$C : \mathbf{r}(u) = 2\cos u\, \mathbf{i} + 2\sin u\, \mathbf{j} + 4\mathbf{k}, \qquad u \in [0, 2\pi].$$
By Stokes's theorem the flux we want is
$$-\int_C \mathbf{v}(\mathbf{r}) \cdot d\mathbf{r} = -\int_C y\, dx + z\, dy + x^2 z^2\, dz$$
$$= -\int_0^{2\pi} \left(-4\sin^2 u + 8\cos u\right)\, du = 4\pi.$$

11. The bounding curve C for S is the bounding curve of the elliptical region $\Omega : \tfrac{1}{4}x^2 + \tfrac{1}{9}y^2 = 1$. Since
$$\nabla \times \mathbf{v} = 2x^2 yz^2 \mathbf{i} - 2xy^2 z^2 \mathbf{j}$$
is zero on the xy-plane, the flux of $\nabla \times \mathbf{v}$ through Ω is zero, the circulation of $\mathbf{v}$ about C is zero, and therefore the flux of $\nabla \times \mathbf{v}$ through S is zero.

13. C bounds the surface

$$S: z = \sqrt{1 - \tfrac{1}{2}(x^2 + y^2)}, \qquad (x, y) \in \Omega$$

with $\Omega : x^2 + (y - \tfrac{1}{2})^2 \leq \tfrac{1}{4}$. Routine calculation shows that $\boldsymbol{\nabla} \times \mathbf{v} = y\mathbf{k}$. The circulation of $\mathbf{v}$ with respect to the upper unit normal $\mathbf{n}$ is given by

$$\iint\limits_{S} (y\mathbf{k} \cdot \mathbf{n})\, d\sigma = \iint\limits_{\Omega} y\, dx dy = \bar{y} A = \frac{1}{2} \left(\frac{\pi}{4} \right) = \frac{1}{8}\pi.$$

$$(18.7.9)$$

If $-\mathbf{n}$ is used, the circulation is $-\tfrac{1}{8}\pi$.   Answer: $\pm\tfrac{1}{8}\pi$.

15. $\boldsymbol{\nabla} \times \mathbf{v} = \mathbf{i} + 2\mathbf{j} + \mathbf{k}$. The paraboloid intersects the plane in a curve C that bounds a flat surface S that projects onto the disc $x^2 + (y - \tfrac{1}{2})^2 = \tfrac{1}{4}$ in the xy-plane. The upper unit normal to S is the vector $\mathbf{n} = \tfrac{1}{2}\sqrt{2}\,(-\mathbf{j} + \mathbf{k})$. The area of the base disc is $\tfrac{1}{4}\pi$. Letting γ be the angle between $\mathbf{n}$ and $\mathbf{k}$, we have $\cos\gamma = \mathbf{n} \cdot \mathbf{k} = \tfrac{1}{2}\sqrt{2}$ and $\sec\gamma = \sqrt{2}$. Therefore the area of S is $\tfrac{1}{4}\sqrt{2}\pi$. The circulation of $\mathbf{v}$ with respect to $\mathbf{n}$ is given by

$$\iint\limits_{S} [(\boldsymbol{\nabla} \times \mathbf{v}) \cdot \mathbf{n}]\, d\sigma = \iint\limits_{S} -\frac{1}{2}\sqrt{2}\, d\sigma = \left(-\frac{1}{2}\sqrt{2} \right) (\text{area of } S) = -\frac{1}{4}\pi.$$

If $-\mathbf{n}$ is used, the circulation is $\tfrac{1}{4}\pi$.   Answer: $\pm\tfrac{1}{4}\pi$.

17. Straightforward calculation shows that

$$\boldsymbol{\nabla} \times (\mathbf{a} \times \mathbf{r}) = \boldsymbol{\nabla} \times [(a_2 z - a_3 y)\,\mathbf{i} + (a_3 x - a_1 z)\,\mathbf{j} + (a_1 y - a_2 x)\mathbf{k}] = 2\mathbf{a}.$$

19. In the plane of C, the curve C bounds some Jordan region that we call Ω. The surface $S \cup \Omega$ is a piecewise–smooth surface that bounds a solid T. Note that $\boldsymbol{\nabla} \times \mathbf{v}$ is continuously differentiable on T.

Thus, by the divergence theorem,

$$\iiint\limits_{T} [\boldsymbol{\nabla} \cdot (\boldsymbol{\nabla} \times \mathbf{v})]\, dx dy dz = \iint\limits_{S \cup \Omega} [(\boldsymbol{\nabla} \times \mathbf{v}) \cdot \mathbf{n}]\, d\sigma$$

where $\mathbf{n}$ is the outer unit normal. Since the divergence of a curl is identically zero, we have

$$\iint\limits_{S \cup \Omega} [(\boldsymbol{\nabla} \times \mathbf{v}) \cdot \mathbf{n}]\, d\sigma = 0.$$

Now $\mathbf{n}$ is $\mathbf{n}_1$ on S and $\mathbf{n}_2$ on Ω. Thus

$$\iint\limits_{S} [(\boldsymbol{\nabla} \times \mathbf{v}) \cdot \mathbf{n}_1]\, d\sigma + \iint\limits_{\Omega} [(\boldsymbol{\nabla} \times \mathbf{v}) \cdot \mathbf{n}_2]\, d\sigma = 0.$$

This gives

$$\iint_S \left[(\nabla \times \mathbf{v}) \cdot \mathbf{n}_1 \right] d\sigma = \iint_\Omega \left[(\nabla \times \mathbf{v}) \cdot (-\mathbf{n}_2) \right] d\sigma = \oint_C \mathbf{v}(\mathbf{r}) \cdot d\mathbf{r}$$

where C is traversed in a positive sense with respect to $-\mathbf{n}_2$ and therefore in a positive sense with respect to $\mathbf{n}_1$. ($-\mathbf{n}_2$ points toward S.)

REVIEW EXERCISES

1. (a) $\mathbf{r}(u) = u\mathbf{i} + u\mathbf{j}, \quad 0 \le u \le 1; \quad \displaystyle\int_C \mathbf{h} \cdot d\mathbf{r} = \int_0^1 (u^3 - u^2)\, du = -\frac{1}{12}$

   (b) $\displaystyle\int_C \mathbf{h} \cdot d\mathbf{r} = \int_0^1 (2u^8 - 3u^7)\, du = -\frac{11}{72}$

3. Since $\mathbf{h}(x, y) = \nabla f$ where $f(x, y) = x^2 y^2 + \frac{1}{2}x^2 - y$,

$$\int_C \mathbf{h}(\mathbf{r}) \cdot d\mathbf{r} = f(2, 4) - f(-1, 2) = \frac{119}{2}$$

   for *any* curve C beginning at $(-1, 2)$ and ending at $(2, 4)$.

5. $\mathbf{h}(x, y, z) = \sin y\,\mathbf{i} + xe^{xy}\,\mathbf{j} + \sin z\,\mathbf{k}; \quad \mathbf{r}(u) = u^2\,\mathbf{i} + u\,\mathbf{j} + u^3\,\mathbf{k}, \quad u \in [0, 3]$

   $x(u) = u^2 \quad y(u) = u \quad z(u) = u^3, \quad x'(u) = 2u, \quad y'(u) = 1, \quad z'(u) = 3u^2$

   $\mathbf{h}(\mathbf{r}(u)) \cdot \mathbf{r}'(u) = 2u \sin u + u^2 e^{u^3} + 3u^2 \sin u^3$

$$\int_C \mathbf{h}(\mathbf{r}) \cdot d\mathbf{r} = \int_0^3 \left(2u \sin u + u^2 e^{u^3} + 3u^2 \sin u^3 \right) du$$

$$= \left[-2u \cos u + 2 \sin u + \tfrac{1}{3} e^{u^3} - \cos u^3 \right]_0^3$$

$$= \tfrac{2}{3} - 6 \cos 3 + 2 \sin 3 + \tfrac{1}{3} e^{27} - \cos 27$$

7. $\mathbf{F}(x, y, z) = xy\,\mathbf{i} + yz\,\mathbf{j} + xz\,\mathbf{k}; \quad \mathbf{r}(u) = u\,\mathbf{i} + u^2\,\mathbf{j} + u^3\,\mathbf{k}.$

   $\mathbf{F}(\mathbf{r}(u)) \cdot \mathbf{r}' = u^3 + 5u^6; \quad W = \displaystyle\int_{-1}^2 (u^3 + 5u^6)\,du = \left[\frac{1}{4}u^4 + \frac{5}{7}u^7 \right]_{-1}^2 = \frac{2685}{28}$

9. A vector equation for the line segment is: $\mathbf{r}(u) = (1 + 2u)\,\mathbf{i} + 4u\,\mathbf{k}, \quad u \in [0, 1]$.

   $\mathbf{F}(\mathbf{r}(u)) \cdot \mathbf{r}' = C\, \dfrac{2 + 20u}{\sqrt{1 + 4u + 20u^2}}; \quad \displaystyle\int_C \mathbf{F} \cdot d\mathbf{r} = C \int_0^1 \frac{(20u + 2)}{\sqrt{1 + 4u + 20u^2}}\, du = 4C$

11. $\dfrac{\partial(ye^{xy} + 2x)}{\partial y} = e^{xy} + xye^{xy} = \dfrac{\partial(xe^{xy} - 2y)}{\partial x} \implies \mathbf{h}$ is a gradient.

(a) $\mathbf{h}(\mathbf{r}(u)) \cdot \mathbf{r}' = 3u^2 e^{u^3} - 4u^3 + 2u; \quad \displaystyle\int_C \mathbf{h} \cdot d\mathbf{r} = \int_0^2 \left(3u^2 e^{u^3} - 4u^3 + 2u\right) du = e^8 - 13$

(b) Let $f(x, y) = e^{xy} + x^2 - y^2$. Then $\nabla f = \mathbf{h}$ and $\displaystyle\int_C \mathbf{h} \cdot d\mathbf{r} = f(2, 4) - f(0, 0) = e^8 - 13$

13. $\mathbf{h}(x, y, z) = \nabla f$ where $f(x, y, z) = x^4 y^3 z^2$.

(a) $\mathbf{h}(\mathbf{r}(u)) = 4u^{15}\,\mathbf{i} + 3u^{14}\,\mathbf{j} + 2u^{13}\,\mathbf{k}; \quad \mathbf{r}'(u) = \mathbf{i} + 2u\,\mathbf{j} + 3u^2\,\mathbf{k}$

$$\int_C \mathbf{h}(\mathbf{r}) \cdot d\mathbf{r} = \int_0^1 16\,u^{15}\,du = 1.$$

(b) $\displaystyle\int_C \mathbf{h}(\mathbf{r}) \cdot d\mathbf{r} = f(\mathbf{r}(1)) - f(\mathbf{r}(0)) = f(1, 1, 1) - f(0, 0, 0) = 1.$

15. (a) $\mathbf{r}(u) = (1 - u)\mathbf{i} + u\mathbf{j}, \quad 0 \le u \le 1.$

$$\int_C 2xy^{1/2}\,dx + yx^{1/2}\,dy = \int_0^1 \left[2(1 - u)u^{1/2}(-1) + u(1 - u)^{1/2}\right] du$$

$$= -2\int_0^1 (1 - u)u^{1/2}\,du + \int_0^1 u(1 - u)^{1/2}\,du$$

$$= -\int_0^1 (1 - u)u^{1/2}\,du = -\frac{4}{15}$$

(b) $\mathbf{r}_1 = \mathbf{i} + u\mathbf{j}, \quad 0 \le u \le 1; \quad \mathbf{r}_2 = (1 - u)\,\mathbf{i} + \mathbf{j}$

$$\int_C 2xy^{1/2}dx + yx^{1/2}dy = \int_0^1 u\,du + \int_0^1 -2(1 - u)\,du = -\frac{1}{2}$$

(c) $\mathbf{r} = \cos u\,\mathbf{i} + \sin u\,\mathbf{j}, \quad 0 \le u \le \pi/2$

$$\int_C 2xy^{1/2}dx + yx^{1/2}dy = \int_0^{\pi/2} \left(-2\sin^{3/2} u \cos u + \cos^{3/2} u \sin u\right) du = -\frac{2}{5}$$

17. $\displaystyle\int_C ye^{xy}\,dx + \cos x\,dy + \left(\frac{xy}{z}\right) dz = \int_0^2 \left(u^2 e^{u^3} + 2u\cos u + 3u^2\right) du$

$$= \left[\tfrac{1}{3}e^{u^3} + 2u\sin u + 2\cos u + u^3\right]_0^2$$

$$= \frac{1}{3}e^8 + \frac{17}{3} + 4\sin 2 + 2\cos 2$$

19. (a) Set $C_1 : \mathbf{r}(u) = u\,\mathbf{i} + u^2\,\mathbf{j}, \quad 0 \le u \le 1; \quad C_2 : \mathbf{r}(u) = (1-u)\,\mathbf{i} + \sqrt{1-u}\,\mathbf{j}, \quad 0 \le u \le 1.$

Then, $C = C_1 + C_2$.

$$\oint_C xy^2\,dx - x^2 y\,dy = \int_{C_1} xy^2\,dx - x^2 y\,dy + \int_{C_2} xy^2\,dx - x^2 y\,dy$$

$$= \int_0^1 (u^5 - 2u^5)\,du + \int_0^1 \left[-(1-u)^2 + \tfrac{1}{2}(1-u)^2 \right]\,du$$

$$= \int_0^1 (-u^5)\,du - \tfrac{1}{2}\int_0^1 (1-u)^2\,du = \left[-\tfrac{1}{6}u^6 + \tfrac{1}{6}(1-u)^3 \right]_0^1 = -\tfrac{1}{3}$$

(b) $P = xy^2; \quad Q = -x^2 y$

$$\oint_C xy^2\,dx - x^2 y\,dy = \int_0^1 \int_{x^2}^{\sqrt{x}} (-4xy)\,dy\,dx = \int_0^1 (2x^2 - 2x^5)\,dx = -\frac{1}{3}$$

21. $P = x - 2y^2; \quad Q = 2xy$

$$\oint_C (x - 2y^2)\,dx + 2xy\,dy = \int_0^2 \int_0^1 6y\,dy\,dx = 6$$

23. $P = \ln(x^2 + y^2); \quad Q = \ln(x^2 + y^2); \quad \dfrac{\partial Q}{\partial x} - \dfrac{\partial P}{\partial y} = \dfrac{2x - 2y}{x^2 + y^2}$

$$\oint_C \ln(x^2 + y^2)\,dx + \ln(x^2 + y^2)\,dy = \iint_\Omega \frac{2x - 2y}{x^2 + y^2}\,dx\,dy$$

$$= \int_0^\pi \int_1^2 \frac{2r\cos\theta - 2r\sin\theta}{r^2}\,r\,dr\,d\theta$$

$$= 2\int_0^\pi \int_1^2 (\cos\theta - \sin\theta)\,dr\,d\theta = -4$$

25. $\displaystyle \oint y^2\,dx = \iint_\Omega -2y\,dx\,dy = \int_0^{2\pi} \int_0^{1+\sin\theta} -2r^2\sin\theta\,dr\,d\theta = \int_0^{2\pi} (-\tfrac{2}{3})(1+\sin\theta)^3 \sin\theta\,d\theta = -\dfrac{5\pi}{2}$

27. $C_1 : \mathbf{r}(u) = -u\,\mathbf{i} + (4 - u^2)\,\mathbf{j}, \; -2 \le u \le 2; \quad C_2 : \mathbf{r}(u) = u\,\mathbf{i}, \; -2 \le u \le 2; \quad C = C_1 \cup C_2$

$$A = \frac{1}{2}\int_C (-y\,dx + x\,dy) = \frac{1}{2}\int_{C_1} (-y\,dx + x\,dy) + \frac{1}{2}\int_{C_2} (-y\,dx + x\,dy)$$

$$= \frac{1}{2}\int_{-2}^2 -(4 - u^2)(-1)\,du - u(-2u)\,du + \frac{1}{2}\int_{-2}^2 0\,du$$

$$= \int_{-2}^2 (4 + u^2)\,du = \frac{32}{3}$$

29. By symmetry, it is sufficient to consider the upper part of the sphere: $z = \sqrt{4 - x^2 - y^2}$

$$\frac{\partial z}{\partial x} = \frac{-x}{\sqrt{4 - x^2 - y^2}}, \qquad \frac{\partial z}{\partial y} = \frac{-y}{\sqrt{4 - x^2 - y^2}}$$

Let Ω be the projection of the sphere onto the xy plane, then

$$S = 2 \iint_\Omega \sqrt{(z_x)^2 + (z_y)^2 + 1} \, dx \, dy = 2 \iint_\Omega \frac{2}{\sqrt{4 - x^2 - y^2}} \, dx \, dy$$

$$= 4 \int_{-\pi/2}^{\pi/2} \int_0^{2\cos\theta} \frac{1}{\sqrt{4 - r^2}} \, r \, dr \, d\theta$$

$$= 4 \int_{-\pi/2}^{\pi/2} \left(2 - 2\sqrt{1 - \cos^2\theta}\right) d\theta = 8(\pi - 2)$$

31. $\dfrac{\partial z}{\partial x} = \dfrac{x}{\sqrt{x^2 + y^2}}, \quad \dfrac{\partial z}{\partial y} = \dfrac{y}{\sqrt{x^2 + y^2}}.$

The projection Ω of the surface onto the xy plane is the disk $x^2 + y^2 \le 9$.

$$S = \iint_\Omega \sqrt{(z_x)^2 + (z_y)^2 + 1} \, dx \, dy = \iint_\Omega \sqrt{2} \, dx \, dy = \int_0^{2\pi} \int_0^3 \sqrt{2} \, r \, dr \, d\theta = 9\pi\sqrt{2}$$

33. $\displaystyle \iint_S yz \, d\sigma = \sqrt{2} \int_0^{2\pi} \int_0^1 r^2 \sin\theta (r \sin\theta + 4) \, dr \, d\theta = \frac{\sqrt{2}\pi}{4}$

35. The cylindrical surface S_1 is parametrized by: $x = u, \; y = 2\cos v, \; z = 2\sin v, \; 0 \le u \le 2, \; 0 \le v \le 2\pi.$

$$\mathbf{N}(u, v) = -2\cos v \, \mathbf{i} - 2\sin v \, \mathbf{j}, \quad \|\|\mathbf{N}(u, v)\|\| = 2$$

$$\iint_{S_1} (x^2 + y^2 + z^2) \, d\sigma = \int_0^2 \int_0^{2\pi} (u^2 + 4) \, 2 \, dv \, du = \frac{128\pi}{3}$$

The disc $S_2 : x = 0, \; y^2 + z^2$ is parametrized by: $x = 0, \; y = u\cos v, \; z = u\sin v, \; 0 \le u \le 2,$ $0 \le v \le 2\pi.$

$$\mathbf{N} = u\,\mathbf{i}, \quad \|\|\mathbf{N}(u, v)\|\| = u; \qquad \iint_{S_2} (x^2 + y^2 + z^2) \, d\sigma = \int_0^2 \int_0^{2\pi} (0 + u^2) \, u \, dv \, du = 8\pi$$

The disc $S_3 : x = 2, \; y^2 + z^2$ is parametrized by: $x = 2, \; y = u\cos v, \; z = u\sin v, \; 0 \le u \le 2,$ $0 \le v \le 2\pi.$

$$\mathbf{N} = u\,\mathbf{i}, \quad \|\|\mathbf{N}(u, v)\|\| = u; \qquad \iint_{S_2} (x^2 + y^2 + z^2) \, d\sigma = \int_0^2 \int_0^{2\pi} (4 + u^2) \, u \, dv \, du = 24\pi$$

Thus, $\displaystyle \iint_S (x^2 + y^2 + z^2) \, d\sigma = \frac{128\pi}{3} + 8\pi + 24\pi = \frac{224\pi}{3}$

37. $\nabla \cdot \mathbf{v} = 4x, \quad \nabla \times \mathbf{v} = 2y\mathbf{k}$

39. $\nabla \cdot \mathbf{v} = 1 + xy, \quad \nabla \times \mathbf{v} = (xz - x)\mathbf{i} - yz\mathbf{j} + z\mathbf{k}$

41. (a) $\nabla \cdot \mathbf{v} = z - x + y$

$$\int_0^1 \int_0^1 \int_0^1 (z - x + y)\,dz\,dy\,dx = \frac{1}{2}$$

(b) at $x = 0, \quad \mathbf{n} = -\mathbf{i}, \mathbf{v} \cdot \mathbf{n} = 0, \displaystyle\int_0^1 \int_0^1 0\,dy\,dz = 0$

at $x = 1, \quad \mathbf{n} = \mathbf{i}, \mathbf{v} \cdot \mathbf{n} = z, \displaystyle\int_0^1 \int_0^1 z\,dy\,dz = 1/2$

at $y = 0, \quad \mathbf{n} = -\mathbf{j}, \mathbf{v} \cdot \mathbf{n} = xy = 0, \displaystyle\int_0^1 \int_0^1 0\,dx\,dz = 0$

at $y = 1, \quad \mathbf{n} = \mathbf{j}, \mathbf{v} \cdot \mathbf{n} = -xy = -x, \displaystyle\int_0^1 \int_0^1 -x\,dx\,dz = -1/2$

at $z = 0, \quad \mathbf{n} = -\mathbf{k}, \mathbf{v} \cdot \mathbf{n} = 0, \displaystyle\int_0^1 \int_0^1 0\,dy\,dx = 0$

at $z = 1, \quad \mathbf{n} = \mathbf{k}, \mathbf{v} \cdot \mathbf{n} = yz, \displaystyle\int_0^1 \int_0^1 y\,dy\,dx = 1/2$

The sum is $1/2$

43. The projection of S onto the xy-plane is: $\quad \Omega : x^2 + y^2 \leq 9$.

$$\iint_S \mathbf{v} \cdot \mathbf{n}\,d\sigma = \iint_\Omega \left(4x^2 + 2xyz + z^2\right) dx\,dy$$

$$= \iint_\Omega \left(4x^2 + 2xy\left[9 - x^2 - y^2\right] + \left[9 - x^2 - y^2\right]^2\right) dx\,dy$$

$$= \int_0^{2\pi} \int_0^3 \left[4r^2 \cos^2 \theta + r^2(9 - r^2)\sin 2\theta + (9 - r^2)^2\right] r\,dr\,d\theta = 324\pi$$

45. (a) $(\nabla \times \mathbf{v}) \cdot \mathbf{n} = (\mathbf{i} + \mathbf{j} + \mathbf{k}) \cdot \left(-\dfrac{1}{2}x\mathbf{i} - \dfrac{1}{2}y\mathbf{j} + \dfrac{\sqrt{4 - x^2 - y^2}}{2}\mathbf{k}\right) = -\dfrac{1}{2}x - \dfrac{1}{2}y + \dfrac{\sqrt{4 - x^2 - y^2}}{2}$

$$\iint_S \left(-\frac{1}{2}x - \frac{1}{2}y + \frac{\sqrt{4-x^2-y^2}}{2}\right) d\sigma = \iint_S \left(-\frac{1}{2}x - \frac{1}{2}y + \frac{\sqrt{4-x^2-y^2}}{2}\right) \frac{2}{\sqrt{4-x^2-y^2}}\, dx\, dy$$

$$= \iint \left(\frac{-x}{\sqrt{4-x^2-y^2}} - \frac{-y}{\sqrt{4-x^2-y^2}} + 1\right) dx\, dy$$

$$= \int_0^{2\pi} \int_0^2 \left(-\frac{r\cos\theta}{\sqrt{4-r^2}} - \frac{r\sin\theta}{\sqrt{4-r^2}} + 1\right) r\, dr\, d\theta = 4\pi$$

(b) $\mathbf{r}(\theta) = 2\cos\theta\,\mathbf{i} + 2\sin\theta\,\mathbf{j}, \quad 0 \le \theta \le 2\pi$

$$\iint_S [(\nabla \times \mathbf{v}) \cdot \mathbf{n}] d\sigma = \oint_C \mathbf{v}(\mathbf{r}) \cdot d\mathbf{r} = \int_0^{2\pi} 4\cos^2\theta\, d\theta = 4\pi$$

CHAPTER 19

SECTION 19.1

1. $y' + xy = xy^3 \implies y^{-3}y' + xy^{-2} = x.$ Let $v = y^{-2}, \quad v' = -2y^{-3}y'.$

$$-\frac{1}{2}v' + xv = x$$
$$v' - 2xv = -2x$$
$$e^{-x^2}v' - 2xe^{-x^2}v = -2xe^{-x^2}$$
$$e^{-x^2}v = e^{-x^2} + C$$
$$v = 1 + Ce^{x^2}$$
$$y^2 = \frac{1}{1 + Ce^{x^2}}.$$

3. $y' - 4y = 2e^x y^{\frac{1}{2}} \implies y^{-\frac{1}{2}}y' - 4y^{\frac{1}{2}} = 2e^x.$ Let $v = y^{\frac{1}{2}}, \quad v' = \frac{1}{2}y^{-\frac{1}{2}}y'.$

$$2v' - 4v = 2e^x$$
$$v' - 2v = e^x$$
$$e^{-2x}v' - 2e^{-2x}v = e^{-x}$$
$$e^{-2x}v = -e^{-x} + C$$
$$v = -e^x + Ce^{2x}$$
$$y = (Ce^{2x} - e^x)^2.$$

5. $(x-2)y' + y = 5(x-2)^2 y^{\frac{1}{2}} \implies y^{-\frac{1}{2}}y' + \frac{1}{x-2}y^{\frac{1}{2}} = 5(x-2).$ Let $v = y^{\frac{1}{2}}, \quad v' = \frac{1}{2}y^{-\frac{1}{2}}y'.$

$$2v' + \frac{1}{x-2}v = 5(x-2)$$
$$v' + \frac{1}{2(x-2)}v = \frac{5}{2}(x-2)$$
$$\sqrt{x-2}\,v' + \frac{1}{2\sqrt{x-2}}v = \frac{5}{2}(x-2)^{\frac{3}{2}}$$
$$\sqrt{x-2}\,v = (x-2)^{\frac{5}{2}} + C$$
$$v = (x-2)^2 + \frac{C}{\sqrt{x-2}}$$
$$y = \left[(x-2)^2 + \frac{C}{\sqrt{x-2}}\right]^2.$$

7. $y' + xy = y^3 e^{x^2} \implies y^{-3}y' + xy^{-2} = e^{x^2}$. Let $v = y^{-2}$, $v' = -2y^{-3}y'$.

$$-\frac{1}{2}v' + xv = e^{x^2}$$

$$v' - 2xv = -2e^{x^2}$$

$$e^{-x^2}v' - 2xe^{-x^2}v = -2$$

$$e^{-x^2}v = -2x + C$$

$$v = -2xe^{x^2} + Ce^{x^2}$$

$$y^{-2} = Ce^{x^2} - 2xe^{x^2}.$$

$C = 4 \implies y^{-2} = 4e^{x^2} - 2xe^{x^2}.$

9. $2x^3 y' - 3x^2 y = y^3 \implies y^{-3}y' - \frac{3}{2x}y^{-2} = \frac{1}{2x^3}$. Let $v = y^{-2}$, $v' = -2y^{-3}y'$.

$$-\frac{1}{2}v' - \frac{3}{2x}v = \frac{1}{2x^3}$$

$$v' + \frac{3}{x}v = -\frac{1}{x^3}$$

$$x^3 v' + 3x^2 v = -1$$

$$x^3 v = -x + C$$

$$v = \frac{C - x}{x^3}$$

$$y^2 = \frac{x^3}{C - x}$$

$1 = \frac{1}{C - x} \implies C = 2 \implies y^2 = \frac{x^3}{2 - x}.$

11. $y' - \frac{y}{x}\ln y = xy \implies \frac{y'}{y} - \frac{1}{x}\ln y = x$. Let $u = \ln y$, $u' = \frac{y'}{y}$.

$$u' - \frac{1}{x}u = x$$

$$\frac{1}{x}u' - \frac{1}{x^2}u = 1$$

$$\frac{1}{x}u = x + C$$

$$u = x^2 + Cx$$

$$\ln y = x^2 + Cx.$$

13. $f(x,y) = \dfrac{x^2+y^2}{2xy};$ $f(tx,ty) = \dfrac{(tx)^2+(ty)^2}{2(tx)(ty)} = \dfrac{t^2(x^2+y^2)}{t^2(2xy)} = \dfrac{x^2+y^2}{2xy} = f(x,y)$

Set $vx = y.$ Then, $v + xv' = y'$ and

$$v + xv' = \frac{x^2+v^2x^2}{2vx^2} = \frac{1+v^2}{2v}$$

$$v - \frac{1+v^2}{2v} + xv' = 0$$

$$v^2 - 1 + 2xvv' = 0$$

$$\frac{1}{x}\,dx + \frac{2v}{v^2-1}\,dv = 0$$

$$\int \frac{1}{x}\,dx + \int \frac{2v}{v^2-1}\,dv = C$$

$$\ln|x| + \ln|v^2-1| = K \quad \text{or} \quad x(v^2-1) = C$$

Replacing v by $y/x,$ we get

$$x\left(\frac{y^2}{x^2} - 1\right) = C \quad \text{or} \quad y^2 - x^2 = Cx$$

15. $f(x,y) = \dfrac{x-y}{x+y};$ $f(tx,ty) = \dfrac{(tx)-(ty)}{tx+ty} = \dfrac{t(x-y)}{t(x+y)} = \dfrac{x-y}{x+y} = f(x,y)$

Set $vx = y.$ Then, $v + xv' = y'$ and

$$v + xv' = \frac{x - vx}{x + vx} = \frac{1-v}{1+v}$$

$$v^2 + 2v - 1 + x(1+v)v' = 0$$

$$\frac{1}{x}\,dx + \frac{1+v}{v^2+2v-1}\,dv = 0$$

$$\int \frac{1}{x}\,dx + \int \frac{1+v}{v^2+2v-1}\,dv = C$$

$$\ln|x| + \tfrac{1}{2}\ln|v^2+2v-1| = K \quad \text{or} \quad x\sqrt{v^2+2v-1} = C$$

Replacing v by $y/x,$ we get

$$x\sqrt{\frac{y^2}{x^2} + 2\frac{y}{x} - 1} = C \quad \text{or} \quad y^2 + 2xy - x^2 = C$$

17. $f(x,y) = \dfrac{x^2 e^{y/x} + y^2}{xy}$; $f(tx, ty) = \dfrac{(tx)^2 - e^{(ty)/(tx)} + (ty)^2}{(tx)(ty)} = \dfrac{t^2 \left(x^2 e^{y/x} + y^2\right)}{t^2 (xy)} = f(x,y)$

Set $vx = y$. Then, $v + xv' = y'$ and

$$v + xv' = \frac{x^2 e^v + v^2 x^2}{vx^2} = \frac{e^v + v^2}{v}$$

$$v^2 + xvv' = e^v + v^2$$

$$-e^v + xvv' = 0$$

$$\frac{1}{x}\, dx = ve^{-v}\, dv$$

$$\int \frac{1}{x}\, dx = \int ve^{-v}\, dv$$

$$\ln|x| = -ve^{-v} - e^{-v} + C$$

Replacing v by y/x, and simplifying, we get

$$y + x = xe^{y/x}(C - \ln|x|)$$

19. $f(x,y) = \dfrac{y}{x} + \sin(y/x)$; $f(tx, ty) = \dfrac{(ty)}{tx} + \sin[(ty/tx)] = \dfrac{y}{x} + \sin(y/x) = f(x,y)$

Set $vx = y$. Then, $v + xv' = y'$ and

$$v + xv' = \frac{vx}{x} + \sin[(vx)/x] = v + \sin v$$

$$xv' = \sin v$$

$$\csc v \, dv = \frac{1}{x}\, dx$$

$$\int \csc v \, dv = \int \frac{1}{x}\, dx$$

$$\ln|\csc v - \cot v| = \ln|x| + K \qquad \text{or} \qquad \csc v - \cot v = Cx$$

Replacing v by y/x, and simplifying, we get

$$1 - \cos(y/x) = Cx\,\sin(y/x)$$

21. The differential equation is homogeneous since

$$f(x,y) = \frac{y^3 - x^3}{xy^2}; \qquad f(tx, ty) = \frac{(ty)^3 - (tx)^3}{(tx)(ty)^2} = \frac{t^3(y^3 - x^3)}{t^3(xy^2)} = \frac{y^3 - x^3}{xy^2} = f(x,y)$$

Set $vx = y$. Then, $v + xv' = y'$ and

$$v + xv' = \frac{(vx)^3 - x^3}{v^2 x^3} = \frac{v^3 - 1}{v^2}$$

$$1 + xv^2 v' = 0$$

$$\frac{1}{x}\, dx + v^2 \, dv = 0$$

$$\int \frac{1}{x}\, dx + \int v^2 \, dv = 0$$

$$\ln|x| + \frac{1}{3}v^3 = C$$

Replacing v by y/x, we get

$$y^3 + 3x^3 \ln|x| = Cx^3$$

Applying the side condition $y(1) = 2$, we have

$$8 + 3\ln 1 = C \quad \Longrightarrow \quad C = 8 \quad \text{and} \quad y^3 + 3x^3 \ln|x| = 8x^3$$

SECTION 19.2

1. $\dfrac{\partial P}{\partial y} = 2xy - 1 = \dfrac{\partial Q}{\partial x};$ the equation is exact on the whole plane.

$\dfrac{\partial f}{\partial x} = xy^2 - y \quad \Longrightarrow \quad f(x,y) = \tfrac{1}{2}x^2y^2 - xy + \varphi(y)$

$\dfrac{\partial f}{\partial y} = x^2y - x + \varphi'(y) = x^2y - x \quad \Longrightarrow \quad \varphi'(y) = 0 \quad \Longrightarrow \quad \varphi(y) = 0 \;\;\text{(omit the constant)}*$

Therefore $f(x,y) = \tfrac{1}{2}x^2y^2 - xy,$ and a one-parameter family of solutions is:

$$\tfrac{1}{2}x^2y^2 - xy = C$$

* We will omit the constant at this step throughout this section.

3. $\dfrac{\partial P}{\partial y} = e^y - e^x = \dfrac{\partial Q}{\partial x};$ the equation is exact on the whole plane.

$\dfrac{\partial f}{\partial x} = e^y - ye^x \quad \Longrightarrow \quad f(x,y) = xe^y - ye^x + \varphi(y)$

$\dfrac{\partial f}{\partial y} = xe^y - e^x + \varphi'(y) = xe^y - e^x \quad \Longrightarrow \quad \varphi'(y) = 0 \quad \Longrightarrow \quad \varphi(y) = 0$

Therefore $f(x,y) = xe^y - ye^x,$ and a one-parameter family of solutions is:

$$xe^y - ye^x = C$$

5. $\dfrac{\partial P}{\partial y} = \dfrac{1}{y} + 2x = \dfrac{\partial Q}{\partial x};$ the equation is exact on the upper half plane.

$\dfrac{\partial f}{\partial x} = \ln y + 2xy \quad \Longrightarrow \quad f(x,y) = x \ln y + x^2y + \varphi(y)$

$\dfrac{\partial f}{\partial y} = \dfrac{x}{y} + x^2 + \varphi'(y) = \dfrac{x}{y} + x^2 \quad \Longrightarrow \quad \varphi'(y) = 0 \quad \Longrightarrow \quad \varphi(y) = 0$

Therefore $f(x,y) = x \ln y + x^2y,$ and a one-parameter family of solutions is:

$$x \ln y + x^2y = C$$

7. $\dfrac{\partial P}{\partial y} = \dfrac{1}{x} = \dfrac{\partial Q}{\partial x};$ the equation is exact on the right half plane.

$\dfrac{\partial f}{\partial x} = \dfrac{y}{x} + 6x \quad \Longrightarrow \quad f(x,y) = y \ln x + 3x^2 + \varphi(y)$

$$\frac{\partial f}{\partial y} = \ln x + \varphi'(y) = \ln x - 2 \quad \Longrightarrow \quad \varphi'(y) = -2 \quad \Longrightarrow \quad \varphi(y) = -2y$$

Therefore $f(x, y) = y \ln x + 3x^2 - 2y$,   and a one-parameter family of solutions is:

$$y \ln x + 3x^2 - 2y = C$$

9. $\dfrac{\partial P}{\partial y} = 3y^2 - 2y \sin x = \dfrac{\partial Q}{\partial x}$;   the equation is exact on the whole plane.

$$\frac{\partial f}{\partial x} = y^3 - y^2 \sin x - x \quad \Longrightarrow \quad f(x, y) = xy^3 + y^2 \cos x - \tfrac{1}{2} x^2 + \varphi(y)$$

$$\frac{\partial f}{\partial y} = 3xy^2 + 2y \cos x + \varphi'(y) = 3xy^2 + 2y \cos x + e^{2y} \quad \Longrightarrow \quad \varphi'(y) = e^{2y} \quad \Longrightarrow \quad \varphi(y) = \tfrac{1}{2} e^{2y}$$

Therefore $f(x, y) = xy^3 + y^2 \cos x - \tfrac{1}{2} x^2 + \tfrac{1}{2} e^{2y}$, and a one-parameter family of solutions is:

$$xy^3 + y^2 \cos x - \tfrac{1}{2} x^2 + \tfrac{1}{2} e^{2y} = C$$

11. (a) Yes: $\dfrac{\partial}{\partial y}[p(x)] = 0 = \dfrac{\partial}{\partial x}[q(y)]$.

(b) For all x, y such that $p(y)q(x) \neq 0$, $\dfrac{1}{p(y)q(x)}$ is an integrating factor.

Multiplying the differential equation by $\dfrac{1}{p(y)q(x)}$, we get

$$\frac{1}{q(x)} + \frac{1}{p(y)} y' = 0$$

which has the form of the differential equation in part (a).

13. $\dfrac{\partial P}{\partial y} = e^{y-x} - 1$ and $\dfrac{\partial Q}{\partial x} = e^{y-x} - xe^{y-x}$;   the equation is not exact.

Since $\dfrac{1}{Q}\left(\dfrac{\partial P}{\partial y} - \dfrac{\partial Q}{\partial x}\right) = \dfrac{1}{xe^{y-x} - 1}\left(xe^{y-x} - 1\right) = 1, \quad \mu(x) = e^{\int dx} = e^x$ is

an integrating factor. Multiplying the given equation by e^x, we get

$$(e^y - ye^x) + (xe^y - e^x)y' = 0$$

This is the equation given in Exercise 3. A one-parameter family of solutions is:

$$xe^y - ye^x = C$$

15. $\dfrac{\partial P}{\partial y} = 6x^2 y + e^y = \dfrac{\partial Q}{\partial x}$;   the equation is exact.

$$\frac{\partial f}{\partial x} = 3x^2 y^2 + x + e^y \quad \Longrightarrow \quad f(x, y) = x^3 y^2 + \tfrac{1}{2} x^2 + xe^y + \varphi(y)$$

$$\frac{\partial f}{\partial y} = 2x^3 y + xe^y + \varphi'(y) = 2x^3 y + y + xe^y \quad \Longrightarrow \quad \varphi'(y) = y \quad \Longrightarrow \quad \varphi(y) = \tfrac{1}{2} y^2$$

Therefore $f(x, y) = x^3 y^2 + \tfrac{1}{2} x^2 + xe^y + \tfrac{1}{2} y^2$,   and a one-parameter family of solutions is:

$$x^3 y^2 + \tfrac{1}{2} x^2 + xe^y + \tfrac{1}{2} y^2 = C$$

17. $\dfrac{\partial P}{\partial y} = 3y^2$ and $\dfrac{\partial Q}{\partial x} = 0;$ the equation is not exact.

Since $\dfrac{1}{Q}\left(\dfrac{\partial P}{\partial y} - \dfrac{\partial Q}{\partial x}\right) = \dfrac{1}{3y^2}(3y^2) = 1,$ $\mu(x) = e^{\int dx} = e^x$ is an

an integrating factor. Multiplying the given equation by $e^x,$ we get

$$\left(y^3 e^x + x e^x + e^x\right) + \left(3y^2 e^x\right) y' = 0$$

$\dfrac{\partial f}{\partial x} = y^3 e^x + x e^x + e^x$ $\implies$ $f(x,y) = y^3 e^x + x e^x + \varphi(y)$

$\dfrac{\partial f}{\partial y} = 3y^2 e^x + \varphi'(y) = 3y^2 e^x$ $\implies$ $\varphi'(y) = 0$ $\implies$ $\varphi(y) = 0$

Therefore $f(x,y) = y^3 e^x + x e^x,$ and a one-parameter family of solutions is:

$$y^3 e^x + x e^x = C$$

19. $\dfrac{\partial P}{\partial y} = 1 = \dfrac{\partial Q}{\partial x};$ the equation is exact.

$\dfrac{\partial f}{\partial x} = x^2 + y$ $\implies$ $f(x,y) = \tfrac{1}{3}x^3 + xy + \varphi(y)$

$\dfrac{\partial f}{\partial y} = x + \varphi'(y) = x + e^y$ $\implies$ $\varphi'(y) = e^y$ $\implies$ $\varphi(y) = e^y$

Therefore $f(x,y) = \tfrac{1}{3}x^3 + xy + e^y,$ and a one-parameter family of solutions is:

$$\tfrac{1}{3}x^3 + xy + e^y = C$$

Setting $x = 1,\ y = 0,$ we get $C = \tfrac{4}{3}$ and

$$\tfrac{1}{3}x^3 + xy + e^y = \tfrac{4}{3}   \text{or}   x^3 + 3xy + 3e^y = 4$$

21. $\dfrac{\partial P}{\partial y} = 4y$ and $\dfrac{\partial Q}{\partial x} = 2y;$ the equation is not exact.

Since $\dfrac{1}{Q}\left(\dfrac{\partial P}{\partial y} - \dfrac{\partial Q}{\partial x}\right) = \dfrac{1}{2xy}(2y) = \dfrac{1}{x},$ $\mu(x) = e^{\int (1/x)\, dx} = e^{\ln x} = x$ is an

integrating factor. Multiplying the given equation by $x,$ we get

$$\left(2xy^2 + x^3 + 2x\right) + \left(2x^2 y\right) y' = 0$$

$\dfrac{\partial f}{\partial y} = 2x^2 y$ $\implies$ $f(x,y) = x^2 y^2 + \varphi(x)$

$\dfrac{\partial f}{\partial x} = 2xy^2 + \varphi'(x) = 2xy^2 + x^3 + 2x$ $\implies$ $\varphi'(x) = x^3 + 2x$ $\implies$ $\varphi = \tfrac{1}{4}x^4 + x^2$

Therefore $f(x,y) = x^2y^2 + \frac{1}{4}x^4 + x^2$,   and a one-parameter family of solutions is:

$$x^2y^2 + \frac{1}{4}x^4 + x^2 = C$$

Setting $x = 1$, $y = 0$, we get $C = \frac{5}{4}$ and

$$x^2y^2 + \frac{1}{4}x^4 + x^2 = \frac{5}{4} \qquad \text{or} \qquad 4x^2y^2 + x^4 + 4x^2 = 5$$

23. $\dfrac{\partial P}{\partial y} = 3y^2$ and $\dfrac{\partial Q}{\partial x} = y^2$;   the equation is not exact.

Since $\dfrac{1}{P}\left(\dfrac{\partial P}{\partial y} - \dfrac{\partial Q}{\partial x}\right) = \dfrac{1}{y^3}\left(2y^2\right) = \dfrac{2}{y}$, $w(y) = e^{-\int (2/y)\,dy} = e^{-2\ln y} = y^{-2}$ is an

integrating factor. Multiplying the given equation by y^{-2}, we get

$$y + \left(y^{-2} + x\right)y' = 0$$

$\dfrac{\partial f}{\partial x} = y \implies f(x,y) = xy + \varphi(y)$

$\dfrac{\partial f}{\partial y} = x + \varphi'(y) = y^{-2} + x \implies \varphi'(y) = y^{-2} \implies \varphi(y) = -\dfrac{1}{y}$

Therefore $f(x,y) = xy - \dfrac{1}{y}$,   and a one-parameter family of solutions is: $xy - \dfrac{1}{y} = C$

Setting $x = -2$, $y = -1$, we get $C = 3$ and the solution $xy - \dfrac{1}{y} = 3$.

25. $\dfrac{\partial P}{\partial y} = -2y\sinh(x - y^2) = \dfrac{\partial Q}{\partial x}$;   the equation is exact.

$\dfrac{\partial f}{\partial x} = \cosh(x - 2y^2) + e^{2x} \implies f(x,y) = \sinh(x - y^2) + \frac{1}{2}e^{2x} + \varphi(y)$

$\dfrac{\partial f}{\partial y} = -2y\cosh(x - y^2) + \varphi'(y) = y - 2y\cosh(x - y^2) \implies \varphi'(y) = y \implies \varphi(y) = \frac{1}{2}y^2$

Therefore $f(x,y) = \sinh(x - y^2) + \frac{1}{2}e^{2x} + \frac{1}{2}y^2$,   and a one-parameter family of solutions is:

$$\sinh(x - y^2) + \frac{1}{2}e^{2x} + \frac{1}{2}y^2 = C$$

Setting $x = 2$, $y = \sqrt{2}$, we get $C = \frac{1}{2}e^4 + 1$ and the solution

$$\sinh(x - y^2) + \frac{1}{2}e^{2x} + \frac{1}{2}y^2 = \frac{1}{2}e^4 + 1$$

27. (a) $\dfrac{\partial P}{\partial y} = 2xy + kx^2$ and $\dfrac{\partial Q}{\partial x} = 2xy + 3x^2 \implies k = 3$.

(b) $\dfrac{\partial P}{\partial y} = e^{2xy} + 2xye^{2xy}$ and $\dfrac{\partial Q}{\partial x} = ke^{2xy} + 2kxye^{2xy} \implies k = 1$.

29. $y' = y^2x^3$;   the equation is separable.

$$y^{-2}\,dy = x^3\,dx \implies -\frac{1}{y} = \frac{1}{4}x^4 + C \implies y = \frac{-4}{x^4 + C}$$

31. $y' + \dfrac{4}{x} y = x^4$;   the equation is linear.

$$H(x) = \int (4/x)\, dx = 4 \ln x = \ln x^4, \quad \text{integrating factor: } e^{\ln x^4} = x^4$$

$$x^4 y' + 4x^3 y = x^8$$

$$\tfrac{d}{dx}\left[x^4 y\right] = x^8$$

$$x^4 y = \tfrac{1}{9} x^9 + C$$

$$y = \tfrac{1}{9} x^5 + C x^{-4}$$

33. $\dfrac{\partial P}{\partial y} = e^{xy} + xye^{xy} = \dfrac{\partial Q}{\partial x}$;   the equation is exact.

$$\frac{\partial f}{\partial x} = ye^{xy} - 2x \quad \Longrightarrow \quad f(x,y) = e^{xy} - x^2 + \varphi(y)$$

$$\frac{\partial f}{\partial y} = xe^{xy} + \varphi'(y) = \frac{2}{y} + xe^{xy} \quad \Longrightarrow \quad \varphi'(y) = \frac{2}{y} \quad \Longrightarrow \quad \varphi(y) = 2 \ln |y|$$

Therefore $f(x,y) = e^{xy} - x^2 + 2 \ln |y|$,   and a one-parameter family of solutions is:

$$e^{xy} - x^2 + 2 \ln |y| = C$$

SECTION 19.3

1. $y' = y \quad \Longrightarrow \quad y = Ce^x$.   Also, $y(0) = 1 \quad \Longrightarrow \quad C = 1$
Thus $y = e^x$ and $y(1) = 2.71828$
(a) 2.48832, relative error= 8.46%.
(b) 2.71825, relative error= 0.001%.

3. (a) 2.59374, relative error= 4.58%.
(b) 2.71828, relative error= 0%.

5. $y' = 2x \quad \Longrightarrow \quad y = x^2 + C$.   Also, $y(2) = 5 \quad \Longrightarrow \quad C = 1$
Thus $y = x^2 + 1$ and $y(1) = 2$.
(a) 1.9, relative error= 5.0%.
(b) 2.0, relative error= 0%.

7. $y' = \dfrac{1}{2y}$
Thus $y = \sqrt{x}$ and $y(2) = \sqrt{2} \simeq 1.41421$.
(a) 1.42052, relative error= −0.45%.
(b) 1.41421, relative error= 0%.

9. (a) 2.65330, relative error= 2.39%.
(b) 2.71828, relative error= 0%.

PROJECT 19.3

1. (a) and (b)

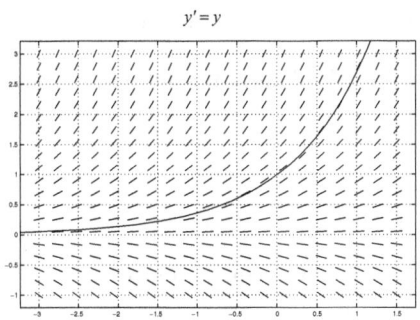

$$y' = y$$

(c) $y - y' = 0$ $H(x) = \int -dx = -x$;   integrating factor: e^{-x}

$$e^{-x}y' - e^{-x}y = 0$$

$$\frac{d}{dx}(e^{-x}y) = 0$$

$$e^{-x}y = C$$

$$y = Ce^x$$

$$y(0) = 1 \implies C = 1. \qquad \text{Thus } y = e^x.$$

3. (a) and (b)

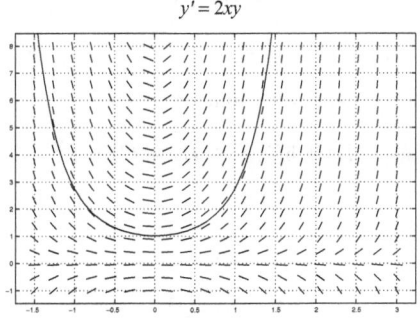

$$y' = 2xy$$

(c) $y' - 2xy = 0$ $H(x) = \int -2x\,dx = -x^2$;   integrating factor: e^{-x^2}

$$e^{-x^2}y' - 2xe^{x^2}y = 0$$

$$\frac{d}{dx}(e^{-x^2}y) = 0$$

$$e^{-x^2}y = C$$

$$y = Ce^{x^2}$$

$$y(0) = 1 \implies C = 1. \qquad \text{Thus } y = e^{x^2}.$$

SECTION 19.4

1. First consider the reduced equation. The characteristic equation is:

$$r^2 + 5r + 6 = (r+2)(r+3) = 0$$

and $u_1(x) = e^{-2x}$, $u_2(x) = e^{-3x}$ are fundamental solutions. A particular solution of the given equation has the form

$$y = Ax + B.$$

The derivatives of y are: $y' = A$, $y'' = 0$.

Substituting y and its derivatives into the given equation gives

$$0 + 5A + 6(Ax + B) = 3x + 4.$$

Thus,

$$6A = 3$$
$$5A + 6B = 4$$

The solution of this pair of equations is: $A = \frac{1}{2}$, $B = \frac{1}{4}$, and $y = \frac{1}{2}x + \frac{1}{4}$.

3. First consider the reduced equation. The characteristic equation is:

$$r^2 + 2r + 5 = 0$$

and $u_1(x) = e^{-x}\cos 2x$, $u_2(x) = e^{-x}\sin 2x$ are fundamental solutions. A particular solution of the given equation has the form

$$y = Ax^2 + Bx + C.$$

The derivatives of y are: $y' = 2Ax + B$, $y'' = 2A$.

Substituting y and its derivatives into the given equation gives

$$2A + 2(2Ax + B) + 5(Ax^2 + Bx + C) = x^2 - 1.$$

Thus,

$$5A = 1$$
$$4A + 5B = 0$$
$$2A + 2B + 5C = -1$$

The solution of this system of equations is: $A = \frac{1}{5}$, $B = -\frac{4}{25}$, $C = -\frac{27}{125}$, and

$$y = \frac{1}{5}x^2 - \frac{4}{25}x - \frac{27}{125}.$$

5. First consider the reduced equation. The characteristic equation is:

$$r^2 + 6r + 9 = (r+3)^2 = 0$$

and $u_1(x) = e^{-3x}$, $u_2(x) = xe^{-3x}$ are fundamental solutions. A particular solution of the given equation has the form

$$y = Ae^{3x}.$$

The derivatives of y are: $y' = 3Ae^{3x}$, $y'' = 9Ae^{3x}$.

Substituting y and its derivatives into the given equation gives

$$9Ae^{3x} + 18Ae^{3x} + 9Ae^{3x} = e^{3x}.$$

Thus, $36A = 1 \implies A = \dfrac{1}{36}$,   and $y = \frac{1}{36}e^{3x}$.

7. First consider the reduced equation. The characteristic equation is:

$$r^2 + 2r + 2 = 0$$

and $u_1(x) = e^{-x}\cos x$, $u_2(x) = e^{-x}\sin x$ are fundamental solutions. A particular solution of the given equation has the form

$$y = Ae^x.$$

The derivatives of y are: $y' = Ae^x$, $y'' = Ae^x$.

Substituting y and its derivatives into the given equation gives

$$Ae^x + 2Ae^x + 2Ae^x = e^x.$$

Thus, $5A = 1 \implies A = \frac{1}{5}$ and $y = \frac{1}{5}e^x$.

9. First consider the reduced equation. The characteristic equation is:

$$r^2 - r - 12 = (r - 4)(r + 3) = 0$$

and $u_1(x) = e^{4x}$, $u_2(x) = e^{-3x}$ are fundamental solutions. A particular solution of the given equation has the form

$$y = A\cos x + B\sin x.$$

The derivatives of y are: $y' = -A\sin x + B\cos x$, $y'' = -A\cos x - B\sin x$.

Substituting y and its derivatives into the given equation gives

$$-A\cos x - B\sin x - (-A\sin x + B\cos x) - 12(A\cos x + B\sin x) = \cos x.$$

Thus,

$$-13A - B = 1$$
$$A - 13B = 0$$

The solution of this system of equations is: $A = -\frac{13}{170}$, $B = -\frac{1}{170}$,   and

$$y = -\frac{13}{170}\cos x - \frac{1}{170}\sin x.$$

is a particular solution of the complete equation.

11. First consider the reduced equation. The characteristic equation is:

$$r^2 + 7r + 6 = (r+6)(r+1) = 0$$

and $u_1(x) = e^{-6x}$, $u_2(x) = e^{-x}$ are fundamental solutions. A particular solution of the given equation has the form

$$y = A\cos 2x + B\sin 2x.$$

The derivatives of y are: $\quad y' = -2A\sin 2x + 2B\cos 2x, \quad y'' = -4A\cos 2x - 4B\sin 2x.$

Substituting y and its derivatives into the given equation gives

$$-4A\cos 2x - 4B\sin 2x + 7(-2A\sin 2x + 2B\cos 2x) + 6(A\cos 2x + B\sin 2x) = 3\cos 2x.$$

Thus,

$$2A + 14B = 3$$
$$-14A + 2B = 0$$

The solution of this system of equations is: $\quad A = \frac{3}{100}, \quad B = \frac{21}{100}$ and

$$y = \frac{3}{10}\cos 2x + \frac{21}{100}\sin 2x.$$

13. First consider the reduced equation. The characteristic equation is:

$$r^2 - 2r + 5 = 0$$

and $u_1(x) = e^x\cos 2x$, $u_2(x) = e^x\sin 2x$ are fundamental solutions. A particular solution of the given equation has the form

$$y = Ae^{-x}\cos 2x + Be^{-x}\sin 2x$$

The derivatives of y are: $\quad y' = -Ae^{-x}\cos 2x - 2Ae^{-x}\sin 2x - Be^{-x}\sin 2x + 2Be^{-x}\cos 2x,$

$y'' = 4Ae^{-x}\sin 2x - 3Ae^{-x}\cos 2x - 4Be^{-x}\cos 2x - 3Be^{-x}\sin 2x.$

Substituting y and its derivatives into the given equation gives

$4Ae^{-x}\sin 2x - 3Ae^{-x}\cos 2x - 4Be^{-x}\cos 2x - 3Be^{-x}\sin 2x -$

$\quad 2\left(-Ae^{-x}\cos 2x - 2Ae^{-x}\sin 2x - Be^{-x}\sin 2x + 2Be^{-x}\cos 2x\right) +$

$\quad 5\left(Ae^{-x}\cos 2x + Be^{-x}\sin 2x\right) = e^{-x}\sin 2x.$

Equating the coefficients of $e^{-x}\cos 2x$ and $e^{-x}\sin 2x$ we get,

$$8A + 4B = 1$$
$$4A - 8B = 0$$

The solution of this system of equations is: $\quad A = \frac{1}{10}, \quad B = \frac{1}{20}$ and

$$y = \frac{1}{10}e^{-x}\cos 2x + \frac{1}{20}e^{-x}\sin 2x.$$

15. First consider the reduced equation. The characteristic equation is:

$$r^2 + 6r + 8 = (r+4)(r+2) = 0$$

and $u_1(x) = e^{-4x}$, $u_2(x) = e^{-2x}$ are fundamental solutions. A particular solution of the given equation has the form

$$y = Axe^{-2x}.$$

The derivatives of y are: $y' = Ae^{-2x} - 2Axe^{-2x}$, $y'' = -4Ae^{-2x} + 4Axe^{-2x}$.

Substituting y and its derivatives into the given equation gives

$$-4Ae^{-2x} + 4Axe^{-2x} + 6\left(Ae^{-2x} - 2Axe^{-2x}\right) + 8Axe^{-2x} = 3e^{-2x}$$

Thus, $2A = 3$ $\implies$ $A = \frac{3}{2}$ and $y = \frac{3}{2}xe^{-2x}$.

17. First consider the reduced equation: $y'' + y = 0$.   The characteristic equation is:

$$r^2 + 1 = 0$$

and $u_1(x) = \cos x$, $u_2(x) = \sin x$ are fundamental solutions. A particular solution of the given equation has the form

$$y = Ae^x.$$

The derivatives of y are: $y' = y'' = Ae^x$.

Substitute y and its derivatives into the given equation:

$$Ae^x + Ae^x = e^x \implies A = \frac{1}{2} \text{ and } y = \frac{1}{2}e^x.$$

The general solution of the given equation is: $y = C_1 \cos x + C_2 \sin x + \frac{1}{2}e^x$.

19. First consider the reduced equation: $y'' - 3y' - 10y = 0$.   The characteristic equation is:

$$r^2 - 3r - 10 = (r - 5)(r + 2) = 0$$

and $u_1(x) = e^{5x}$, $u_2(x) = e^{-2x}$ are fundamental solutions. A particular solution of the given equation has the form

$$y = Ax + B.$$

The derivatives of y are: $y' = A$, $y'' = 0$.

Substitute y and its derivatives into the given equation:

$$-3A - 10(Ax + B) = -x - 1 \implies A = \frac{1}{10}, \quad B = \frac{7}{100} \text{ and } y = \frac{1}{10}x + \frac{7}{100}$$

The general solution of the given equation is:

$$y = C_1 e^{5x} + C_2 e^{-2x} + \frac{1}{10}x + \frac{7}{100}$$

21. First consider the reduced equation: $y'' + 3y' - 4y = 0$.   The characteristic equation is:

$$r^2 + 3r - 4 = (r + 4)(r - 1) = 0$$

and $\quad u_1(x) = e^x, \quad u_2(x) = e^{-4x} \quad$ are fundamental solutions. A particular solution of the given equation has the form

$$y = Axe^{-4x}.$$

The derivatives of y are: $\quad y' = Ae^{-4x} - 4Axe^{-4x}, \quad y'' = -8Ae^{-4x} + 16Axe^{-4x}.$

Substitute y and its derivatives into the given equation:

$$-8Ae^{-4x} + 16Axe^{-4x} + 3\left(Ae^{-4x} - 4Axe^{-4x}\right) - 4Axe^{-4x} = e^{-4x}.$$

This implies $\quad -5A = 1, \quad$ so $\quad A = -\frac{1}{5} \quad$ and $\quad y = -\frac{1}{5}xe^{-4x}.$

The general solution of the given equation is: $\quad y = C_1 e^x + C_2 e^{-4x} - \frac{1}{5}xe^{-4x}.$

23. First consider the reduced equation: $\quad y'' + y' - 2y = 0. \quad$ The characteristic equation is:

$$r^2 + r - 2 = (r+2)(r-1) = 0$$

and $\quad u_1(x) = e^{-2x}, \quad u_2(x) = e^x \quad$ are fundamental solutions. A particular solution of the given equation has the form

$$y = x(A + Bx)e^x.$$

The derivatives of y are:

$$y' = (A + (2B+A)x + Bx^2)e^x, \quad y'' = (2A + 2B + (4B+A)x + Bx^2)e^x.$$

Substitute y and its derivatives into the given equation:

$$(2A + 2B + (4B+A)x + Bx^2 + A + (2B+A)x + Bx^2 - 2Ax - 2Bx^2)e^x = 3xe^x.$$

This implies $\quad A = -\frac{1}{3} \; , B = \frac{1}{2} \quad$ so $\quad y = x(-\frac{1}{3} + \frac{1}{2}x)e^x.$

The general solution of the given equation is: $\quad y = C_1 e^{-2x} + C_2 e^x - \frac{1}{3}xe^x + \frac{1}{2}x^2 e^x.$

25. Let $y_1(x)$ be a solution of $\quad y'' + ay' + by = \phi_1(x), \quad$ let $y_2(x)$ be a solution of $\quad y'' + ay' + by = \phi_2(x),$ and let $\quad z = y_1 + y_2. \quad$ Then

$$z'' + az' + bz = (y_1'' + y_2'') + a(y_1' + y_2') + b(y_1 + y_2)$$
$$= (y_1'' + ay_1' + by_1) + (y_2'' + ay_2' + y_2) = \phi_1 + \phi_2.$$

27. First consider the reduced equation: $\quad y'' + 4y' + 3y = 0. \quad$ The characteristic equation is:

$$r^2 + 4r + 3 = (r+3)(r+1) = 0$$

and $\quad u_1(x) = e^{-3x}, \quad u_2(x) = e^{-x} \quad$ are fundamental solutions. Since $\quad \cosh x = \frac{1}{2}\left(e^x + e^{-x}\right), \quad$ a particular solution of the given equation has the form

$$y = Ae^x + Bxe^{-x}$$

The derivatives of y are: $\quad y' = Ae^x + Be^{-x} - Bxe^{-x} \quad y'' = Ae^x - 2Be^{-x} + Bxe^{-x}.$

Substitute y and its derivatives into the given equation:

$$Ae^x - 2Be^{-x} + Bxe^{-x} + 4\left(Ae^x + Be^{-x} - Bxe^{-x}\right) + 3\left(Ae^x + Bxe^{-x}\right) = \tfrac{1}{2}\left(e^x + e^{-x}\right).$$

Equating coefficients, we get $A = \tfrac{1}{16}$, $B = \tfrac{1}{4}$,   and so $y = \tfrac{1}{16} e^x + \tfrac{1}{4} xe^{-x}$.

The general solution of the given equation is: $y = C_1 e^{-3x} + C_2 e^{-x} + \tfrac{1}{16} e^x + \tfrac{1}{4} xe^{-x}$.

29. First consider the reduced equation $y'' - 2y' + y = 0$.   The characteristic equation is:

$$r^2 - 2r + 1 = (r - 1)^2 = 0$$

and $u_1(x) = e^x$, $u_2(x) = xe^x$ are fundamental solutions. Their Wronskian is given by

$$W = u_1 u_2' - u_2 u_1' = e^x(e^x + xe^x) - xe^x(e^x) = e^{2x}$$

Using variation of parameters, a particular solution of the given equation will have the form

$$y = u_1 z_1 + u_2 z_2,$$

where

$$z_1 = -\int \frac{xe^x(xe^x \cos x)}{e^{2x}}\, dx = -\int x^2 \cos x\, dx = -x^2 \sin x - 2x \cos x + 2 \sin x,$$

$$z_2 = \int \frac{e^x(xe^x \cos x)}{e^{2x}}\, dx = \int x \cos x\, dx = x \sin x + \cos x$$

Therefore,

$$y = e^x\left(-x^2 \sin x - 2x \cos x + 2 \sin x\right) + xe^x\left(x \sin x + \cos x\right) = 2e^x \sin x - xe^x \cos x.$$

31. First consider the reduced equation $y'' - 4y' + 4y = 0$.   The characteristic equation is:

$$r^2 - 4r + 4 = (r - 2)^2 = 0$$

and $u_1(x) = e^{2x}$, $u_2(x) = xe^{2x}$ are fundamental solutions. Their Wronskian is given by

$$W = u_1 u_2' - u_2 u_1' = e^{2x}\left(e^{2x} + 2xe^{2x}\right) - xe^{2x}(2e^{2x}) = e^{4x}.$$

Using variation of parameters, a particular solution of the given equation will have the form

$$y = u_1 z_1 + u_2 z_2,$$

where

$$z_1 = -\int \frac{xe^{2x}\left(\tfrac{1}{3} x^{-1} e^{2x}\right)}{e^{4x}}\, dx = -\frac{1}{3}\int dx = -\tfrac{1}{3} x,$$

$$z_2 = \int \frac{e^{2x}\left(\tfrac{1}{3} x^{-1} e^{2x}\right)}{e^{4x}}\, dx = \frac{1}{3}\int \frac{1}{x}\, dx = \tfrac{1}{3} \ln|x|.$$

Therefore,

$$y = e^{2x}\left(-\tfrac{1}{3} x\right) + xe^{2x}\left(\tfrac{1}{3} \ln|x|\right) = -\tfrac{1}{3} xe^{2x} + \tfrac{1}{3} x \ln|x| e^{2x}.$$

Note: Since $u = -\frac{1}{3}xe^{2x}$ is a solution of the reduced equation,

$$y = \frac{1}{3}x\ln|x|\,e^{2x}$$

is also a particular solution of the given equation.

33. First consider the reduced equation $y'' + 4y' + 4y = 0$. The characteristic equation is:

$$r^2 + 4r + 4 = (r+2)^2 = 0$$

and $u_1(x) = e^{-2x}$, $u_2(x) = xe^{-2x}$ are fundamental solutions. Their Wronskian is given by

$$W = u_1 u_2' - u_2 u_1' = e^{-2x}\left(e^{-2x} - 2xe^{2x}\right) - xe^{-2x}(-2e^{-2x}) = e^{-4x}.$$

Using variation of parameters, a particular solution of the given equation will have the form

$$y = u_1 z_1 + u_2 z_2,$$

where

$$z_1 = -\int \frac{xe^{-2x}\left(x^{-2}e^{-2x}\right)}{e^{-4x}}\,dx = -\int \frac{1}{x}\,dx = -\ln|x|$$

$$z_2 = \int \frac{e^{-2x}\left(x^{-2}e^{-2x}\right)}{e^{-4x}}\,dx = \int \frac{1}{x^2}\,dx = -\frac{1}{x}$$

Therefore,

$$y = e^{-2x}\left(-\ln|x|\right) + xe^{-2x}\left(-\frac{1}{x}\right) = -e^{-2x}\ln|x| - e^{-2x}.$$

Note: Since $u = -e^{-2x}$ is a solution of the reduced equation, we can take

$$y = -\ln|x|\,e^{2x}.$$

35. First consider the reduced equation $y'' - 2y' + 2y = 0$. The characteristic equation is:

$$r^2 - 2r + 2 = 0$$

and $u_1(x) = e^x\cos x$, $u_2(x) = e^x\sin x$ are fundamental solutions. Their Wronskian is given by

$$W = e^x\cos x\left[e^x\sin x + e^x\cos x\right] - e^x\sin x\left[e^x\cos x - e^x\sin x\right] = e^{2x}$$

Using variation of parameters, a particular solution of the given equation will have the form

$$y = u_1 z_1 + u_2 z_2,$$

where

$$z_1 = -\int \frac{e^x\sin x \cdot e^x\sec x}{e^{2x}}\,dx = -\int \tan x\,dx = -\ln|\sec x| = \ln|\cos x|$$

$$z_2 = \int \frac{e^x\cos x \cdot e^x\sec x}{e^{2x}}\,dx = \int dx = x$$

Therefore,

$$y = e^x\cos x\left(\ln|\cos x|\right) + e^x\sin x(x) = e^x\cos x\ln|\cos x| + xe^x\sin x.$$

37. Assume that the forcing function $F(t) = F_0$ (constant). Then the differential equation has a particular solution of the form $i = A$. The derivatives of i are: $i' = i'' = 0$. Substituting i and its derivatives into the equation, we get

$$\frac{1}{C}A = F_0 \implies A = CF_0 \implies i = CF_0.$$

The characteristic equation for the reduced equation is:

$$Lr^2 + Rr + \frac{1}{C} = 0 \implies r_1, r_2 = \frac{-R \pm \sqrt{R^2 - 4L/C}}{2L} = \frac{-R\sqrt{C} \pm \sqrt{CR^2 - 4L}}{2L\sqrt{C}}$$

(a) If $CR^2 = 4L$, then the characteristic equation has only one root: $r = -R/2L$, and $u_1 = e^{-(R/2L)t}$, $u_2 = te^{-(R/2L)t}$ are fundamental solutions.

The general solution of the given equation is:

$$i(t) = C_1 e^{-(R/2L)t} + C_2 t e^{-(R/2L)t} + CF_0$$

and its derivative is:

$$i'(t) = -C_1(R/2L)e^{-(R/2L)t} + C_2 e^{-(R/2L)t} - C_2(R/2L)t e^{-(R/2L)t}.$$

Applying the side conditions $i(0) = 0$, $i'(0) = F_0/L$, we get

$$C_1 + CF_0 = 0$$

$$(-R/2L)C_1 + C_2 = F_0/L$$

The solution is $C_1 = -CF_0$, $C_2 = \frac{F_0}{2L}(2 - RC)$.

The current in this case is:

$$i(t) = -CF_0 e^{-(R/2L)t} + \frac{F_0}{2L}(2 - RC)t e^{-(R/2L)t} + CF_0.$$

(b) If $CR^2 - 4L < 0$ then the characteristic equation has complex roots:

$$r_1 = -R/2L \pm i\beta, \quad \text{where} \quad \beta = \sqrt{\frac{4L - CR^2}{4CL^2}} \quad (\text{here } i^2 = -1)$$

and fundamental solutions are: $u_1 = e^{-(R/2L)t}\cos\beta t$, $u_2 = e^{-(R/2L)t}\sin\beta t$.

The general solution of the given differential equation is:

$$i(t) = e^{-(R/2L)t}\left(C_1 \cos\beta t + C_2 \sin\beta t\right) + CF_0$$

and its derivative is:

$$i'(t) = (-R/2L)e^{-(R/2L)t}\left(C_1 \cos\beta t + C_2 \sin\beta t\right) + \beta e^{-(R/2L)t}\left(-C_1 \sin\beta t + C_2 \cos\beta t\right).$$

Applying the side conditions $i(0) = 0$, $i'(0) = F_0/L$, we get

$$C_1 + CF_0 = 0$$

$$(-R/2L)C_1 + \beta C_2 = F_0/L$$

The solution is $C_1 = -CF_0$, $C_2 = \frac{F_0}{2L\beta}(2 - RC)$.

The current in this case is:

$$i(t) = e^{-(R/2L)t}\left(\frac{F_0}{2L\beta}(2 - RC)\sin\beta t - CF_0 \cos\beta t\right) + CF_0.$$

39. (a) Let $y_1(x) = \sin\left(\ln x^2\right)$. Then

$$y_1' = \left(\frac{2}{x}\right)\cos\left(\ln x^2\right) \quad \text{and} \quad y_1'' = -\left(\frac{4}{x^2}\right)\sin\left(\ln x^2\right) - \left(\frac{2}{x^2}\right)\cos\left(\ln x^2\right)$$

Substituting y_1 and its derivatives into the differential equation, we have

$$x^2\left[-\left(\frac{4}{x^2}\right)\sin\left(\ln x^2\right) - \left(\frac{2}{x^2}\right)\cos\left(\ln x^2\right)\right] + x\left[\left(\frac{2}{x}\right)\cos\left(\ln x^2\right)\right] + 4\sin\left(\ln x^2\right) = 0$$

The verification that y_2 is a solution is done in exactly the same way.

The Wronskian of y_1 and y_2 is:

$$W(x) = y_1 y_2' - y_2 y_1'$$

$$= \sin\left(\ln x^2\right)\left[-\left(\frac{2}{x}\right)\sin\left(\ln x^2\right)\right] - \cos\left(\ln x^2\right)\left[\left(\frac{2}{x}\right)\cos\left(\ln x^2\right)\right]$$

$$= -\frac{2}{x} \neq 0 \text{ on } (0,\infty)$$

(b) To use the method of variation of parameters as described in the text, we first re-write the equation in the form

$$y'' + x^{-1}y' + 4x^{-2}y = x^{-2}\sin(\ln x).$$

Then, a particular solution of the equation will have the form $y = y_1 z_1 + y_2 z_2$, where

$$z_1 = -\int \frac{\cos(\ln x^2)x^{-2}\sin(\ln x)}{-2/x}\,dx$$

$$= \tfrac{1}{2}\int \cos(2\ln x)x^{-1}\sin(\ln x)\,dx$$

$$= \tfrac{1}{2}\int \cos 2u \,\sin u\,du \qquad (u = \ln x)$$

$$= \tfrac{1}{2}\int (2\cos^2 u - 1)\sin u\,du$$

$$= -\tfrac{1}{3}\cos^3 u + \tfrac{1}{2}\sin u$$

and

$$z_2 = \int \frac{\sin(\ln x^2)x^{-2}\sin(\ln x)}{-2/x}\,dx$$

$$= -\tfrac{1}{2}\int \sin(2\ln x)x^{-1}\sin(\ln x)\,dx$$

$$= -\tfrac{1}{2}\int \sin 2u\,\sin u\,du \qquad (u = \ln x)$$

$$= -\int \sin^2 u\,\cos u\,du$$

$$= -\tfrac{1}{3}\sin^3 u$$

Thus, $y = \sin 2u\left(-\tfrac{1}{3}\cos^3 u + \tfrac{1}{2}\sin u\right) - \cos 2u\left(\tfrac{1}{3}\sin^3 u\right)$ which simplifies to:

$$y = \tfrac{1}{3}\sin u = \tfrac{1}{3}\sin(\ln x).$$

SECTION 19.5

1. The equation of motion is of the form
$$x(t) = A\sin(\omega t + \phi_0).$$
The period is $T = 2\pi/\omega = \pi/4$. Therefore $\omega = 8$. Thus
$$x(t) = A\sin(8t + \phi_0) \text{ and } v(t) = 8A\cos(8t + \phi_0).$$
Since $x(0) = 1$ and $v(0) = 0$, we have
$$1 = A\sin\phi_0 \quad \text{and} \quad 0 = 8A\cos\phi_0.$$
These equations are satisfied by taking $A = 1$ and $\phi_0 = \pi/2$.

Therefore the equation of motion reads
$$x(t) = \sin\left(8t + \tfrac{1}{2}\pi\right).$$
The amplitude is 1 and the frequency is $8/2\pi = 4/\pi$.

3. We can write the equation of motion as
$$x(t) = A\sin\left(\frac{2\pi}{T}t\right).$$
Differentiation gives
$$v(t) = \frac{2\pi A}{T}\cos\left(\frac{2\pi}{T}t\right).$$
The object passes through the origin whenever $\sin[(2\pi/T)] = 0$.

Then $\cos[(2\pi/T)t] = \pm 1$ and $v = \pm 2\pi A/T$.

5. In this case $\phi_0 = 0$ and, measuring t in seconds, $T = 6$.

Therefore $\omega = 2\pi/6 = \pi/3$ and we have
$$x(t) = A\sin\left(\frac{\pi}{3}t\right), \quad v(t) = \frac{\pi A}{3}\cos\left(\frac{\pi}{3}t\right).$$
Since $v(0) = 5$, we have $\pi A/3 = 5$ and therefore $A = 15/\pi$.

The equation of motion can be written
$$x(t) = (15/\pi)\sin\left(\tfrac{1}{3}\pi t\right)$$

7. $x(t) = x_0\sin\left(\sqrt{k/m}\,t + \tfrac{1}{2}\pi\right)$

9. The equation of motion for the bob reads
$$x(t) = x_0\sin\left(t\sqrt{k/m} + \tfrac{1}{2}\pi\right). \qquad \text{(Exercise 7)}$$
Since $v(t) = \sqrt{k/m}\,x_0\cos\left(\sqrt{k/m}\,t + \tfrac{1}{2}\pi\right)$, the maximum speed is $\sqrt{k/m}\,x_0$.

The bob takes on half of that speed where $\left| \cos\left(\sqrt{k/m}\, t + \frac{1}{2}\pi \right) \right| = \frac{1}{2}$. Therefore

$$\left| \sin\left(\sqrt{k/m}\, t + \frac{1}{2}\pi \right) \right| = \sqrt{1 - \frac{1}{4}} = \frac{1}{2}\sqrt{3} \quad \text{and} \quad x(t) = \pm\frac{1}{2}\sqrt{3}\, x_0.$$

11. $\text{KE} = \frac{1}{2}m[v(t)]^2 = \frac{1}{2}m(k/m)x_0{}^2 \cos^2\left(\sqrt{k/m}\, t + \frac{1}{2}\pi \right)$

$$= \frac{1}{4}kx_0{}^2 \left[1 + \cos\left(2\sqrt{k/m}\, t + \pi \right) \right].$$

$$\text{Average KE} = \frac{1}{2\pi\sqrt{m/k}} \int_0^{2\pi\sqrt{m/k}} \frac{1}{4}kx_0{}^2 \left[1 + \cos\left(2\sqrt{k/m}\, t + \pi \right) \right]\, dt$$

$$= \frac{1}{4}kx_0{}^2.$$

13. Setting $y(t) = x(t) - 2$, we can write $x''(t) = 8 - 4x(t)$ as $y''(t) + 4y(t) = 0$.

This is simple harmonic motion about the point $y = 0$; that is, about the point $x = 2$. The equation of motion is of the form

$$y(t) = A\sin(2t + \phi_0).$$

The condition $x(0) = 0$ implies $y(0) = -2$ and thus

$(*)$ $\qquad\qquad\qquad A\sin\phi_0 = -2$

Since $y'(t) = x'(t)$ and $y'(t) = 2A\cos(2t + \phi_0)$, the condition $x'(0) = 0$ gives $y'(0) = 0$, and thus

$(**)$ $\qquad\qquad\qquad 2A\cos\phi_0 = 0.$

Equations $(*)$ and $(**)$ are satisfied by $A = 2$, $\phi_0 = \frac{3}{2}\pi$. The equation of motion can therefore be written

$$y(t) = 2\sin\left(2t + \frac{3}{2}\pi \right).$$

The amplitude is 2 and the period is π.

15. (a) Take the downward direction as positive. We begin by analyzing the forces on the buoy at a general position x cm beyond equilibrium. First there is the weight of the buoy: $F_1 = mg$. This is a downward force. Next there is the buoyancy force equal to the weight of the fluid displaced; this force is in the opposite direction: $F_2 = -\pi r^2 (L + x)\rho$. We are neglecting friction so the total force is

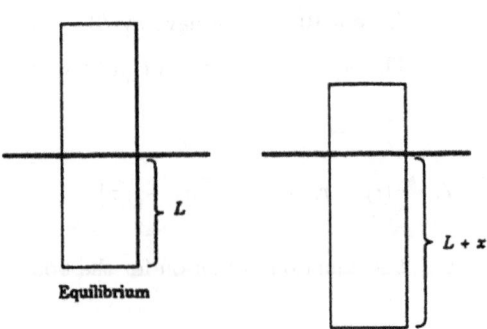

$$F = F_1 + F_2 = mg - \pi r^2 (L + x)\rho = \left(mg - \pi r^2 L\rho \right) - \pi r^2 x\rho.$$

We are assuming at the equilibrium point that the forces (weight of buoy and buoyant force of fluid) are in balance:

$$mg - \pi r^2 L\rho = 0.$$

Thus,

$$F = -\pi r^2 x \rho.$$

By Newton's

$$F = ma \qquad \text{(force = mass × acceleration)}$$

we have

$$ma = -\pi r^2 x \rho \qquad \text{and thus} \qquad a + \frac{\pi r^2 \rho}{m} x = 0.$$

Thus, at each time t,

$$x''(t) + \frac{\pi r^2 \rho}{m} x(t) = 0.$$

(b) The usual procedure shows that

$$x(t) = x_0 \sin\left(r\sqrt{\pi\rho/m}\, t + \tfrac{1}{2}\pi\right).$$

The amplitude A is x_0 and the period T is $(2/r)\sqrt{m\pi/\rho}$.

17. From (19.5.4), we have

$$x(t) = Ae^{(-c/2m)t} \sin(\omega t + \phi_0) = \frac{A}{e^{(c/2m)t}} \sin(\omega t + \phi_0) \quad \text{where} \quad \omega = \frac{\sqrt{4km - \omega^2}}{2m}.$$

If c increases, then both the amplitude, $\left|\dfrac{A}{e^{(c/2m)t}}\right|$ and the frequency $\dfrac{\omega}{2\pi}$ decrease.

19. Set $x(t) = 0$ in (19.5.6). The result is:

$$C_1 e^{(-c/2m)t} + C_2 t e^{(-c/2m)t} = 0 \implies C_1 + C_2 t = 0 \implies t = -C_1/C_2$$

Thus, there is at most one value of t at which $x(t) = 0$.

The motion changes directions when $x'(t) = 0$:

$$x'(t) = -C_1(c/2m)e^{(-c/2m)t} + C_2 e^{(-c/2m)t} - C_2(c/2m)t e^{(-c/2m)t}.$$

Now,

$$x'(t) = 0 \implies -C_1(c/2m) + C_2 - C_2 t(c/2m) = 0 \implies t = \frac{C_2 - C_1(c/2m)}{C_2(c/2m)}$$

and again we conclude that there is at most one value of t at which $x'(t) = 0$.

21. $x(t) = A\sin(\omega t + \phi_0) + \dfrac{F_0/m}{\omega^2 - \gamma^2} \cos(\gamma t)$

If $\omega/\gamma = m/n$ is rational, then $2\pi m/\omega = 2\pi n/\gamma$ is a period.

23. The characteristic equation is

$$r^2 + 2\alpha r + \omega^2 = 0; \qquad \text{the roots are} \quad r_1, \; r_2 = -\alpha \pm \sqrt{\alpha^2 - \omega^2}$$

Since $0 < \alpha < \omega$, $\alpha^2 < \omega^2$ and the roots are complex. Thus, $u_1(t) = e^{-\alpha t} \cos \beta t$, $\quad u_2(t) = e^{-\alpha t} \sin \beta t$, where $\beta = \sqrt{\omega^2 - \alpha^2}$ are fundamental solutions, and the general solution is:

$$x(t) = e^{-\alpha t}(C_1 \cos \beta t + C_2 \sin \beta t); \quad \beta = \sqrt{\alpha^2 - \omega^2}$$

25. Set $\omega = \gamma$ in the particular solution x_p given in Exercise 24. Then we have

$$x_p = \frac{F_0}{2\alpha\gamma m} \sin \gamma t$$

As $\quad c = 2\alpha m \to 0^+$, $\quad$ the amplitude $\quad \left| \frac{F_0}{2\alpha\gamma m} \right| \to \infty$

27. $\left(\omega^2 - \gamma^2 \right)^2 + 4\alpha^2 \gamma^2 = \omega^4 + \gamma^4 + 2\gamma^2(2\alpha^2 - \omega^2)$ increases as γ increases.

REVIEW EXERCISES

1. The equation is linear: $\quad H(x) = \displaystyle\int 1 dx = x \implies e^{H(x)} = e^x$

$$\frac{d}{dx}(e^x y) = 2e^{-x} \implies e^x y = -2e^{-x} + C; \quad \text{the solution is: } y = -2e^{-2x} + Ce^{-x}$$

3. The equation is separable:

$$\frac{y}{y^2 + 1} \, dy = \frac{1}{\cos^2 x} \, dx = \sec^2 x \, dx$$

$$\frac{1}{2} \ln(y^2 + 1) = \tan x + C$$

The solution is: $\quad \ln(1 + y^2) = 2 \tan x + C$

5. The equation can be written $\quad y' - \dfrac{2}{x} y = \dfrac{1}{x^2} y^2 \quad$ a Bernoulli equation.

$$y^{-2} y' - \frac{2}{x} y^{-1} = \frac{1}{x^2}$$

Let $v = y^{-1}$. Then $v' = -y^{-2} y'$, and we get the linear equation

$$v' + \frac{2}{x} v = -\frac{1}{x^2}.$$

Integrating factor: $\quad H(x) = \displaystyle\int (2/x) \, dx = \ln x^2$ and $e^{H(x)} = x^2$.

$$x^2 v' + 2xv = -1$$

$$x^2 v = -x + C$$

$$v = -\frac{1}{x} + \frac{C}{x^2} = \frac{C - x}{x^2}$$

The solution for the original equation is $y = \dfrac{x^2}{C - x}$

7. Since $\dfrac{\partial(y\sin x + xy\cos x)}{\partial y} = \sin x + x\cos x = \dfrac{\partial(x\sin x + y^2)}{\partial x}$,   the equation is exact.

$$f(x,y) = \int (y\sin x + xy\cos x)\,dx = xy\sin x + \phi(y).$$

$\frac{\partial f}{\partial y} = x\sin x + \phi'(y) = y^2 + x\sin x \Longrightarrow \phi'(y) = y^2 \quad \Longrightarrow \quad \phi(y) = \frac{1}{3}y^3$

The solution is $x\sin x + \frac{1}{3}y^3 = C$

9. The equation is separable:

$$\frac{1+y}{y}\,dy = (x^2 - 1)\,dx$$

$$\ln|y| + y = \frac{1}{3}x^3 - x + C$$

11. The equation can be written as $y' + \dfrac{2}{x}y = x^2$, a linear equation.

$H(x) = \ln x^2; \quad e^{H(x)} = x^2.$

$$x^2 y' + 2xy = x^4$$

$$x^2 y = \frac{1}{5}x^5 + C$$

$$y = \frac{1}{5}x^3 + Cx^{-2}$$

13. The differential equation is homogeneous.

Set $v = y/x$. Then $y = vx$ and $\dfrac{dy}{dx} = v + x\dfrac{dv}{dx}$.

$$v + x\frac{dv}{dx} = \frac{x^2 + x^2 v^2}{2x^2 v} = \frac{1 + v^2}{2v}$$

$$x\frac{dv}{dx} = \frac{1 + v^2}{2v} - v = \frac{1 - v^2}{2v}$$

$$\frac{2v}{1 - v^2}\,dv = \frac{1}{x}\,dx$$

$$-\ln|1 - v^2| = \ln|x| + C$$

$$1 - v^2 = \frac{C}{x}$$

Replacing v by y/x, we get $x^2 - y^2 = Cx$.

Applying the initial condition $y(1) = 2$ gives $C = -3$. The solution of the initial-value problem is:
$x^2 + 3x - y^2 = 0$.

15. Since $\dfrac{\partial(x + y)^2}{\partial y} = 2x + 2y = \dfrac{\partial(2xy + x^2 - 1)}{\partial x}$,   the equation is exact.

$$f(x,y) = \int (x + y)^2\,dx = \int (x^2 + 2xy + y^2)\,dx = \frac{1}{3}x^3 + x^2 y + xy^2 + \phi(y).$$

$\dfrac{\partial f}{\partial y} = x^2 + 2xy + \phi'(y) = 2xy + x^2 - 1 \Longrightarrow \phi'(y) = -1 \quad \Longrightarrow \quad \phi(y) = -y$

The general solution is: $\frac{1}{3}x^3 + x^2 y + xy^2 - y = C$

Applying the initial condition $y(1) = 1$ gives $C = 4/3$. The solution of the initial-value problem is:

$\frac{1}{3}x^3 + x^2 y + xy^2 - y = 4/3$

17. The equation is a Bernoulli equation; rewrite it as: $y^{-2}y' + xy^{-1} = x$.

Set $v = y^{-1}$. Then $v' = -y^{-2}y'$, and we have

$$v' - xv = -x,$$

a linear equation. $H(x) = \int(-x)\,dx = -\frac{1}{2}x^2$ and $e^{H(x)} = e^{-x^2/2}$

$$e^{-x^2/2}v' - xe^{-x^2/2}v = -xe^{-x^2/2}$$

$$e^{-x^2/2}v = e^{-x^2/2} + C$$

$$v = 1 + Ce^{x^2/2}$$

$$y = \frac{1}{1 + Ce^{x^2/2}}$$

Applying the initial condition $y(0) = 2$ gives $C = -1/2$. The solution of the initial-value problem
is: $y = \dfrac{2}{2 - e^{x^2/2}}$.

19. The characteristic equation is: $r^2 - 2r + 2 = 0$. The roots are: $r_1,\ r_2 = 1 \pm i$.
The general solution is:

$$e^x(C_1 \cos x + C_2 \sin x).$$

21. The characteristic equation for the reduced equation is: $r^2 - r - 2 = 0$. The roots are: $r = 2, -1$.
Use undetermined coefficients to find a particular solution of the nonhomogeneous equation:

$$z = A\cos 2x + B\sin 2x$$
$$z' = -2A\sin 2x + 2B\cos 2x$$
$$z'' = -4A\cos 2x - 4B\sin 2x$$

Substituting $z,\ z',\ z''$ into the differential equation yields the pair of equations:

$$-6A - 2B = 0,\ 2A - 6B = 1 \quad \Longrightarrow \quad A = \frac{1}{20},\ B = -\frac{3}{20}.$$

The general solution is: $y = C_1 e^{2x} + C_2 e^{-x} + \frac{1}{20}\cos 2x - \frac{3}{20}\sin 2x$

23. The characteristic equation for the reduced equation is: $r^2 - 6r + 9 = 0$. The roots are: $r_1 = r_2 = 3$.
Use undetermined coefficients to find a particular solution of the nonhomogeneous equation.

Since $z = e^{3x}$ and $z = xe^{3x}$ are solutions of the reduced equation, set $z = Ax^2e^{3x}$.

$$z = Ax^2e^{3x}$$
$$z' = 2Axe^{3x} + 3Ax^2e^{3x}$$
$$z'' = 2Ae^{3x} + 12Axe^{3x} + 9Ax^2e3x$$

Substituting z, z', z'' into the differential equation gives:

$$2A = 3 \quad \Longrightarrow \quad A = \frac{3}{2}.$$

The general solution is: $y = C_1e^{3x} + C_2xe^{3x} + \frac{3}{2}x^2e^{3x}$.

25. The characteristic equation for the reduced equation is: $r^2 - 2r + 1 = 0$. The roots are: $r_1 = r_2 = 1$. Use variation of parameters to find a particular solution of the nonhomogeneous equation.

Set $u_1 = e^x$ and $u_2 = xe^x$. Then their Wronskian is $W(x) = e^{2x}$.

$$z_1 = -\int \frac{xe^x(1/x)e^x}{e^{2x}}\, dx = -\int dx = -x \qquad z_2 = \int \frac{e^x(1/x)e^x}{e^{2x}}\, dx = \int (1/x)\, dx = \ln x$$
$$y_p = -xe^x + xe^x \ln x$$

The general solution of the equation is: $y = C_1e^x + C_2xe^x + xe^x \ln x, \ x > 0$

27. The characteristic equation for the reduced equation is: $r^2 + 4r + 4 = 0$. The roots are: $r_1 = r_2 = -2$. Use undetermined coefficients to find a particular solution of the nonhomogeneous equation:

$$z = Ax^2e^{-2x} + Be^{2x}$$
$$z' = 2Axe^{-2x} - 2Ax^2e^{-2x} + 2Be^{2x}$$
$$z'' = 2Ae^{-2x} - 8Axe^{-2x} + 4Ax^2e^{-2x} + 4Be^{2x}$$

Substituting z, z', z'' into the differential equation yields the equations:

$$2A = 4, \ 16B = 2 \quad \Longrightarrow \quad A = 2, \ B = \frac{1}{8}.$$

The general solution is: $y = C_1e^{-2x} + C_2xe^{-2x} + \frac{1}{8}e^{2x} + 2x^2e^{-2x}$

29. First find the general solution of the differential equation.

The characteristic equation for the reduced equation is: $r^2 + r = 0$. The roots are: $r_1 = -1$, $r_2 = 0$.

Use undetermined coefficients to find a particular solution of the nonhomogeneous equation:

Set $z = Ax^2 + Bx$

$$z = Ax^2 + Bx$$
$$z' = 2Ax + B$$
$$z'' = 2A$$

Substituting z, z', z'' into the differential equation yields the equations:

$$2A = 1, \ A + B = 0 \quad \Longrightarrow \quad A = 1/2, \ B = -1.$$

The general solution of the differential equation is: $y = C_1 e^{-x} + C_2 + \frac{1}{2}x^2 - x$

Applying the initial conditions $y(0) = 1$, $y'(0) = 0$, we get the pair of equations

$$C_1 + C_2 = 1, \quad -C_1 - 1 = 0, \quad \Longrightarrow \quad C_1 = -1, \ C_2 = 2.$$

The solution of the initial-value problem is: $\quad y = 2 - e^{-x} + \frac{1}{2}x^2 - x$

31. First find the general solution of the differential equation.

The characteristic equation for the reduced equation is: $r^2 - 5r + 6 = 0$. The roots are: $r_1 = 2$, $r_2 = 3$.

Use undetermined coefficients to find a particular solution of the nonhomogeneous equation:

Set $\ z = Axe^{2x}$

$$z = Axe^{2x}$$
$$z' = Ae^{2x} + 2Axe^{2x}$$
$$z'' = 4Ae^{2x} + 4Axe^{2x}$$

Substituting z, z', z'' into the differential equation gives: $-A = 10$, $A = -10$.

The general solution of the differential equation is: $y = C_1 e^{2x} + C_2 e^{3x} - 10xe^{2x}$

Applying the initial conditions $y(0) = 1$, $y'(0) = 1$, we get $C_1 = -8$, $C_2 = 9$.

The solution of the initial-value problem is: $\quad y = 9e^{3x} - 8e2x - 10xe^{2x}$

33. Assume $x(t) = A\sin(wt + \phi_0)$.

From $T = 2\pi/\omega = \pi/2$, $\omega = 4$ and $x(t) = A\sin(4t + \phi_0)$

$x(0) = 2 \quad \Longrightarrow \quad A\sin(\phi)0) = 2; \quad x'(0) = 0 \quad \Longrightarrow \quad 4A\cos(\phi_0) = 0 \quad \Longrightarrow \quad \phi_0 = \dfrac{\pi}{2}$ and $A = 2$.

Therefore,

$$x(t) = 2\sin(4t + \pi/2); \quad \text{amplitude } A = 2; \quad \text{frequency } 2/\pi.$$

35. Assume that the downward direction is positive. Then

$$4x''(t) = -64x(t) + 8\sin 4t, \quad x(0) = -\frac{1}{2}, \quad x'(0) = 0$$

This equation can be written as

$$x'' + 16x = 2\sin 4t$$

The characteristic equation for the reduced equation is: $r^2 + 16 = 0$ and the roots are $r = \pm 4i$.

Use undetermined coefficients to find a particular solution of the nonhomogeneous equation:

Set $\ z = At\cos 4t + Bt\sin 4t$

$$z = At\cos 4t + Bt\sin 4t$$
$$z' = A\cos 4t - 4At\sin 4t + B\sin 4t + 4Bt\cos 4t$$
$$z'' = -8A\sin 4t - 16At\cos 4t + 8B\cos 4t - 16Bt\sin 4t$$

Substituting z, z', z'' into the differential equation yields the equations, we get $A = -\frac{1}{4}$, $B = 0$.

The general solution of the differential equation is:

$$x(t) = C_1 \cos 4t + C_2 \sin 4t - \frac{1}{4}t \cos 4t.$$

Applying the initial conditions $x(0) = -1/2$, $x'(0) = 0$, we get $C_1 = -1/2$, $C_2 = 1/16$.

The equation of motion is:

$$x(t) = -\frac{1}{2}\cos 4t - \frac{1}{4}t \cos 4t + \frac{1}{16}\sin 4t$$

NOTES

NOTES

NOTES

NOTES

NOTES

NOTES

NOTES

NOTES

NOTES

NOTES

NOTES

NOTES

NOTES

NOTES

NOTES

NOTES

NOTES

NOTES

NOTES

NOTES

NOTES

NOTES

NOTES

NOTES

NOTES

NOTES

NOTES

NOTES

NOTES

NOTES

NOTES

NOTES

NOTES